Ship Structures

Ship Structures

Editors

Joško Parunov
Yordan Garbatov

MDPI • Basel • Beijing • Wuhan • Barcelona • Belgrade • Manchester • Tokyo • Cluj • Tianjin

Editors
Joško Parunov
University of Zagreb
Croatia

Yordan Garbatov
Universidade de Lisboa
Portugal

Editorial Office
MDPI
St. Alban-Anlage 66
4052 Basel, Switzerland

This is a reprint of articles from the Special Issue published online in the open access journal *Journal of Marine Science and Engineering* (ISSN 2077-1312) (available at: https://www.mdpi.com/journal/jmse/special_issues/ship_structures).

For citation purposes, cite each article independently as indicated on the article page online and as indicated below:

LastName, A.A.; LastName, B.B.; LastName, C.C. Article Title. *Journal Name* **Year**, *Volume Number*, Page Range.

ISBN 978-3-0365-4129-7 (Hbk)
ISBN 978-3-0365-4130-3 (PDF)

Cover image courtesy of Brodosplit JSC

Contents

Journal of
Marine Science and Engineering

Editorial

Ship Structures

Joško Parunov [1,*] and Yordan Garbatov [2,*]

[1] Faculty of Mechanical Engineering and Naval Architecture, University of Zagreb, 10000 Zagreb, Croatia
[2] Centre for Marine Technology and Ocean Engineering (CENTEC), Instituto Superior Técnico, Universidade de Lisboa, 1049-001 Lisbon, Portugal
* Correspondence: jparunov@fsb.hr or josko.parunov@fsb.hr (J.P.); yordan.garbatov@tecnico.ulisboa.pt (Y.G.)

Citation: Parunov, J.; Garbatov, Y. Ship Structures. *J. Mar. Sci. Eng.* **2022**, *10*, 374. https://doi.org/10.3390/jmse10030374

Received: 1 March 2022
Accepted: 1 March 2022
Published: 6 March 2022

Publisher's Note: MDPI stays neutral with regard to jurisdictional claims in published maps and institutional affiliations.

1. Introduction

This book contains fifteen recent research studies in the broad field of ship structural design, analysis and degradation, where two studies deal with corrosion degradation in ship structures, while the remaining contributions belong to three major steps in the efficient design and analysis of ship structures, i.e., modelling of the environment and environmental loads, analysis of a ship structural response, and the definition of criteria for different failure modes of ship structures.

More specifically, the first two works are related to the probabilistic description of environmental parameters relevant to the analysis of ship structures [1,2], while the other two cover modelling of the environmental loads [3,4].

A ship structural response is divided into static, quasi-static and dynamic, and one of the contributions belongs to the latter group, i.e., vibratory ship structural response [5].

The criteria that ship structures should satisfy are identified according to three different limit states, i.e., ultimate limit state (ULS), serviceability limit state (SLS) and accidental limit state (ALS). The ULS represents serious failure of ship structures, which has an immediate impact on the ship's structural safety. Such a failure is caused by an overload of the structure, resulting in either a large spread of plasticity, due to the material yielding, or elastic–plastic buckling collapse, caused by compressive loads. The ULS may be associated with the structural failure of different components of ship structures, such as the plates and stiffened panels, or to the whole ship hull girder. Four studies are dedicated to different aspects of the ultimate strength of ship structures [6–9].

The SLS represents the failure modes that disturb the normal functioning of the ship's structural components, although the structural safety is not immediately put in danger. Excessive structural vibration and fatigue failures of the ship's structural details are typical examples of the SLS. One of the studies presented deals with the vibratory response, while the other two analyze the fatigue capacity of the ship's structural components [10,11].

The ALS usually refers to the residual strength of the ship's structures after a marine accident, i.e., collision or grounding, and two studies are dedicated to these problems [12,13].

Besides fatigue structural failure, corrosion degradation is another important effect that appears once the coating protection has failed during the ship's service life. Understanding, modelling, analyzing and mitigating the corrosion phenomenon are important issues in the design and maintenance of the ship, as the ship structures should have satisfactory safety, even at the end of their lifetime. Two studies deal with the modelling and analysis of corrosion degradation in ship structures [14,15].

In the next section, a brief overview of the papers published in the book is presented, ordered according to this Introduction.

2. Paper Details

Wind and waves present the main causes of environmental loading on seagoing ships and offshore structures. A comprehensive description of the wind and wave climate in the Adriatic Sea is presented in [1]. The conditional modelling approach, i.e., the marginal distribution of significant wave height and the conditional distribution of peak period and wind speed, is used in the study. The model parameters are fitted and presented, while extreme significant wave heights are evaluated for 20-, 50- and 100-year return periods. The results are shown for 39 uniformly spaced locations across the offshore Adriatic Sea, which can serve as an input for almost all kinds of analyses of ships and offshore structures.

Uncertainties in the prediction of extreme waves are presently a concern for the research community, having important practical implications in the design and operation of ships and offshore structures. Uncertainty, caused by the choice of probability distributions and fitting techniques in the analysis of extreme values of wave heights, is discussed in [2]. Methods and fitting techniques are tested on a long-term database for a location in the Adriatic Sea. The variability of the results and trends of extreme wave height estimates for long return periods are presented, and the limitations of certain methods and techniques are pointed out.

The estimation of wind loads on exposed parts of ships and offshore structures represents a challenging task because of its implications for various aspects of exploitation. An extended method for estimating the wind loads on container ships is presented in [3], using the Generalized Regression Neural Network (GRNN). The obtained results are in favorable agreement with the experimental measurements in the wind tunnel, as well as with the computational fluid dynamics (CFD) simulation performed.

The computational method for calculating the extreme vertical wave-induced bending moments on a passenger ship along the shipping route, based on the hindcast database, is presented in [4]. An operability analysis is performed to account for the ship's operational restrictions. Conducting an integrated operability analysis and extreme wave load computation is the main objective of the study. The method is employed to analyze a passenger ship sailing across the Adriatic Sea.

The vibratory structural response of connecting beams, using the component mode synthesis method, is improved in [5]. Polynomials that combine simple and fixed supports have been proposed to satisfy the boundary conditions at a junction. A comparison with the finite element results shows that they have good agreement with the method for practical purposes. The method is useful for the vibration analysis of many local structures in a ship, such as equipment supports.

The ultimate structural capacity assessment, particularly of the longitudinal strength, is crucial to ensure the safety of ships, crews, the marine environment, and the cargoes carried. A review of the references dealing with the collapse strength of the plates, stiffened panels and the entire hull is provided in [6]. The main contributions of the paper are the conclusions and suggestions about potential future research in the field.

The ultimate strength of rectangular plates subjected to cyclic load reversals is studied in [7]. The finite element solution is implemented to estimate the load-carrying capacity, accounting for the influence of the initial imperfections, plate thicknesses and aspect ratio. It was found that the uni-modal initial imperfection has a significant impact on the ultimate capacity reduction.

Residual stresses and the initial imperfections caused by the welding process are important for the prediction of the ultimate capacity of plates and stiffened panels. It is, therefore, challenging to predict these effects by using welding simulations when designing the ship. Recognizing that welding simulations coupled with structural analysis could be rather complex and time consuming, a simplified procedure for the welding simulations of a rectangular plate is developed in [8]. The collapse strength of the welded plates is then compared to the residual stress-free case, both for tension and compression, and including the initial imperfections. Based on the comparison of the results with codes and most of the established formulations, conclusions and recommendations about the applicability of the proposed method are provided.

A finite element method for predicting welding-induced distortion in multi-pass welding is developed in [9]. The study focuses on the extraction of the equivalent plastic strain and heat-affected zone (HAZ) width through 3D thermal elastic–plastic analysis (TEPA) for each welding pass. The predicted welding distortion is compared with the measured experimental data, indicating that good agreement can be obtained.

The fatigue strength of T-type specimens under two-step repeating variable amplitude loading conditions is studied experimentally in [10]. Discussions about non-linear fatigue damage accumulation, and the effect of the interaction between the high- and low-amplitude loadings on the fatigue life were carried out, and relevant conclusions were obtained, according to the series of fatigue tests.

Hatch corners, especially those in large bulk carriers, are among the most important details regarding the potential appearance of fatigue cracks. Small-scale fatigue experiments of hatch corners are, therefore, performed in [11], aiming to investigate the effect of stress relaxation using shot peening treatment. The results show that the fatigue lives of the shot-peened ship hatch corner specimens are always longer than those of the untreated specimens.

The limit state function for the assessment of the longitudinal strength of damaged ships under combined bending moments is studied in [12]. As the limit state function cannot be obtained directly, the results for the residual strength are used for fitting the approximate limit state function. Various fitting methods have been proposed, and it is concluded that the weighted piecewise fitting method is the best choice for all the investigated cases.

The possibility of crack propagation during the towing period of a damaged ship, and the consequent reduction in the residual longitudinal strength, are investigated in [13]. Fluctuating wave-induced stresses are considered to be the main cause of crack propagation, which is studied using linear elastic fracture mechanics. The proposed method can be considered as part of the emergency response procedure during the salvage of a damaged ship.

Using the probabilistic model to estimate the percentage of corrosion depth for the inner bottom plates of ageing bulk carriers is proposed in [14]. The ratio, considered as the random variable, of the corrosion rate and the average initial inner bottom plate's thickness are studied. Three three-parameter distributions for estimating the cumulative probability distribution and the probability density function of the random variable are used. The statistical and empirical results are presented in numerical and graphical form, and conclusions are provided.

The research study presented in [15] deals with the mechanical behavior of butt-welded specimens made of shipbuilding steel, exposed to a natural marine environment for a relatively long period of up to 3 years. Relative mass change, due to corrosion degradation over time, is observed, along with the calculated corrosion rates. In addition, the corroded surfaces of specimens are inspected using optical and scanning electron microscopy. It is concluded that the sea splash generally has the most negative impact on the corrosion rate, while pit depths are generally the greatest in the heat-affected zone area of the specimen.

Author Contributions: Conceptualization, J.P. and Y.G.; writing—original draft preparation, J.P. and Y.G.; writing—review and editing, J.P. and Y.G.; supervision, J.P. and Y.G.; project administration, J.P. and Y.G.; funding acquisition, J.P. and Y.G. All authors have read and agreed to the published version of the manuscript.

Funding: Joško Parunov was supported by the Croatian Science Foundation, under the project MODUS IP-2019-04-2085, and Yordan Garbatov was supported by the Portuguese Foundation for Science and Technology (Fundação para a Ciência e Tecnologia—FCT), under contract UIDB/UIDP/00134/2020.

Institutional Review Board Statement: Not applicable.

Informed Consent Statement: Not applicable.

Data Availability Statement: Not applicable.

Acknowledgments: The editors wish to express sincere gratitude to all the authors and reviewers.

Conflicts of Interest: The authors declare no conflict of interest.

References

1. Katalinić, M.; Parunov, J. Comprehensive Wind and Wave Statistics and Extreme Values for Design and Analysis of Marine Structures in the Adriatic Sea. *J. Mar. Sci. Eng.* **2021**, *9*, 522. [CrossRef]
2. Katalinić, M.; Parunov, J. Uncertainties of Estimating Extreme Significant Wave Height for Engineering Applications Depending on the Approach and Fitting Technique—Adriatic Sea Case Study. *J. Mar. Sci. Eng.* **2020**, *8*, 259. [CrossRef]
3. Prpić-Oršić, J.; Valčić, M.; Čarija, Z. A Hybrid Wind Load Estimation Method for Container Ship Based on Computational Fluid Dynamics and Neural Networks. *J. Mar. Sci. Eng.* **2020**, *8*, 539. [CrossRef]
4. Petranović, T.; Mikulić, A.; Katalinić, M.; Ćorak, M.; Parunov, J. Method for Prediction of Extreme Wave Loads Based on Ship Operability Analysis Using Hindcast Wave Database. *J. Mar. Sci. Eng.* **2021**, *9*, 1002. [CrossRef]
5. Park, J.H.; Yoon, D.Y. A Proposal of Mode Polynomials for Efficient Use of Component Mode Synthesis and Methodology to Simplify the Calculation of the Connecting Beams. *J. Mar. Sci. Eng.* **2021**, *9*, 20. [CrossRef]
6. Tekgoz, M.; Garbatov, Y. Collapse Strength of Intact Ship Structures. *J. Mar. Sci. Eng.* **2021**, *9*, 1079. [CrossRef]
7. Tekgoz, M.; Garbatov, Y. Strength Assessment of Rectangular Plates Subjected to Extreme Cyclic Load Reversals. *J. Mar. Sci. Eng.* **2020**, *8*, 65. [CrossRef]
8. Gordo, J.M. Effect of Residual Stresses on the Elastoplastic Behavior of Welded Steel Plates. *J. Mar. Sci. Eng.* **2020**, *8*, 702. [CrossRef]
9. Lee, J.; Perrera, D.; Chung, H. Multi-Pass Welding Distortion Analysis Using Layered Shell Elements Based on Inherent Strain. *J. Mar. Sci. Eng.* **2021**, *9*, 632. [CrossRef]
10. Gan, J.; Sun, D.; Deng, H.; Wang, Z.; Wang, X.; Yao, L.; Wu, W. Fatigue Characteristic of Designed T-Type Specimen under Two-Step Repeating Variable Amplitude Load with Low-Amplitude Load below the Fatigue Limit. *J. Mar. Sci. Eng.* **2021**, *9*, 107. [CrossRef]
11. Gan, J.; Gao, Z.; Wang, Y.; Wang, Z.; Wu, W. Small–Scale Experimental Investigation of Fatigue Performance Improvement of Ship Hatch Corner with Shot Peening Treatments by Considering Residual Stress Relaxation. *J. Mar. Sci. Eng.* **2021**, *9*, 419. [CrossRef]
12. Zhu, Z.; Ren, H.; Wang, X.; Zhao, N.; Li, C. Methods for Fitting the Limit State Function of the Residual Strength of Damaged Ships. *J. Mar. Sci. Eng.* **2022**, *10*, 102. [CrossRef]
13. Gledić, I.; Mikulić, A.; Parunov, J. Improvement of the Ship Emergency Response Procedure in Case of Collision Accident Considering Crack Propagation during Salvage Period. *J. Mar. Sci. Eng.* **2021**, *9*, 737. [CrossRef]
14. Ivošević, Š.; Meštrović, R.; Kovač, N. A Probabilistic Method for Estimating the Percentage of Corrosion Depth on the Inner Bottom Plates of Aging Bulk Carriers. *J. Mar. Sci. Eng.* **2020**, *8*, 442. [CrossRef]
15. Vukelic, G.; Vizentin, G.; Brnic, J.; Brcic, M.; Sedmak, F. Long-Term Marine Environment Exposure Effect on Butt-Welded Shipbuilding Steel. *J. Mar. Sci. Eng.* **2021**, *9*, 491. [CrossRef]

Journal of
*Marine Science
and Engineering*

Article

Comprehensive Wind and Wave Statistics and Extreme Values for Design and Analysis of Marine Structures in the Adriatic Sea

Marko Katalinić [1] and Joško Parunov [2,*]

[1] Faculty of Maritime Studies, University of Split, 21000 Split, Croatia; marko.katalinic@pfst.hr
[2] Faculty of Mechanical Engineering and Naval Architecture, University of Zagreb, 10000 Zagreb, Croatia
* Correspondence: josko.parunov@fsb.hr; Tel.: +385-(0)1-6168-226

Abstract: Wind and waves present the main causes of environmental loading on seagoing ships and offshore structures. Thus, its detailed understanding can improve the design and maintenance of these structures. Wind and wave statistical models are developed based on the WorldWaves database for the Adriatic Sea: for the entire Adriatic Sea as a whole, divided into three regions and for 39 uniformly spaced locations across the offshore Adriatic. Model parameters are fitted and presented for each case, following the conditional modelling approach, i.e., the marginal distribution of significant wave height and conditional distribution of peak period and wind speed. Extreme significant wave heights were evaluated for 20-, 50- and 100-year return periods. The presented data provide a consistent and comprehensive description of metocean (wind and wave) climate in the Adriatic Sea that can serve as input for almost all kind of analyses of ships and offshore structures.

Keywords: offshore Adriatic Sea; significant wave height; peak period; wind speed; extreme value; conditional modelling; joint distributions

Citation: Katalinić, M.; Parunov, J. Comprehensive Wind and Wave Statistics and Extreme Values for Design and Analysis of Marine Structures in the Adriatic Sea. *J. Mar. Sci. Eng.* **2021**, *9*, 522. https://doi.org/10.3390/jmse9050522

Academic Editor: Wei-Bo Chen

Received: 19 April 2021
Accepted: 7 May 2021
Published: 12 May 2021

Publisher's Note: MDPI stays neutral with regard to jurisdictional claims in published maps and institutional affiliations.

1. Introduction

Wave (re-analysis) databases, comprising numerical wave model hindcasts and assimilated altimetry satellite data, present a comprehensive state-of-the-art source for analysis of metocean data that serve as input for the design and assessment of oceangoing vessels and offshore structures [1]. The WorldWaves database was used in this study to derive wind and wave statistics for the Adriatic Sea, which is a semi-enclosed sea basin with specific wind–wave climate. The basin is analyzed as a whole, divided into three regions and at 39 evenly spaced locations. Procedures and concepts applied in the study resulted with environmental wind and wave models that provide a detailed and structured insight into the wind–wave climate of this basin. Joint probability distributions of significant wave height and peak periods and distribution of wind speed to significant wave height are developed together with extreme wave height estimation for 20, 50 and 100 year return periods.

The obtained results are useful for the design, operation planning, maintenance and life-time extension of marine-related engineering objects in the Adriatic. Specific calculations such as: extreme sea states analysis for the design of marine structures [2], coupled aero- and hydro-dynamic analysis of floating offshore wind turbines [3], long-term fatigue calculations [4], structural reliability [5] and mooring analyses [6], can benefit from the input of the developed data. Wave statistics in the Adriatic is also used for exploring wave energy potential [7,8] and for planning and evaluating the performance of high-requirement service vessels such as the coastal patrol boat developed for the Adriatic [9]. Safety analysis of a fishing vessel due to roll in a seaway [10] and organization of special marine operations [11] are other examples where accurate wave statistics for the Adriatic is indispensable.

In the second half of 20th century, wave statistics in the Adriatic was based on visual observations collected from merchant and meteorological vessels and published in the wave climatology atlas [12]. Wave statistics was rather roughly graphically presented in the form of wave roses that were later digitalized and studied in terms of extreme values [13]. These data, however, suffer from known inaccuracies of visual wave observations and lack of extreme events due to heavy weather avoidance by ships [14]. Later, measurements have been performed from four floating buoys installed along the west coast of the Adriatic [15]. However, these data were recorded and are available only for limited time of 5 years, with some interruptions due to failure and maintenance. One of the most comprehensive wave-data sources in general are the measurements from the Aqua Alta oceanographic tower, located near Venice in the north part of the Adriatic [16]. Although 38 years of continues uninterrupted measurements are available, they refer only to one specific location and cannot be used for the whole Adriatic Sea.

The present study represents the first complete wave and wind statistics for the whole Adriatic Sea, representing progress compared to previous studies [17]. The way the data are presented enables its application to almost all purposes related to marine structural design and analysis, as reviewed in the preceding paragraphs. Site specific design (e.g., for offshore structures) might require spatial interpolation of presented data; however, it can be directly applicable for the analysis of ships, as they are not confined to a specific location. Particular attention is paid to the accuracy of calculated extreme values, to avoid too conservative results that could lead to non-economical and over-dimensioned structures [18].

The paper is organized as follows. Initially, the WorldWaves database, used as the underlying material for the study, is described in Section 2 along with preliminary data structuring such as the development of sea-state and wind contingency tables, wave roses visualizations and example validations against existing buoy data. In Section 3 the theoretical background of applied methods is presented for the development of the joint probability distribution models. In Section 4, all model parameters and extreme values are presented graphically, both regionally and per individual location analyzed. Section 5 gives a discussion on the results and Section 6 presents the conclusion. At the end of the paper, Appendix A provides basic data, such as location coordinates and regional subdivision and Appendix B gives detailed tabulated results that are graphically presented in Section 4.

2. Data

The underlying data used for the analysis of wave and wind climate analysis in the Adriatic Sea were extracted from the WorldWaves (WW) database. The database represents numerical wave model hindcasts with assimilated available satellite altimetry measurements [19,20]. It includes 39 locations, evenly distributed across the Adriatic Sea with $0.5° \times 0.5°$ (lat.-long.) spacing, in the period from September 1992 to January 2016. The underlying numerical wave model WAM (Wave Modelling) is run at the ECMWF (European Centre for Medium-Range Weather Forecasts) which acts as a European meteorological institute providing numerical atmospheric and ocean forecasts, archiving data and improving forecasting models. WAM is extensively validated in the literature [21,22]. Satellite altimetry measurements are, in general, validated by in-situ measurements made with wave buoys and are considered an empirical source of data for larger domains but lack in continuity as they are confined by individual satellite tracks and overflight times. WW includes satellite altimetry data from satellite missions taking measurements over the Mediterranean (i.e., the Adriatic): European Remote Sensing Satellites (ERS-1 and ERS-2), Ocean Topography Experiment (TOPEX), Geosat Follow-On, Jason and Environmental Satellite (Envisat).

WW Data Subdivision, Preparation and Preliminary Considerations

A total of 39 locations were available within the WW database for the Adriatic Sea. The locations were analyzed: individually, grouped into regions (southern, central and northern Adriatic, according to DHMZ–Croatian Meteorological and Hydrological Service official subdivision) and joint together for the basin as a whole. Location, their numbering and regional subdivision are presented in Figure 1, and exact geographic coordinates are given in Appendix A Table A1.

Figure 1. Studied locations in the Adriatic Sea as available from the WW database.

At each location, 12 physical wave and wind parameters are available at 6-h intervals (four per day) as presented in Appendix A Table A2 and for each location there are total of 34,460 lines of records.

Maximum recorded significant wave heights, along with accompanying parameters, were extracted and are presented in Appendix A Table A3. The single highest significant wave height in the database is recorded at location 9 (E14.5°–N44.0°) 16.11.2002, reading H_s = 6.72 m during southeast wind (local names *jugo/scirocco*). For comparison, the single highest wave measured until now along the east coast reads H_{max} = 10.87 m off the city of Dubrovnik on 12.11.2019, associated with the significant wave height of H_s = 4.75 m. The highest significant wave height so far is measured from the gas platform in the north Adriatic and reads 7.5 m [23].

Visualization of H_s time series for a one-year period, presented in Figure 2, confirms expected higher variability during winter months.

An example validation of WW data against available in-situ wave buoy measurement data from the Italian RON project [15] is shown for H_s and T_p on locations nearby for one winter month period, in Figures 3 and 4.

The WW and RON locations compared in Figures 3 and 4 are about 26 km apart. The time series shows a good match, especially for significant wave heights. General H_s trends match well and peaks coincide. Variations between extremes in Figure 3 can be accounted to distance between the compared locations (with influence of the coastline and surrounding orography), to physical and numerical limitations and settings of the numerical model and measurement buoy properties. Deviations between wave period Tp records are slightly larger than for H_s. Buoy data for wave periods show local "jumps" which could suggest that the buoy data possibly need additional filtering.

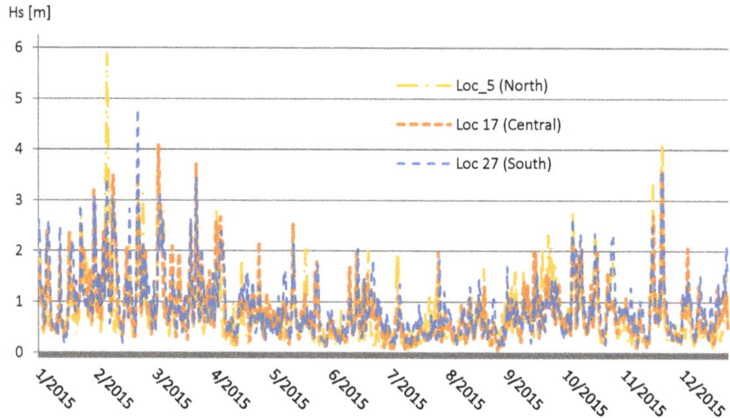

Figure 2. H_s time series during 2015 at three locations, in North, Central and South Adriatic.

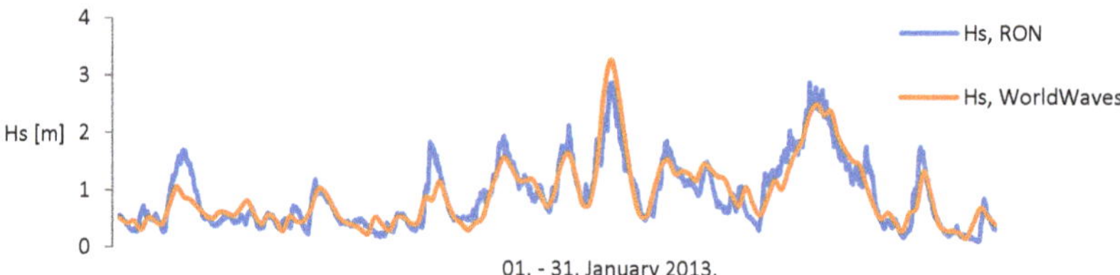

Figure 3. H_s time series from WW database (location 4) and RON buoy "Ancona" (13°43′10″ E–43°49′26″ N).

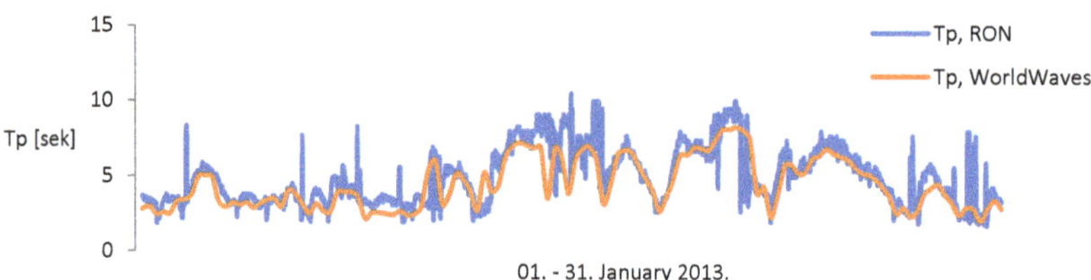

Figure 4. Tp time series from WW database (location 4) and RON buoy "Ancona" (13°43′10″ E–43°49′26″ N).

To prepare the data for analysis, the following frequency of occurrence tables were extracted from the WW database:

1. Sea state tables (H_s–Tp), for the following:
    - The Adriatic Sea with all location merged as presented in Table 1;
    - Adriatic regions as presented in Appendix A (North, Table A4; Central, Table A5; and South, Table A6);
    - Each of the 39 locations individually.
2. Wind speed to significant wave height (u_w–H_s), for the following:
    - The Adriatic Sea with all location merged as presented in Table 2;
    - Adriatic regions (North, Central and South) as presented in Appendix A (North, Table A7; Central, Table A8; and South, Table A9);
    - Each of the 39 locations individually.

Table 1. Sea state table–entire Adriatic Sea basin, period September 1992–January 2016.

T_p/H_s	0.0–0.5	0.5–1.0	1.0–1.5	1.5–2.0	2.0–2.5	2.5–3.0	3.0–3.5	3.5–4.0	4.0–4.5	4.5–5.0	5.0–5.5	5.5–6.0	6.0–6.5	Sum
0–1	0	0	0	0	0	0	0	0	0	0	0	0	0	0
1–2	0	0	0	0	0	0	0	0	0	0	0	0	0	0
2–3	221,186	65,034	754	7	0	0	0	0	0	0	0	0	0	286,981
3–4	149,331	225,790	32,459	1233	39	1	1	0	0	0	0	0	0	408,854
4–5	36,859	130,876	80,830	19,669	1732	109	7	1	0	0	0	0	0	270,083
5–6	17,747	52,597	54,473	38,089	15,605	3194	469	54	3	0	0	0	0	182,231
6–7	10,379	24,193	20,917	17,297	14,389	8836	3255	731	183	32	5	2	1	100,220
7–8	3704	8591	8993	6373	4838	3810	3141	1879	701	215	54	18	10	42,327
8–9	1863	3459	3515	2727	1930	1361	956	677	470	197	78	13	8	17,254
9–10	1174	1055	1115	819	654	495	336	231	156	92	48	26	3	6204
10–11	376	434	420	294	217	120	98	64	40	22	6	2	1	2094
11–12	363	105	107	81	91	31	26	16	10	9	5	2	1	847
12–13	434	39	23	10	8	10	7	7	2	2	1	1	0	544
13–14	0	0	0	0	0	0	0	0	0	0	0	0	0	0
Sum	443,416	512,173	203,606	86,599	39,503	17,967	8296	3660	1565	569	197	64	24	1,317,639

Note. Total of 73% of sea states are less than 1 m.

Table 2. Wind speed—significant wave height occurrence—entire Adriatic, Sept 1992–Jan 2016.

u_w/H_s	0–0.25	0.25–0.5	0.5–0.75	0.75–1	1–1.25	1.25–1.5	1.5–1.75	1.75–2	2–2.25	2.25–2.5	2.5–2.75	2.75–3	3–3.25	3.25–3.5	Sum
0–1	20,070	33,291	10,955	3152	1198	462	187	81	29	26	4	3	1	0	69,459
1–2	32,250	68,242	25,570	7861	2803	1134	453	200	83	23	9	7	2	2	138,639
2–3	27,155	84,153	38,583	12,391	4471	1707	703	313	131	61	23	17	5	3	169,716
3–4	14,064	81,193	52,240	18,013	6548	2608	1067	437	209	86	47	24	7	9	176,552
4–5	5139	54,153	66,294	24,730	9238	3666	1504	614	306	104	70	23	26	8	165,875
5–6	1666	16,761	68,671	35,231	12,912	5099	2161	990	425	209	90	48	16	6	144,285
6–7	393	3321	37,421	45,709	18,845	7541	3211	1419	628	295	132	63	38	17	119,033
7–8	103	964	8754	37,098	26,470	11,372	4919	2082	954	452	232	97	39	26	93,562
8–9	33	310	1519	13,596	25,426	16,301	7712	3415	1534	731	317	160	57	30	71,141
9–10	11	77	416	2830	12,104	16,639	10,722	5572	2490	1196	502	235	96	75	52,965
10–11	6	23	129	625	3206	8879	10,611	7563	3966	1962	815	426	175	112	38,498
11–12	4	16	44	168	770	2781	5704	6793	5306	3051	1556	700	328	155	27,376
12–13	0	10	33	51	161	779	1995	3477	4147	3606	2245	1206	585	249	18,544
13–14	0	6	9	31	66	207	585	1191	2063	2588	2283	1684	975	545	12,233
14–15	0	2	8	15	28	88	192	397	709	1194	1425	1365	1087	698	7208
15–16	0	0	4	9	11	32	69	141	243	398	604	724	823	709	3767
Sum	100,894	342,522	310,650	201,510	124,257	79,295	51,795	34,685	23,223	15,982	10,354	6782	4260	2644	13,088

Additionally, the well-known directionality of higher wave height associated with S-SE winds (*jugo/scirocco*) and NE winds (*bura/bora*), as the Adriatic basin specificities due to the surrounding orography, is confirmed by wave roses. Wave roses for the Adriatic as a whole are presented in Figure 5.

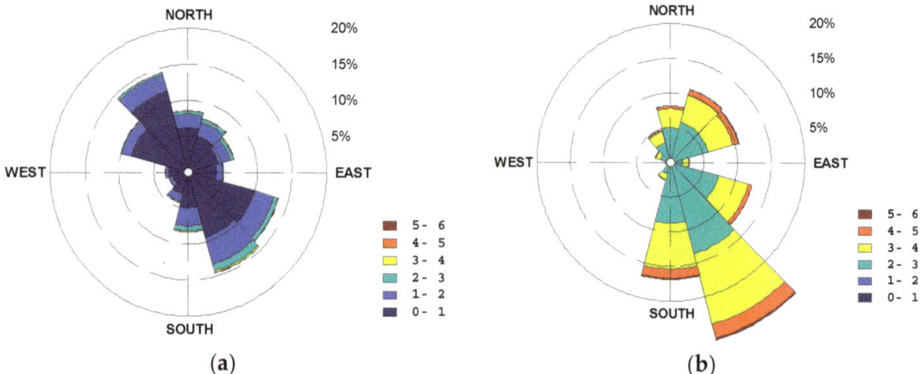

Figure 5. Wave roses for the entire Adriatic basin: (**a**) all sea states and (**b**) sea states with $H_s > 2.5$ m.

There is a noticeable amount of smaller N–NW waves accounted to the same direction (*maestral/maestrale*) wind—a typical daily coastal circulation, caused by temperature oscillation between land and sea during summer months, which is important to the leisure nautical sector due to its predictability and mild character. Regional wave roses are available in Appendix A Figures A1 and A2 and suggest higher dominance of S–SE waves in South and Central Adriatic and NE high wave dominance in the North Adriatic region.

3. Methods

The WW data were analyzed in accordance with Det Norske Veritas (DNV) classification society recommendations for determining environmental conditions and loads on marine structures [2]. A joint distribution is applied, consisting of a marginal three-parameter Weibull distribution of significant wave heights and a conditional log-normal distribution for peak periods. For wind speeds, a conditional distribution is given as function of significant wave height and described by a two-parameter Weibull distribution.

3.1. Joint Distribution of Significant Wave Height and Peak Periods

Based on sea state tables, both regional and per individual location, a joint probability distribution model was derived with the aim to optimize the parameters that will later be used to determine extreme sea states for return periods longer than the database span and to provide a consistent approach for determination of loads for fatigue and strength analyses of ships and offshore structures.

CMA approach (Conditional Modelling Approach) is applied for modelling the joint distribution of significant wave height and peak (spectral) period. In general, this method defines the probability density function (PDF) by a marginal distribution and a set of conditional probability densities, each of which is modelled by a parametric function whose parameters are optimized by mathematical fitting techniques to represent data from the WW database in the best possible way. The CMA approach of fitting joint distribution to the wave data was first proposed by Bitner-Gregersen and Haver [24] and discussed in detail in Reference [25]. It is currently an integral part of standardized engineering procedures (e.g., Reference [2]) and scientific practice (e.g., Reference [26]).

3.1.1. Marginal Distribution of Significant Wave Height

The proposed CMA model uses the three-parameter Weibull distribution to describe a PDF of significant wave height H_s as first proposed in Reference [27].

$$f_{H_S}(\hat{H}_s) = \frac{\beta_{H_s}}{\alpha_{H_s}} \left(\frac{\hat{H}_s - \gamma_{H_s}}{\alpha_{H_s}} \right)^{\beta_{H_s}-1} exp\left\{ -\left(\frac{\hat{H}_s - \gamma_{H_s}}{\alpha_{H_s}} \right)^{\beta_{H_s}} \right\} \tag{1}$$

where α_{H_s} is the scaling parameter, β_{H_s} is the shape parameter and γ_{H_s} is the location parameter. The parameters are optimized on linearized scale with the least square method (LSM). The cumulative 3-parameter Weibull probability density function (CDF) then follows:

$$F(H_S) = P(\hat{H}_s < H_S) = 1 - exp\left[-\frac{(\hat{H}_s - \gamma_{H_s})}{\alpha_{H_s}} \right]^{\beta_{H_s}}, \qquad H_S, \hat{H}_s \geq \gamma_{H_s} \tag{2}$$

where $P(\hat{H}_s < H_S)$ represents the probability that a certain random significant wave height $\hat{H}_s$ will take on a value less than H_S. Probability of exceedance is than given as

$$Q(\hat{H}_s) = 1 - P(\hat{H}_s) \tag{3}$$

When a large number of observations are available, as in the WW database, a common approach is to sort the data into $H_{s,i}$ bins, as presented in the sea-state tables, e.g., Table 1. The empirical probability of exceeding of each bin is then determined by the usual expression [13]:

$$Q(H_{s,i}) = \frac{\sum_{j=1}^{i} f_j}{N+1} \tag{4}$$

where $\sum_{j=1}^{i} f_j$ represents the cumulative frequency of all values equal to or greater than $H_{s,i}$, while N is the total number of observations (the sum of all observations as shown in the sea state tables).

The theoretical probability $P(\hat{H}_{s,i})$ that $H_{s,i}$ will not be exceeded can be determined according to Equation (2). To fit the theoretical distribution to the empirical points, calculated from the database, the distribution parameters α_{H_s} and β_{H_s} were optimized using the least squares method on a linearized double-log scale.

$$ln\left(-ln\left(F\left((\hat{H}_s)\right)\right)\right) = \beta_{H_s} * ln\left((\hat{H}_s - \gamma_{H_s})\right) - \alpha_{H_s} * ln(\beta_{H_s}) \tag{5}$$

Shape and scale parameters are determined from the linearized model coefficients (y_1–slope and y_0–ordinate intersection) according to the following:

$$\alpha_{H_s} = y_1 \tag{6}$$

$$\beta_{H_s} = e^{-\frac{y_0}{y_1}} \tag{7}$$

The choice of the third parameter—the proper threshold parameter γ_{H_s}—whilst permitting some data points to lay below is an anomaly that is sometimes dealt with by discarding (censoring) smallest empirical CDF data points prior to fitting the theoretical CDF. For example, suggestions given in Reference [5], in order to formalize such an approach, recommended discarding the data points corresponding to the probability level $F = 0.2$ and below, but simultaneously argued that such criterion does not work equally well on all datasets.

The threshold parameter γ_{H_s} in this paper is chosen without discarding any data and in accordance with the procedure given in Reference [28]. The procedure concept is to test different values of γ_{H_s} which then obviously affects the quality of the fit. Since the expected relationship in Equation (5) is expected to be linear the assumption is that the optimal threshold parameter γ_{Hs} will provide the best possible approximation to a linear model. This argument is formalized by applying an optimization algorithm to maximize the coefficient of determination R^2, as a statistical measure of accuracy, of a linear regression on the transformed variables $ln\left(\left(\hat{H}_s - \gamma_{Hs}\right)\right.$ and $ln\left(-ln\left(F\left(\left(\hat{H}_s\right)\right)\right)\right)$ across all possible threshold values.

It was also noticed that the choice of the bin size used to calculate the empirical probability of exceedance in Equation (4) has a significant influence on the theoretical model because it determines the resolution of the empirical CDF data points on which the theoretical model is fitted. A too-coarse resolution loses precision and increases error, while too fine resolution results in sporadic empty bins in the upper *Hs* range thus creating false points of the empirical cumulative density function that would influence the results and respective evaluated extremes. After initial testing of model behavior on the analyzed dataset, the bin size was determined by defining 30 equally spaced bins between the minimum and the maximum recorded value on each location separately.

3.1.2. Conditional Distribution of Peak Periods Depending on Significant Wave Height

The conditional distribution of peak period T_p based on H_s is modelled by the log-normal PDF, as first proposed by Bitner-Gregersen and Haver in Reference [24] (see also Reference [29]):

$$f_{T_P|H_S}\left(\hat{T}_p|\hat{H}_s\right) = \frac{1}{\sigma\,\hat{T}_p\,\sqrt{2\,\pi}}exp\left\{-\frac{\left(ln\hat{T}_p - \mu\right)^2}{2\sigma^2}\right\} \tag{8}$$

where the distribution parameters μ is the mean logarithmic value of the variable $\mu = E\left[lnT_p\right]$, and σ is the standard deviation of the variable logarithmic value $\sigma = std\left[lnT_p\right]$ [16]. Both parameters are, easy to calculate statistical quantities, dependent on H_s. Since H_s is divided into bins, according to Table 1, the change of σ and μ calculated for each bin is modelled as follows:

$$\mu = a_0 lnH_s + a_1 \tag{9}$$

$$\sigma = a_2 * lnH_s + a_3 \tag{10}$$

Coefficients a_0, a_1, a_2 and a_3 are calculated for each location individually, regionally and for the entire Adriatic Sea.

3.2. Joint Distribution of Significant Wave Height and Wind Speed

Like joint distribution of significant wave height and peak period, a statistical model for joint distribution of significant wave height and wind speed is developed. The wind speed u_w is defined as a variable dependent on the significant wave height, H_s. Although physically inversing the cause and consequence, such description, together with results from Section 3.1, presents a complete metocean description dependent on a single variable, i.e., H_s. Wind speed and direction data are an integral part of the underlying WW database and refer to the data used by ECMWF to force WAM model. Wind speeds are given at 10 m above sea level with an assumed duration of 6 h (as for the corresponding sea state).

The conditional distribution of the wind speed as a function of the significant wave height can be described by the two-parameter Weibull distribution as proposed by Bitner-Gregersen and Haver [24] and included in DNV RP C-205 [2]:

$$f_{U|H_S}(u_w|H_s) = k\frac{u_w^{k-1}}{U_c^k}exp\left[-\left(\frac{u_w}{U_c}\right)^k\right] \tag{11}$$

where the scale parameter U_c and the shape parameter k are estimated from the available data, using the following model:

$$k = c_1 + c_2 H_s^{c3} \quad U_c = c_4 + c_5 H_s^{c6} \tag{12}$$

3.3. Extremes Values of Significant Wave Height for Long Return Periods

Extreme H_s values, for return periods (RP) longer than the scope of the WW database, can been evaluated based on using the three-parameter Weibull distribution fit described in Section 3.1.1. The fitted distribution upper tail is extrapolated to theoretical probability of exceedance $Q(H_s^{RP})$ of a certain H_s value and return period RP. Probability of exceedance $Q(H_s^{RP})$ is generally determined as follows:

$$Q\left(H_s^{RP}\right) = \frac{T_{REG}}{N} * \frac{1}{RP} \tag{13}$$

where T_{REG} is the duration of uninterrupted observations within the database (23.5 year), and N is the total number of data records. Once the 3-parameter Weibull CDF (as given in Equation (2)) is fitted to data, the significant wave height that will be exceeded once for certain return period can be determined as its inverse:

$$H_s^{RP} = \alpha_{Hs} * \left(-ln\left(Q\left(H_s^{RP}\right)\right)\right)^{1/\beta_{Hs}} + \gamma_{Hs} \tag{14}$$

4. Results

Within this section, models parameters, as described in Section 3, fitted to data are presented graphically for brevity and the same tabulated data are presented in Appendix B.

4.1. Parameters of the Joint Distribution of Significant Wave Height and Peak Wave Periods

The three-parameter Weibull distribution parameters were fitted for each of the 39 location, for data merged according to the regional subdivision and for the all location merged, i.e., the entire Adriatic Sea. An example fitting on a linearized scale, as per Equation (5), is presented in Figure 6 for location 9, where the maximum H_s was recorded within the database.

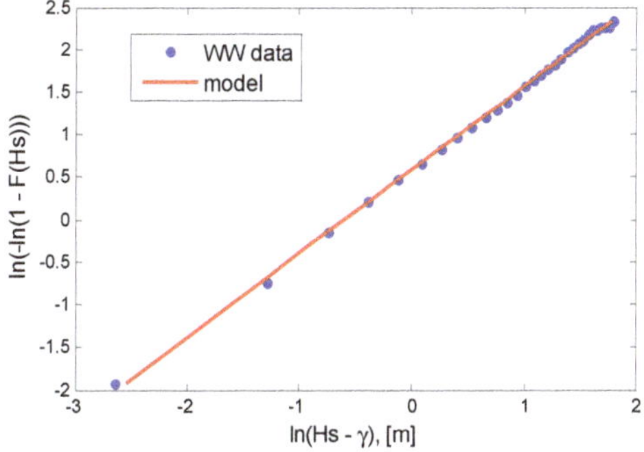

Figure 6. Parameter Weibull parameter fit for marginal distribution of *Hs*. (E 14.5–N 44.0).

The fit presented in Figure 6 shows an example validation of the model. All 39 location fits were visually inspected, and the calculated coefficient of determination, R^2, ranging between $R^2_{min} = 0.9971$ and $R^2_{max} = 0.9996$, confirms that the model is appropriate. Likewise, an example fit is presented in Figure 7 of the log-normal distribution fit for a conditional distribution of peak periods dependent on significant wave height.

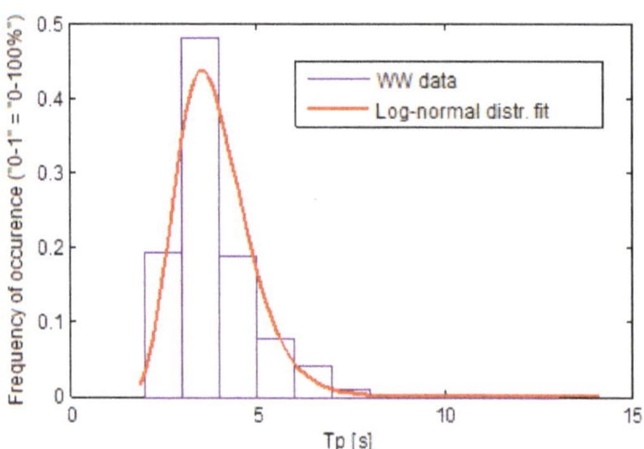

Figure 7. Log-normal conditional distribution fit of *Tp* on *Hs*, e.g., H_s = 0.5–1.0 m; E 14.5–N 44.0.

The sea state, H_s and T_p description are completed by determining the mean and the standard deviation values that define the log-normal distribution for the entire bin range of H_s, as presented in Figure 8.

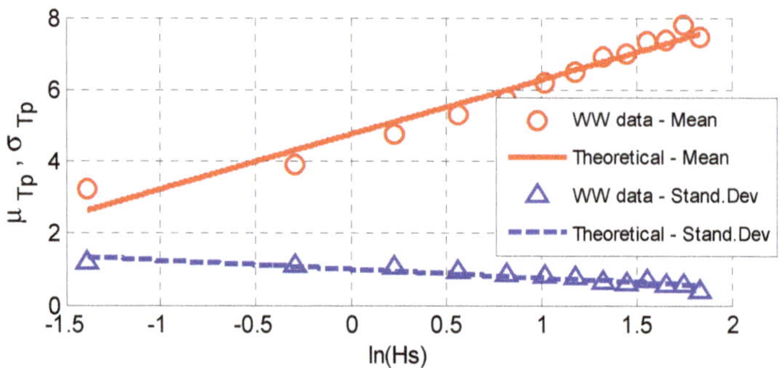

Figure 8. Mean and standard deviation for calculation of *Tp* distribution across *Hs* bin range; E 14.5–N 44.0.

The model parameters for the joint distribution of significant wave height and peak period are finally presented in Table 3 for regions, and Figures 9 and 10 for individual locations.

Table 3. Model parameters for joint distribution of *Hs* and *Tp*. Adriatic regional subdivision.

Region	α_{Hs}	β_{Hs}	γ_{Hs}	a_0	a_1	a_2	a_3
Adriatic Sea	0.553	0.987	0.071	1.7862	5.1201	−0.2003	1.0634
North Adriatic	0.527	0.955	0.076	1.6161	4.8126	−0.2876	1.0508
Central Adriatic	0.659	1.091	0.068	1.8676	5.0563	−0.1995	0.9285
South Adriatic	0.762	1.166	0.091	1.9353	5.2776	−0.1492	1.0775

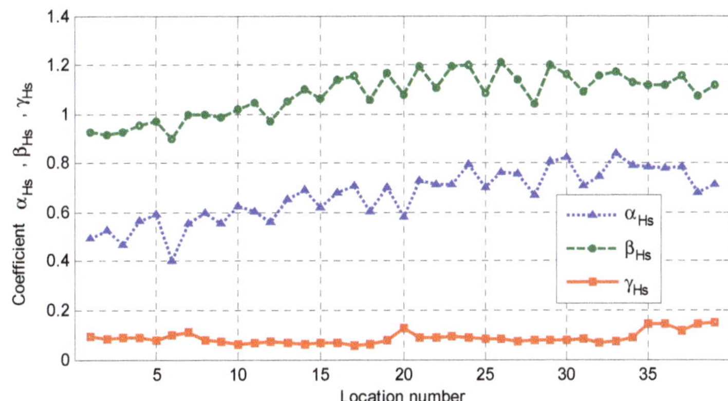

Figure 9. Model parameters for marginal distribution of significant wave height per location.

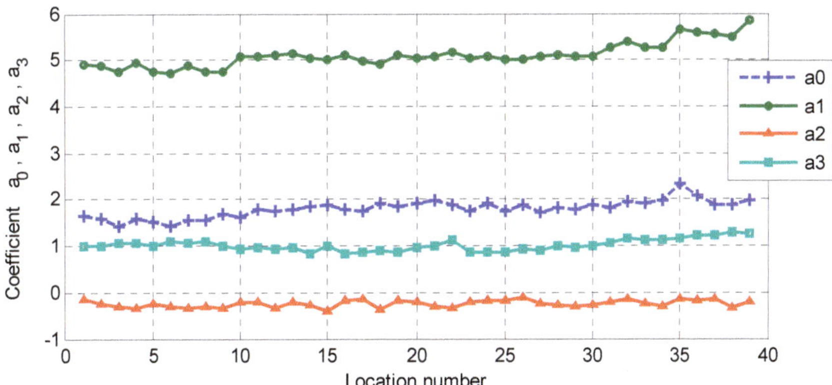

Figure 10. Model parameters for mean and standard deviation across *Hs* bin range to model log-normal distribution of peak period *Tp*-per location.

For precision, and to enable repeatability and practical usage, numerical parameters for individual locations are given in Appendix B Table A10.

4.2. Parameters of the Joint Distribution of Wind Speed and Significant Wave Height

For H_s bins, the theoretical distribution of wind speeds proposed by the model in Equation (11) is derived. The model parameters k and U_c are optimized using the nonlinear least squares method. An example fit of the two-parameter Weibull distribution of wind speed for bin H_s = 2.25–2.5 m for the entire Adriatic is shown in Figure 11.

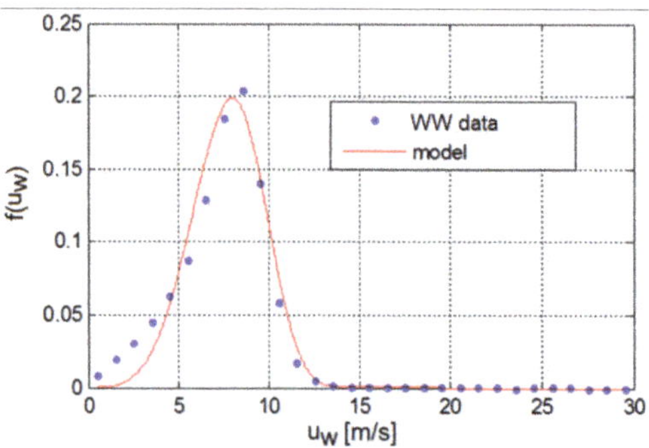

Figure 11. Fitting the model to WW data; entire Adriatic; Hs,i = 2.25–2.5 m.

By repeating the same procedure for each $H_{s,i}$ bin, a model of fit parameters k and U_c is obtained. Their results for the entire Adriatic are presented in Figure 12 as a function of H_s.

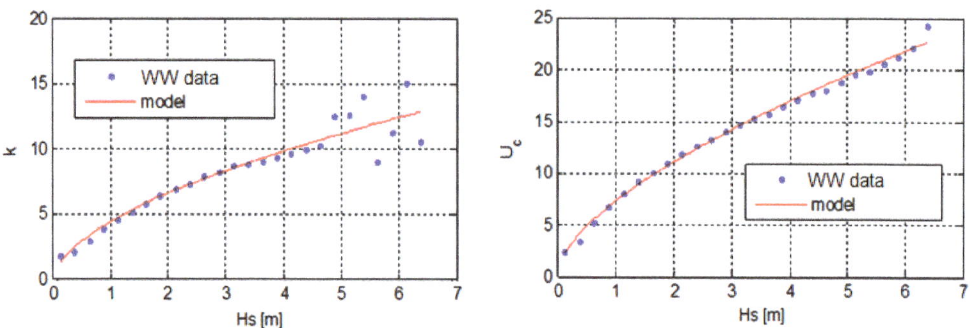

Figure 12. Distribution of k and Uc for Adriatic as a whole.

A data scattering of WW data points for the shape parameter k can be seen for higher Hs values. This feature is even more pronounced having a more detailed look for certain locations and is due to small amount of data at high Hs. Poor agreement of the statistical model at higher values also has a negative effect on poorer agreement of the model with data at lower values. It was found that more than 99.5% of the recorded data are usually below significant wave height of 3.25–3.75 m. In order to achieve a better agreement data corresponding to the highest 0.5% Hs were discarded, both for individual locations analysis and grouped data, having in mind that such filtering makes the model acceptable for fatigue or seakeeping considerations of offshore structures but not for the extreme value analysis.

The model parameters for the joint distribution of wind speed and significant wave height are finally presented in Table 4 for regions, and Figures 13 and 14 for individual locations.

Table 4. Model parameters for joint distribution of Hs and u_w. Adriatic regional subdivision.

Region	c_1	c_2	c_3	c_4	c_5	c_6
Adriatic Sea	1.6533	2.2548	1.1557	−0.3633	7.7434	0.5817
North Adriatic	1.6910	1.8402	1.2730	−0.0925	7.4936	0.6245
Central Adriatic	1.4848	2.5383	1.0783	−0.2775	7.7063	0.5978
South Adriatic	1.5796	2.5238	1.0601	−0.2048	7.5145	0.5886

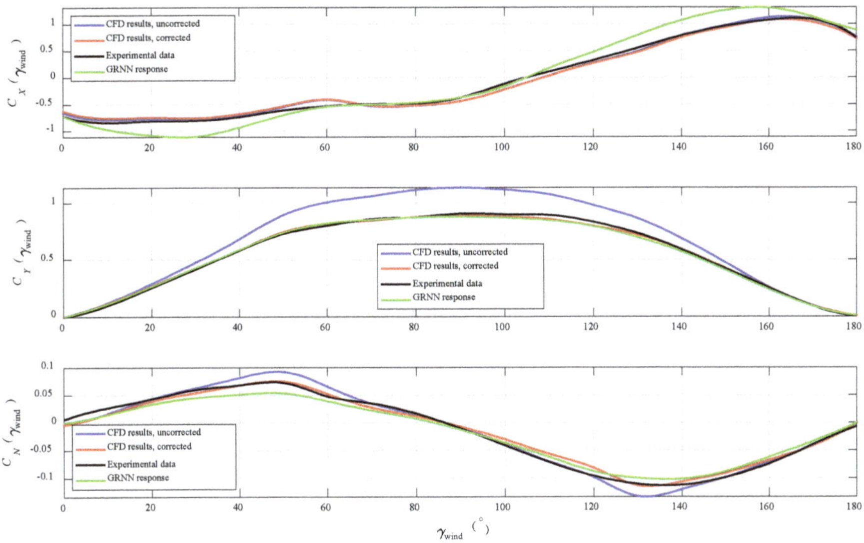

Figure 9. Wind loads coefficients for configuration #3.

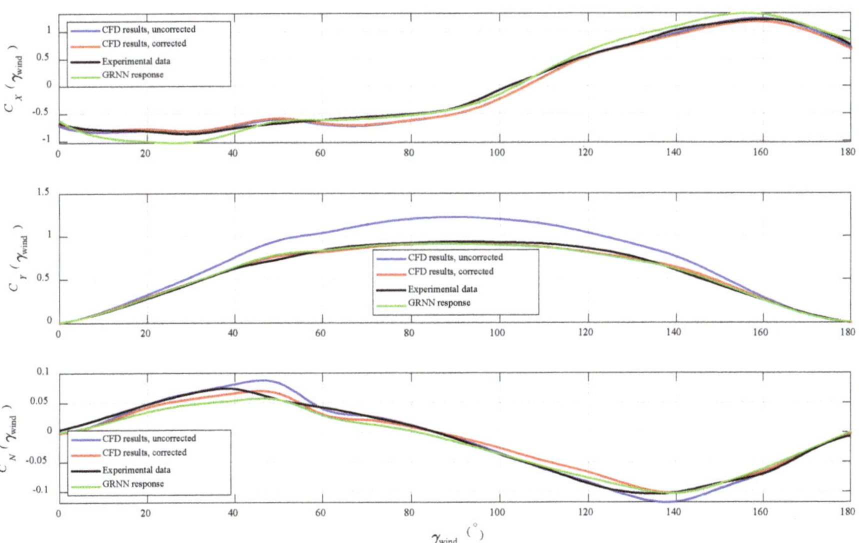

Figure 10. Wind loads coefficients for configuration #4.

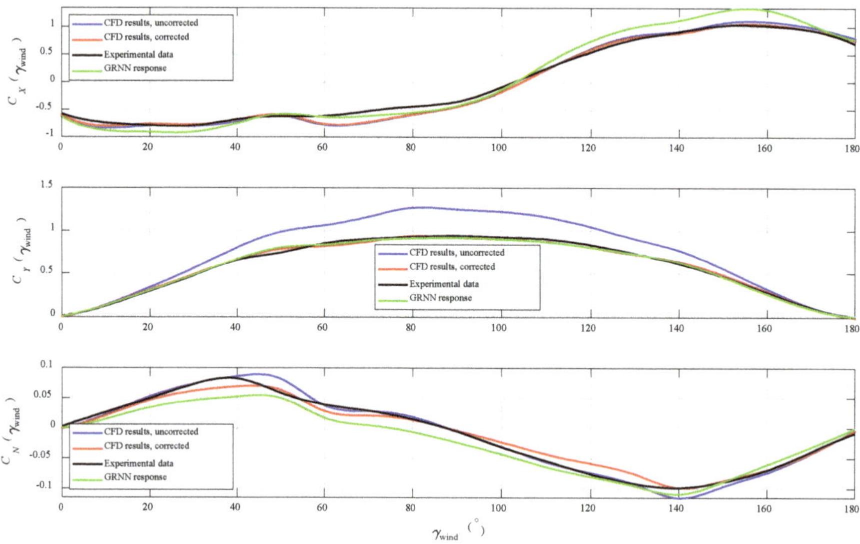

Figure 11. Wind loads coefficients for configuration #5.

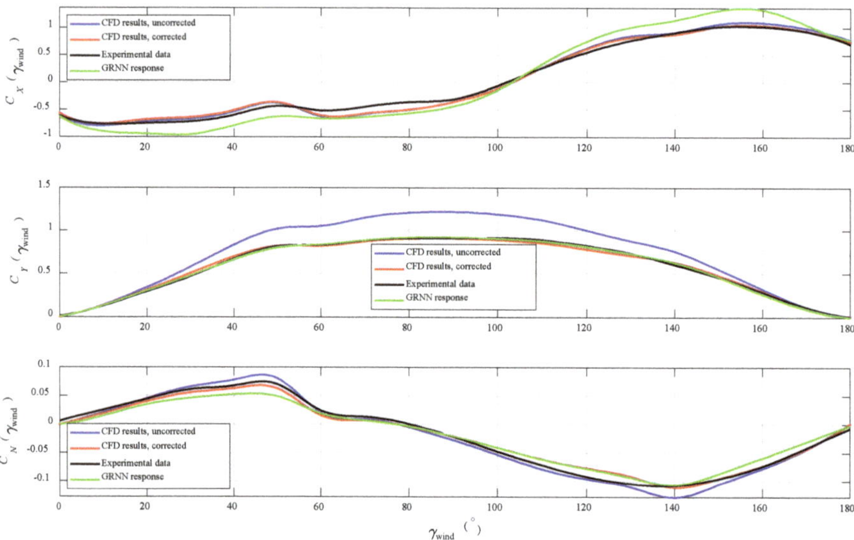

Figure 12. Wind loads coefficients for configuration #6.

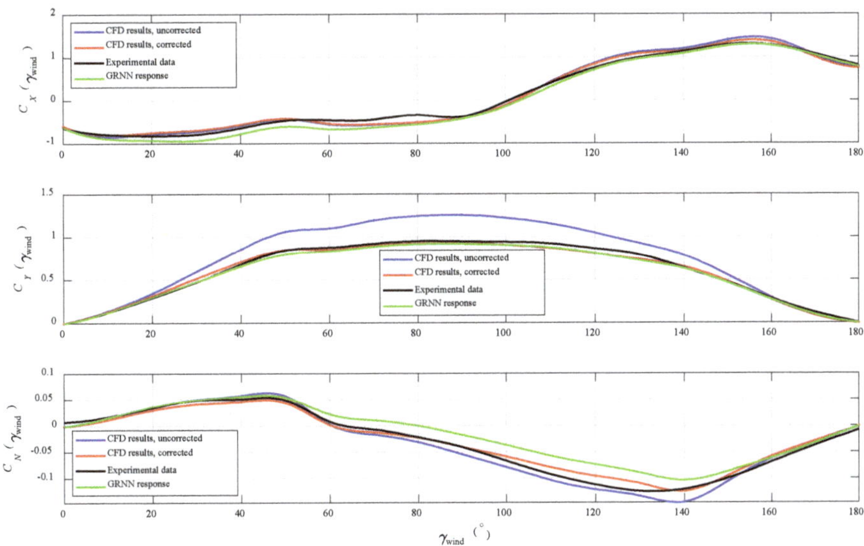

Figure 13. Wind loads coefficients for configuration #7.

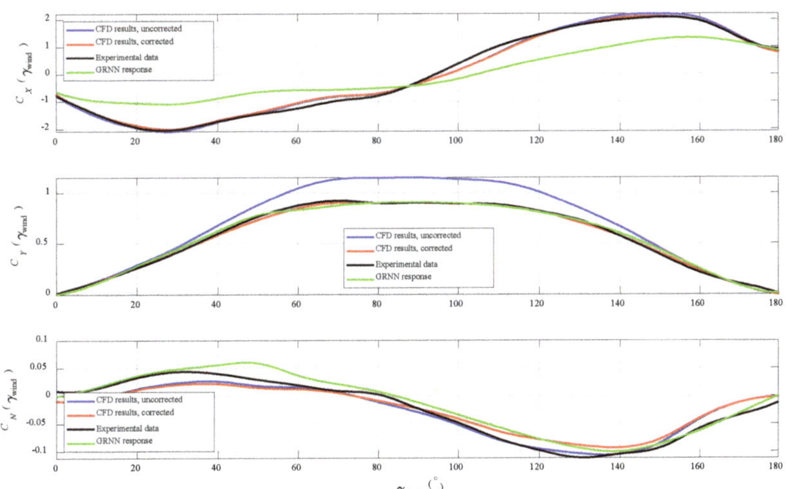

Figure 14. Wind loads coefficients for configuration #8.

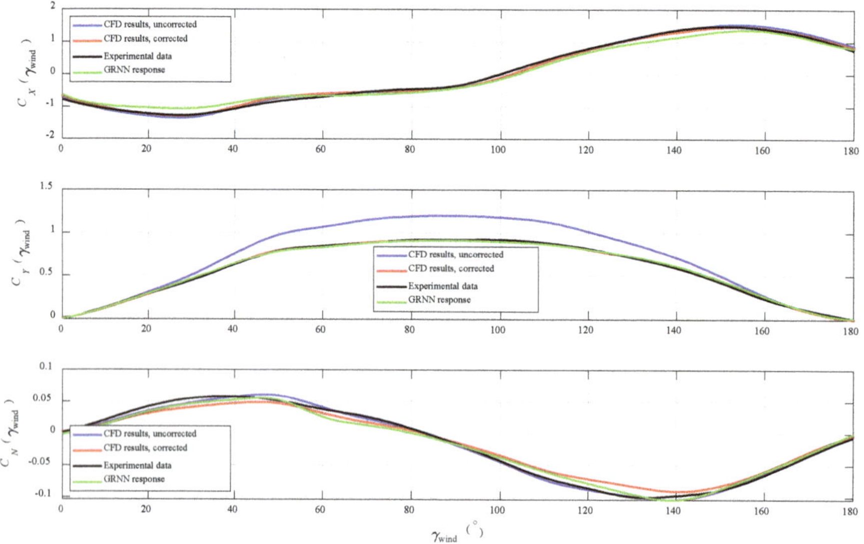

Figure 15. Wind loads coefficients for configuration #9.

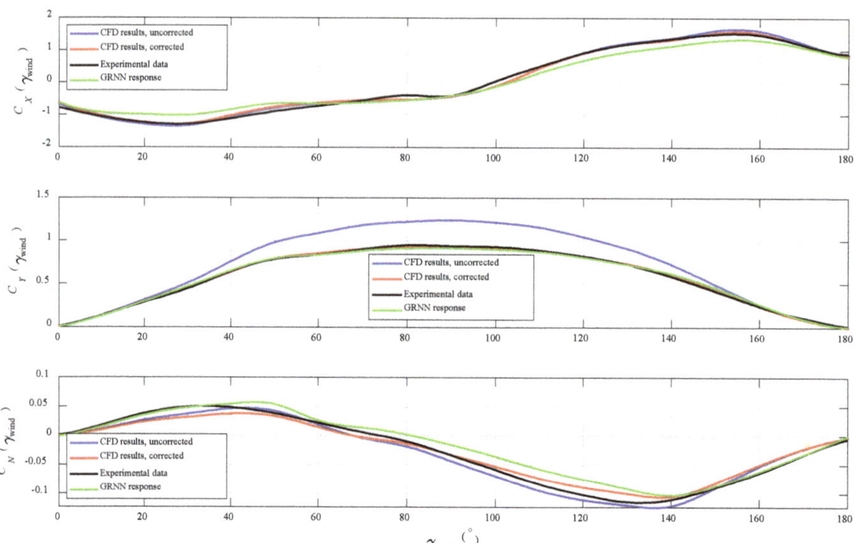

Figure 16. Wind loads coefficients for configuration #10.

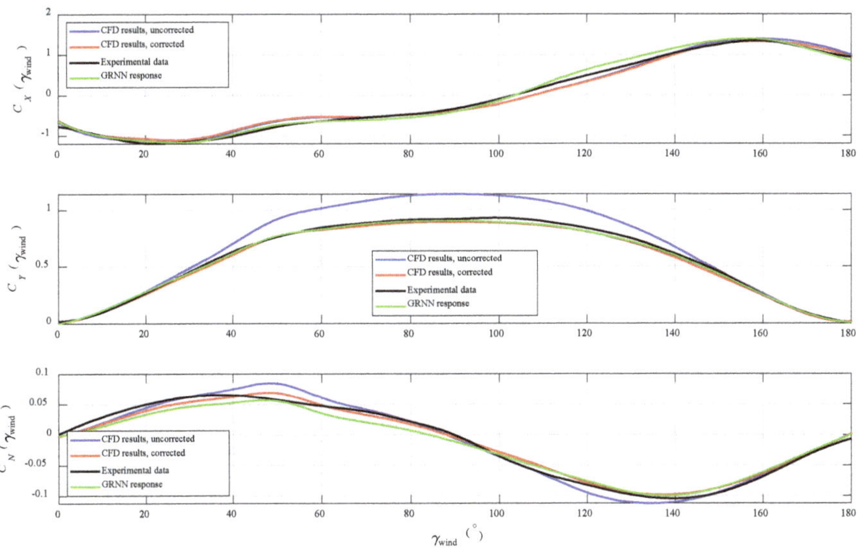

Figure 17. Wind loads coefficients for configuration #11.

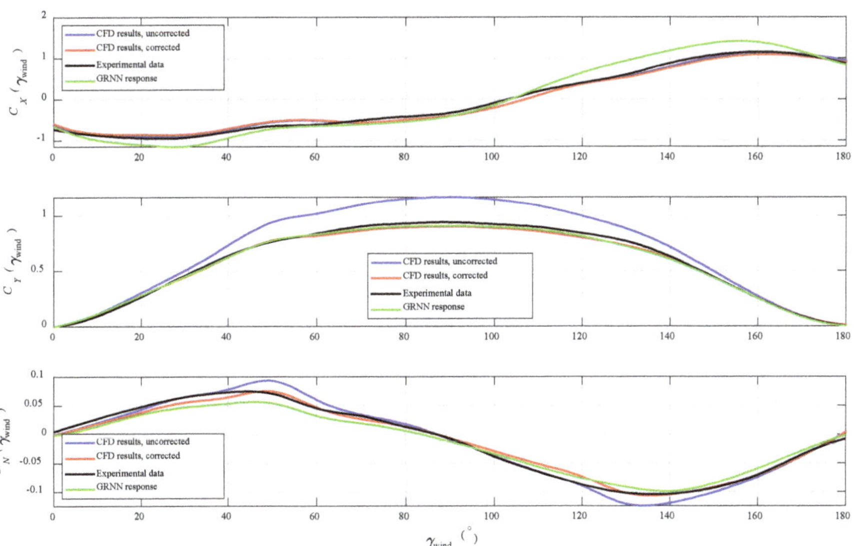

Figure 18. Wind loads coefficients for configuration #12.

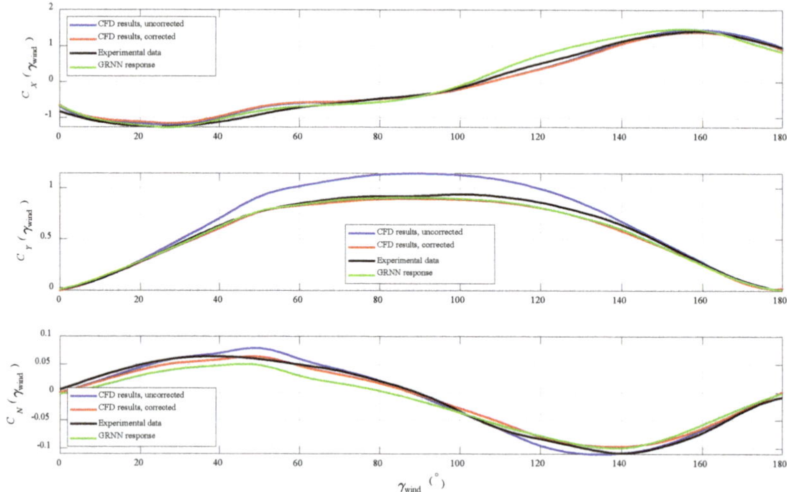

Figure 19. Wind loads coefficients for configuration #13.

Table 2. Comparison of obtained results in terms of mean values (μ_i) and standard deviations (σ_i) of absolute differences of wind load coefficients in cases A (CFD vs. data) and B (GRNN vs. CFD).

t	A: CFD Results vs. Experimental Data			B: GRNN Responses vs. CFD Results		
	$\mu_X(C_{X,t}^{CFD}-C_{X,t}^{data})$ $\sigma_X(C_{X,t}^{CFD}-C_{X,t}^{data})$	$\mu_Y(C_{Y,t}^{CFD}-C_{Y,t}^{data})$ $\sigma_Y(C_{Y,t}^{CFD}-C_{Y,t}^{data})$	$\mu_N(C_{N,t}^{CFD}-C_{N,t}^{data})$ $\sigma_N(C_{N,t}^{CFD}-C_{N,t}^{data})$	$\mu_X(C_{X,t}^{GRNN}-C_{X,t}^{CFD})$ $\sigma_X(C_{X,t}^{GRNN}-C_{X,t}^{CFD})$	$\mu_Y(C_{Y,t}^{GRNN}-C_{Y,t}^{CFD})$ $\sigma_Y(C_{Y,t}^{GRNN}-C_{Y,t}^{CFD})$	$\mu_N(C_{N,t}^{GRNN}-C_{N,t}^{CFD})$ $\sigma_N(C_{N,t}^{GRNN}-C_{N,t}^{CFD})$
1	0.006830 0.077113	0.000001 0.022939	0.001013 0.007878	−0.018669 0.083724	−0.013219 0.008449	0.001119 0.004969
2	0.028897 0.070867	−0.031025 0.029837	0.000250 0.006619	0.019692 0.156943	0.036904 0.021176	0.005489 0.006890
3	−0.001519 0.067631	−0.002002 0.016848	0.001721 0.007465	0.021038 0.217758	−0.005834 0.007852	−0.003374 0.008509
4	−0.040542 0.069472	−0.001895 0.024665	0.001351 0.008477	0.039521 0.126639	−0.001958 0.011143	−0.005018 0.004919
5	−0.027537 0.071166	0.000302 0.021155	0.000559 0.008445	0.066050 0.124721	−0.003438 0.013194	−0.011163 0.009887
6	0.004378 0.066045	0.000507 0.021938	0.000101 0.007114	−0.011767 0.190404	−0.001516 0.020347	−0.000170 0.007039
7	0.009061 0.079468	−0.009399 0.028647	0.002226 0.007970	−0.095680 0.076000	−0.013755 0.017702	0.013060 0.009611
8	0.015437 0.112929	−0.004332 0.014446	0.001051 0.016156	−0.032795 0.642980	0.008992 0.020995	0.009923 0.017630
9	−0.000038 0.061642	−0.000605 0.010249	0.001022 0.008962	−0.020133 0.106340	0.001654 0.008042	−0.002510 0.005232
10	0.006541 0.072194	−0.000360 0.012062	0.000443 0.009953	−0.028850 0.157382	−0.001015 0.014933	0.008996 0.009849
11	0.008620 0.090098	−0.021369 0.014297	0.000902 0.006874	0.020806 0.128550	0.009962 0.006351	−0.004635 0.005230
12	−0.012284 0.079605	−0.015912 0.020094	0.000107 0.007031	0.042008 0.231174	0.003056 0.007483	−0.002915 0.008500
13	0.025503 0.103171	−0.026480 0.019025	0.001303 0.007147	0.035129 0.163835	0.008007 0.006242	−0.006211 0.007112

5. Discussion

An enhanced methodology for estimating the wind loads on container ships is presented. Ship frontal and lateral projected areas, that is, their associated closed contours, are represented with elliptic Fourier descriptors. EFDs of closed contours and wind load data derived from CFD analysis are used for GRNN training. As it can be seen in the previous sections, particularly in Chapter 4, very limited number of container configurations were available for conducting this analysis. Therefore, natural choice was to select the so-called leave-one-out cross-validation technique. With this approach, training was performed the number of times that is equal to the number of different container configurations that were analyzed. During this analysis, with each consecutive training, a different container configuration was leaved out for testing, while all other were used for training. The validation of CFD model is based on 13 different container configurations of 9000+ container ship for which the experimental results are provided by Andersen [11]. As indicated, the GRNN was tested on before mentioned 13 different container configurations in a way that the network is trained with the set of data obtained for 12 configurations and the remaining one is used for testing. Taking into account limited input data used for network training, the results show good agreement with the experimental data obtained in wind tunnels.

This approach takes into account all aspects of the variability of the above water frontal and lateral ship profile but with limitation related to the main frontal projection that cannot capture the variability of longitudinal cross sections. However, it is very suitable for the assessment of wind loads on container ships wherever a wind load data for similar ships with various container configurations are available from wind tunnels or are obtained by CFD analysis. In this case, whereas the validity of the CFD analysis has been proven on experimental data, all configurations of interest can be analysed with CFD analysis and larger database can be created. In this way, the cheaper and faster calculation can fill the gap between ship shapes for which calculations or experiments are performed. It is reasonable to assume that the network trained with larger training data set will provide results that are more accurate. Moreover, for the estimation of real wind-induced forces and moment acting on a container ship in open sea-like conditions, the network should be trained with uncorrected values of wind loads coefficients. In other words, the important focus should be put on the initial wind conditions during CFD simulations, because GRNN will be trained with these CFD results and consequently will be trained for some specific wind conditions. Regardless of the conditions under which it was trained, in actual applications of wind loads estimation, trained GRNN requires only simple data such as wind speed and direction, which can be obtained using the wind sensor (anemometer) and areas of frontal and lateral projections that can be easily determined. In other words, it does not require additional wind tunnel testing or CFD simulations, even in a case of completely new configurations that were not used in the training phase. However, if tunnel testing or CFD simulations can be performed and wind load coefficients can be obtained for these "new configurations," GRNN can be easily and quickly retrained and thus providing higher accuracy and reliability for future estimations.

Further research should be aimed toward uncertainties and reliability analysis, as well to increasing the number of container configurations by means of verified CFD analysis in order to enlarge the available database for neural network training. The selection of neural network type and number of harmonics could also affect the results, so they should be further analyzed as well. The next direction for further research will aim towards the development of a pseudo 3D approach to wind loads estimation on a container ship. This approach will be based on EFDs that are used for ship frontal and lateral closed contour representation of ship cross sections. In this way, all aspects of the 3D variability of the above-water frontal and lateral ship profile will be taken into account.

Author Contributions: Conceptualization, J.P.-O., Z.Č. and M.V.; methodology, J.P.-O., Z.Č. and M.V.; software, Z.Č. and M.V.; validation, J.P.-O., Z.Č. and M.V.; formal analysis, Z.Č. and M.V.; investigation, J.P.-O. and M.V.; resources J.P.-O. and M.V.; data curation, Z.Č. and M.V.; writing, original draft preparation, J.P.-O.; writing, review and editing, J.P.-O., Z.Č. and M.V.; visualization, Z.Č. and M.V.; supervision, J.P.-O.; project administration, J.P.-O. and M.V.; funding acquisition, J.P.-O. All authors have read and agreed to the published version of the manuscript.

Funding: This work was fully supported by the Croatian Science Foundation under the project IP-2018-01-3739. This work was also supported by the University of Rijeka (project no. uniri-tehnic-18-18 1146 and uniri-tehnic-18-266 6469).

J. Mar. Sci. Eng. **2020**, *8*, 539

Conflicts of Interest: The authors declare no conflict of interest.

Abbreviations

The following abbreviations are used in this manuscript:

CAD	Computer-Aided Design
CFD	Computational Fluid Dynamics
EFD	Elliptic Fourier Descriptors
GRNN	Generalized Regression Neural Network
LNG	Liquefied Natural Gas
NED	North-East-Down
NN	Neural Network
RANS	Reynolds-averaged Navier–Stokes

References

1. Isherwood, R.M. Wind resistance of merchant ships. *R. Inst. Nav. Archit.* **1972**, *114*, 327–338.
2. Gould, R.W.F. *The Estimation of Wind Loads on Ship Superstructures*; Monograph, No. 8; The Royal Institution of Naval Architects: London, UK, 1982; p. 34.
3. Blendermann, W. *Schiffsform und Windlast-Korrelations-und Regressionanalyse von Windkanalmes-Sungen am Modell*; Report No. 533; Institut fur Schiffbau der Universitat Hamburg: Hamburg, Germany, 1993; 132p.
4. Blendermann, W. Parameter identification of wind loads on ships. *J. Wind Eng. Ind. Aerodyn.* **1994**, *51*, 339–351. [CrossRef]
5. Blendermann, W. Estimation of wind loads on ships in wind with a strong gradient. In Proceedings of the 14th International Conference on Offshore Mechanics and Arctic Engineering (OMAE), New York, NY, USA, 18–22 June 1995; ASME: New York, NY, USA, 1995; Volume 1-A, pp. 271–277.
6. Blendermann, W. Floating docks: Prediction of wind loads in extreme winds. *Schiff Hafen* **1996**, *48*, 67–70.
7. Haddara, M.R.; Guedes Soares, C. Wind loads on marine structures. *Mar. Struct.* **1999**, *12*, 199–209. [CrossRef]
8. Brizzolara, S.; Rizzuto, E. Wind heeling moments on very large ships. Some insights through CFD results. In Proceedings of the 9th International Conference on Stability of Ships and Ocean Vehicles, Rio de Janeiro, Brazil, 25–29 September 2006; pp. 781–793.
9. Wnek, A.D.; Guedes Soares, C. CFD assessment of the wind loads on an LNG carrier and floating platform models. *Ocean. Eng.* **2015**, *97*, 30–36. [CrossRef]
10. Janssen, W.D.; Blocken, B.; van Wijhe, H.J. CFD simulations of wind loads on a container ship: Validation and impact of geometrical simplifications. *J. Wind. Eng. Ind. Aerodyn.* **2017**, *166*, 106–116. [CrossRef]
11. Andersen, I.M.V. Wind loads on post-panamax containership. *Ocean Eng.* **2013**, *58*, 115–134. [CrossRef]
12. Valčić, M.; Prpić-Oršić, J. Hybrid method for estimating wind loads on ships based on elliptic Fourier analysis and radial basis neural networks. *Ocean Eng.* **2016**, *122*, 227–240. [CrossRef]
13. Prpić-Oršić, J.; Valčić, M.; Vučinić, D. Application of pattern recognition method for estimating wind loads on ships and marine objects. In *Materialwissenschaft und Werkstofftechnik—Material Science and Engineering Technology*; WILEY-VCH Verlag GmbH & Co. KGaA: Weinheim, Germany, 2017; Volume 48, pp. 387–400.
14. Valčić, M.; Prpić-Oršić, J.; Vučinić, D. Application of pattern recognition method for estimating wind loads on ships and marine objects. In *Visual Computing—Advancing Engineering Practice*; Series Lecture Notes in Mechanical Engineering; Springer: Singapore, 2019; 38p.
15. Prpić-Oršić, J.; Valčić, M. Sensitivity analysis of wind load estimation method based on elliptic Fourier descriptors. In Proceedings of the Maritime Technology and Engineering—MARTECH 2016, Lisabon, Portugal, 4–6 July 2016; pp. 151–160.
16. Specht, D.F. A General Regression Neural Network. *IEEE Trans. Neural Netw.* **1991**, *2*, 568–576. [CrossRef] [PubMed]
17. Freeman, H. Computer Processing of Line-Drawing Images. *Comput. Surv.* **1974**, *6*, 57–97. [CrossRef]
18. Granlund, G.H. Fourier Preprocessing for Hand Print Character Recognition. *IEEE Trans. Comput.* **1972**, *21*, 195–201. [CrossRef]
19. Pavlidis, T. *Algorithms for Graphics and Image Processing*; Springer: Berlin/Heidelberg, Germany, 1982.

J. Mar. Sci. Eng. **2020**, *8*, 539

20. Nixon, M.S.; Aguado, A.S. *Feature Extraction and Image Processing*, 2nd ed.; Academic Press (Elsevier): London, UK, 2008.
21. Kuhl, F.P.; Giardina, C.R. Elliptic Fourier Features of a Closed Contour. *Comput. Graph. Image Process.* **1982**, *18*, 236–258. [CrossRef]
22. Fossen, T.I. *Handbook of Marine Craft Hydrodynamics and Motion Control*; John Wiley & Sons Ltd.: Chichester, UK, 2011.
23. Norwegian Maritime Directory. Regulations for Mobile Offshore Units. 1997. Available online: http://www.sjofartsdir.no/en/Legislation_and_International_Relations/Translated_Norwegian_legislation/Regulations-for-Mobile-Offshore-Units (accessed on 20 June 2020).
24. Wieringa, J. Updating the Davenport roughness classification. *J. Wind Eng. Ind. Aerodyn.* **1992**, *41*, 357–368. [CrossRef]
25. Blocken, B.; Carmeliet, J.; Stathopoulos, T. CFD evaluation of the wind speed conditions in passages between buildings—Effect of wall-function roughness modifications on the atmospheric boundary layer flow. *J. Wind Eng. Ind. Aerodyn.* **2007**, *95*, 941–962. [CrossRef]
26. Ferziger, J.H.; Peric, M. *Computational Methods for Fluid Dynamics*; Springer Science & Business Media: Berlin/Heidelberg, Germany, 2012.

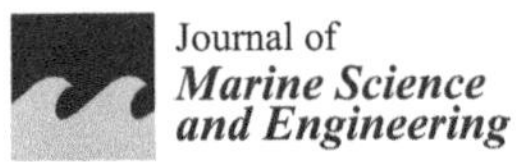

Journal of
Marine Science and Engineering

MDPI

Article

Method for Prediction of Extreme Wave Loads Based on Ship Operability Analysis Using Hindcast Wave Database

Tamara Petranović [1], Antonio Mikulić [1], Marko Katalinić [2], Maro Ćorak [3] and Joško Parunov [1,*]

[1] Faculty of Mechanical Engineering and Naval Architecture, University of Zagreb, 10000 Zagreb, Croatia; tamara.petranovic@fsb.hr (T.P.); antonio.mikulic@fsb.hr (A.M.)
[2] Faculty of Maritime Studies, University of Split, 21000 Split, Croatia; marko.katalinic@pfst.hr
[3] Maritime Department, University of Dubrovnik, 20000 Dubrovnik, Croatia; mcorak@unidu.hr
* Correspondence: josko.parunov@fsb.hr

Abstract: The method for the prediction of extreme vertical wave bending moments on a passenger ship based on the hindcast database along the shipping route is presented. Operability analysis is performed to identify sea states when the ship is not able to normally operate and which are likely to be avoided. Closed-form expressions are used for the calculation of transfer functions of ship motions and loads. Multiple operability criteria are used and compared to the corresponding limiting values. The most probable extreme wave bending moments for the short-term sea states at discrete locations along the shipping route are calculated, and annual maximum extreme values are determined. Gumbel probability distribution is then fitted to the annual extreme values, and wave bending moments corresponding to a return period of 20 years are determined for discrete locations. The system reliability approach is used to calculate combined extreme vertical wave bending moment along the shipping route. The method is employed on the example of a passenger ship sailing across the Adriatic Sea (Split, Croatia, to Ancona, Italy). The contribution of the study is the method for the extreme values of wave loads using the hindcast wave database and accounting for ship operational restrictions.

Keywords: seakeeping operability criteria; operability analysis; wave bending moment; transfer functions; hindcast wave database

Citation: Petranović, T.; Mikulić, A.; Katalinić, M.; Ćorak, M.; Parunov, J. Method for Prediction of Extreme Wave Loads Based on Ship Operability Analysis Using Hindcast Wave Database. *J. Mar. Sci. Eng.* **2021**, *9*, 1002. https://doi.org/10.3390/jmse9091002

Academic Editor: Decheng Wan

Received: 29 August 2021
Accepted: 10 September 2021
Published: 14 September 2021

Publisher's Note: MDPI stays neutral with regard to jurisdictional claims in published maps and institutional affiliations.

1. Introduction

The extreme wave loads on ships are usually determined by weighting short-term sea state responses by their probabilities of occurrences. When linear seakeeping computations are performed, the transfer functions of ship responses are combined with the wave spectra to define short-term responses. The sea state statistics contained in the wave scatter diagram are then used to compute the long-term distribution from which design value can be derived. Extreme value adopted as the design value of global wave load is usually calculated from the long-term distribution for exceeding probability of 10^{-8}. The procedure recommended by the International Association of Classification Societies (IACS) for the computation of long-term global wave loads on ships is given in [1].

The recommended procedure assumes that the wave heading angles are uniformly distributed in all sea states. The possibility that the shipmaster changes heading angle in extreme sea conditions is not considered [2]. Such maneuvers are normally taken to reduce excessive ship responses in heavy weather, most often the ship rolling motion [3]. Consequently, the probability of beam seas is reduced, which is ignored in common computational procedure, but could affect the extreme wave loads [1]. The effect of the heavy weather avoidance is also neglected, although shipmasters in practice tend to avoid the most severe sea states, e.g., by changing ship course or by waiting for favorable weather forecast [4]. The avoidance of heavy weather is partially considered by accepting the wave atlas Global Wave Statistics (GWS) as the source of the wave data [5]. Namely, data in the GWS are collected by the observations from merchant ships that sail along standard trading routes, excluding thus implicitly the most severe weather conditions [3]. Another difficulty represents the fact that weather routing is not the same for all ship types. For example, container ships experience frequent weather routing while oil tankers and bulk carriers rarely change their planned course [6]. Wave statistics in the GWS are subjected to the uncertainties of visual wave observations and wave directionality, as equal probability of waves from all directions is assumed [3].

Alternative methods for the computation of extreme wave loads on ships are considered nowadays to account rationally for mentioned environmental and operational uncertainties. Thus, IACS recently redefined the vertical hull girder loads for container ships using the direct calculation approach [7]. The routing factor has been introduced to reduce the design loads, which were found larger compared to the rule formulation. The improvement of the wave data accuracy is considered by using hindcast numerical wave databases [8,9]. An example of the application of the hindcast database for vessel response prediction and fatigue damage estimation is presented in [10]. To efficiently employ the wave hindcast database for the computation of wave loads on ships, it is essential to account appropriately for weather routing effects. One possible approach is to compare the hindcast database to the visual observations [11].

Although hindcast databases have already been used in ship seakeeping and structural analysis (e.g., [8–10]), incorporation of the operability analysis in the calculation of the extreme wave loads on ships is still missing. That aspect is crucial to obtain realistic estimates of extreme wave loads. In the present study, a method is proposed to calculate extreme wave loads that a ship encounters along a specific route using the hindcast database and accounting for operational restrictions. A case study of a passenger ship sailing from Croatian port Split to Italian port Ancona and vice versa in the Adriatic Sea is presented. All sea states along the shipping route contained in the hindcast wave database for the period from January 1997 to January 2020 are considered in four available locations along the route. The most probable short-term extreme vertical wave bending moments are determined for each sea state that the ship could encounter. The probability distribution of annual extreme values is then determined for each location. System probability is determined by combining extreme wave bending moments along the route, and bending moments for large return periods are calculated. The effect of heavy weather avoidance is accounted for by ship operability analysis, considering different operability criteria. Wave loads are calculated for three cases:

1. Without any operational restrictions, i.e., for all sea states that ship could potentially encounter along the route;
2. Full operational restrictions, i.e., the case when sea states not satisfying any of the operational criteria are avoided;
3. Partial operability restrictions, i.e., the case when speed is considerably reduced for sea states not satisfying any of the operational criteria.

The focus of the present study is on the probabilistic method for the computation of extreme wave loads. Therefore, a simplified but efficient approach for the calculation of wave-induced responses is used, employing closed-form expressions formulated by Jensen et al. [12]. Seakeeping limiting values for operational criteria are taken specifically for the passenger ship [13].

The paper is organized as follows: The ship and shipping route used in the analysis are described in Section 2. Operability analysis is described in Section 3. The method for the prediction of extreme wave bending moments is given in Section 4, while the results of the analysis are provided in Section 5. Discussion of the results is given in Section 6, and conclusions of the study are provided at the end of the paper. In Appendix A, histograms of annual maximum vertical wave bending moments without and with operational restrictions are given, and in Appendix B, histograms of annual maximum vertical wave bending moments for a case with reduced speed are presented.

2. Ship and Shipping Route

2.1. Ship Particulars

The passenger ship (Figure 1) used for the seakeeping operability analysis is similar to the ferry currently used for coastal and international lines in the Adriatic Sea. The main particulars of the ship that correspond to the features of the vessel within the "larger ferry category" are provided in Table 1.

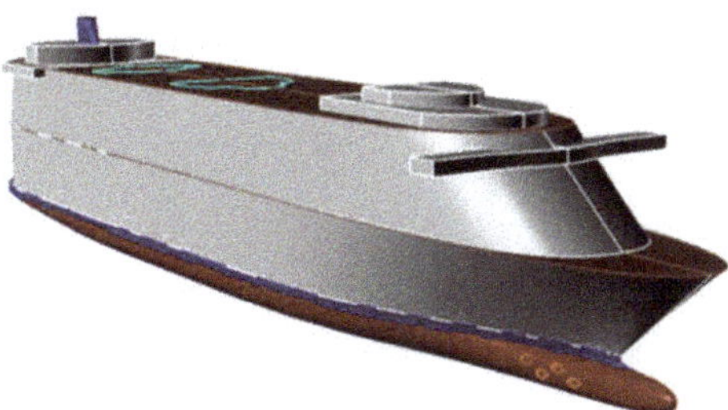

Figure 1. 3D model of the ship used in the study (Adopted from [14]).

Table 1. Main particulars of the ship (Adopted from [14]).

Particular	Value	Measuring Unit	Description
L_{OA}	114	m	Length overall
L_{PP}	103.2	m	Length between perpendiculars
B	18.7	m	Width
T	5	m	Draft
Δ	6565	t	Displacement
C_B	0.617	-	Block coefficient
GM	1.94	m	Metacentric height
v_{max}	17	kn	Maximum speed

2.2. Shiping Route

The passenger ship used in the study is intended to sail between the Croatian port Split and Italian port Ancona across the Adriatic Sea. The heading from Split to Ancona is 275° in nautical coordinates, and that from Ancona to Split is 95° (0° is toward the north). Operability criteria and loads of the ship are determined in four points near the shipping route, with geographical coordinates given in Table 2. The sailing route of the passenger ship, together with four considered points, is shown in Figure 2.

Table 2. Locations of points in the Adriatic Sea (Adopted from [15]).

Point in the Adriatic Sea	Latitude (°)	Longitude (°)
Point 1	43.5	15.5
Point 2	43.5	15
Point 3	43.5	14.5
Point 4	43.5	14

Figure 2. Points near the Split–Ancona route (Adopted from [15]).

3. Operability Analysis

3.1. Transfer Functions for Motions

Wave-induced ship motions are estimated by using closed-form, semi-analytical expressions [12]. The procedure requires only the main ship dimensions as input data: ship length, width, draft, block coefficient, and speed. Closed-form equations for the motion in the vertical plane are derived according to the linear strip theory by neglecting the coupling terms between heave and pitch and assuming a constant sectional added mass, equal to the displaced water. Semi-analytical expressions for transfer functions of heave and pitch are available in [12].

The transfer function of roll is defined in [12] as:

$$\Phi_\varphi = \frac{|M|}{\sqrt{\left[-\overline{\omega}^2(T_N/2\pi)^2 + 1\right]^2 C_{44}^2 + \overline{\omega}^2 B_{44}^2}} \tag{1}$$

where $\overline{\omega}$ is the encounter frequency, $|M|$ is the amplitude of the roll excitation moment, B_{44} is the hydrodynamic roll damping, C_{44} is the restoring moment coefficient, and T_N is the natural roll period. The viscous roll damping needs to be included in the calculation, and in the present study, 20% of critical damping is used.

3.2. Wave Data

WorldWaves hindcast wave database, containing numerical wave simulation results, calibrated by satellite data over 39 evenly scattered points in the Adriatic Sea for a 23-year period (Figure 3) is employed. Historical information about sea states in four locations near the sailing route is used for this study (location numbers 10, 12, 15, and 18 in Figure 3). Each point contains 12 physical parameters measured every 6 h from January 1997 to January 2020, which makes a total of 33,600 recorded logs. For the purposes of this study, the significant wave height, the peak wave period, the mean wave directions, and time and date records are used [16].

Table 3 shows the maximum recorded significant wave heights in four considered points together with the dates of recordings. It is interesting to notice that maximum significant wave height at three locations is recorded on the same date, indicating a high spatial correlation among extreme sea states on this route.

Figure 3. Thirty-nine evenly distributed locations in the Adriatic Sea (Adopted from [14]).

Table 3. Maximum significant wave heights in 4 locations near the sailing route.

Point in the Adriatic Sea (Figure 2)	Max. H_S (m)	Date
1	5.89	29 October 2018
2	6.18	29 October 2018
3	5.98	29 October 2018
4	5.65	11 November 2013

3.3. JONSWAP Wave Spectrum

The Joint North Sea Wave Project (JONSWAP) wave spectrum, which is the adjustment of the Pierson–Moskowitz spectrum for the existence of a limited fetch and wind persistence, is employed in this study for the semi-enclosed basin of the Adriatic Sea. Adequacy of the JONSWAP wave spectrum formulation for the Adriatic Sea is shown in [16].

The JONSWAP wave spectrum used in the present study is defined by Det Norske Veritas [17]:

$$S_J(\omega) = A_\gamma S_{PM}(\omega)\gamma^{exp\left(-0.5\left(\frac{\omega-\omega_p}{\sigma\omega_p}\right)^2\right)},$$ (2)

where γ is a non-dimensional peak shape parameter, $A_\gamma = 1 - 0.287\ln(\gamma)$ is a normalizing factor, $S_{PM}(\omega)$ is the Pierson–Moskowitz spectrum, $\omega_p = 2\pi/T_p$ is the angular spectral peak frequency, T_p is the peak wave period, and σ is the spectral width parameter.

The non-dimensional peak shape parameter reads:

$$-\gamma = 5 \; for \; \frac{T_p}{\sqrt{H_s}} \leq 3.6;$$
$$\gamma = exp\left(5.75 - 1.15\frac{T_p}{\sqrt{H_s}}\right) \; for \; 3.6 < \frac{T_p}{\sqrt{H_s}} < 5;$$
$$\gamma = 1 \; for \; 5 \leq \frac{T_p}{\sqrt{H_s}}.$$ (3)

3.4. Operability Criteria and Limits

Seakeeping operability criteria considered in this paper are the root mean square (RMS) of roll, the RMS of pitch, the RMS of the vertical acceleration at the forward perpendicular (FP), the probability of slamming, the probability of green water, the probability of propeller emergence, and the weighted average motion sickness incidence (WAMSI). The operability analysis is performed to estimate the percentage of time during which the ship may not be able to sail on the given route and with a given speed. Limiting values of selected operability criteria are shown in Table 4. Limiting values employed are universal for all ship types, except limiting values of the roll and the vertical acceleration at FP, which are specific for passenger ships [13].

Table 4. Operability criteria limiting values.

Operability Criterion	Limiting Value
RMS of roll	2.5°
RMS of pitch	1.5°
RMS of vertical acceleration at FP	0.05 g
Probability of slamming	0.03
Probability of green water	0.05
Probability of propeller emergence	0.25
WAMSI	20% in 4 h

Expressions for calculating operability criteria using response spectra and corresponding spectral moments are provided, e.g., in [18].

The passenger comfort criterion used in the present study is the WAMSI, given by:

$$\text{WAMSI} = \frac{\int \text{MSI}\, W\, dx}{\int W\, dx} \tag{4}$$

where MSI is the motion sickness incidence and W is the weighting function. Integrals in Equation (4) are determined along the length of the passenger deck. The weighting function represents the distribution of the people on the passenger deck, taken equal to 1 in the present study, which means that equally weighted subjects are evenly distributed along the passenger deck. The MSI is defined as the percentage of subjects who experienced nausea within a certain amount of time calculated according to the expressions provided in [19]:

$$\text{MSI} = 100\left[0.5 + erf\left(\frac{log_{10}\left(\frac{0.798\sqrt{m_{0_v}}}{g}\right) - \mu_{MSI}}{0.4}\right)\right], \tag{5}$$

where:

$$\mu_{MSI} = -0.819 + 2.32\left(log_{10}\left(\sqrt{\frac{m_{0_v}}{m_{2_u}}}\right)\right)^2, \tag{6}$$

where m_{0_v} and m_{2_u} are the zeroth moment of the response spectrum of vertical acceleration and the second moment of the response spectrum of vertical motion.

Recently, Overall Motion Sickness Incidence (OMSI) has been introduced as a measure of passenger comfort, defined as the mean MSI over the length and breadth of the passenger deck [20], which is not considered in the present study.

4. Method for the Prediction of Extreme Wave Loads

The transfer function of vertical wave-induced bending moment (VWBM) at midship is calculated according to the closed-form expression [12]:

$$\Phi_M = \kappa \frac{1 - kT}{(k_e L_{pp})^2}\left[1 - \cos\left(\frac{k_e L_{pp}}{2}\right) - \frac{k_e L_{pp}}{4}\sin\left(\frac{k_e L_{pp}}{2}\right)\right] \cdot F_V(F_n) \cdot F_C(C_B) \cdot \sqrt[3]{|\cos\beta|} \cdot \rho g B_0 L_{pp}^2 \tag{7}$$

where k_e represents the effective wave number, β is the heading angle, κ is the Smith correction factor, $F_V(F_n)$ is the speed correction factor, $F_C(C_B)$ is the correction factor for the block coefficient, $\rho = 1025\ \text{kg/m}^3$ is the sea density, and B_0 is the measured maximum ship breadth at the waterline. Details on how to calculate terms appearing in Equation (7) are given in [12].

The most probable extreme value (MPEV) of VWBM in the short-term sea state depends on the sea state duration $T_d = 21,600$ s , which corresponds to wave data recording at 6-hour intervals:

$$\text{MPEV(VWBM)} = \sqrt{m_{0_M} \cdot 2ln\left(\frac{T_d}{T_z}\right)}. \tag{8}$$

T_z is the average zero-crossing period of the response, and it is defined as:

$$T_z = 2\pi \sqrt{\frac{m_{0_M}}{m_{2_M}}} \tag{9}$$

where m_{0_M} and m_{2_M} are the zeroth and second moments of the response spectrum of VWBM [21].

For a period of the availability of wave data (January 1997 to January 2020), annual maximum VWBMs are extracted for all sea states in four locations along the route. Histograms are fitted with Gumbel distribution, which is commonly used as the extreme value distribution function. Parameters of Gumbel distribution are estimated using the method of moments [22,23].

The probability of exceedance of the long-term extreme VWBM can be written as:

$$Q = \frac{1}{RP} \tag{10}$$

and the probability of non-exceedance is:

$$F = 1 - Q. \tag{11}$$

where RP is the return period in years [23].

To determine the resulting extreme value distribution along the shipping route, extreme value distributions at individual locations are to be statistically combined. A certain level of VWBM is exceeded along the whole route if exceeded in any of the locations along the route. This can be modeled by considering individual locations as members in a series probabilistic system. First-order upper and lower bounds on the extreme value distribution along the route can be set by considering members of the series system as statistically independent or fully correlated, respectively. The concept was initially proposed by Mansour and Preston considering wave zones in GWS along the route [24]. If one looks at the data given in Table 3, it appears that extreme values at different locations along the route occur at the same time, so the correlation of sea states between locations could be quite large.

If sea states along the location are assumed statistically independent, the system probability of non-exceedance of the extreme vertical bending moment along the sailing route $P_{ne,SI}$ reads:

$$P_{ne,SI} = \prod_{i=1}^{n}(P_{ne_i})^{k_i}, \tag{12}$$

while the system probability of non-exceedance of extreme vertical bending moment for fully correlated sea states along the route $P_{ne,FC}$ is [25]:

$$P_{ne,FC} = \min_{i}(P_{ne_i})^{k_i}. \tag{13}$$

k_i appearing in Equations (12) and (13) represents the fraction of time that the ship spends in each of n wave locations. If the time spent in port is neglected, then $\sum_{i=1}^{n} k_i = 1$. It is assumed in this study that the ship spends an equal amount of time at each of four locations; i.e., $n = 4$ and $k_i = 0.25$. If probability distributions are different for two opposite traveling directions at the same location, then $n = 8$ and $k_i = 0.125$.

Combined probability distributions along the route, given by Equations (12) and (13), can be calculated numerically. Combined probability distributions are plotted on the VWBM $- P_{ne}$ diagram, and VWBMs for a 20-year return period are obtained.

5. Results

5.1. Operability Plot

The polar plot in Figure 4 represents the limit values of calculated operability criteria for different wave heading angles and significant wave heights. Only a maximum ship speed of 17 knots is considered in the operability analysis. Values in the polar plot are estimated for the modal spectral frequency ω_m, which is in the Adriatic Sea given as [17]:

$$\omega_m = 0.52 + \frac{1.4}{0.7 + H_s}. \tag{14}$$

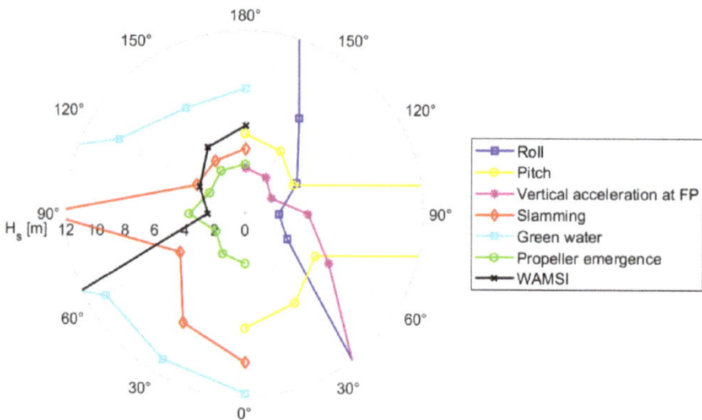

Figure 4. Polar plot (head sea is 180°) (Adopted from [15]).

When the ship sails from Ancona to Split, during the most frequent south-eastern wind, waves encounter the ship at 220°, which is equal to 140° due to symmetry. From Figure 4, one may notice that the limiting vertical acceleration at the FP is exceeded firstly at the significant wave height $H_s = 2.75\ m$. When the ship sails from Split to Ancona, waves encounter the ship at 40°, and it can be seen from Figure 4 that the limit value of the propeller emergence is exceeded firstly at significant wave height $H_s = 2.75\ m$.

The frequencies of exceedance of operability criteria limits for all sea states contained in the hindcasted wave database are presented in Table 5. Frequencies of exceedance for each criterion are obtained by dividing the number of exceedances by the total number of sea states. The total frequency of exceedance when at least one criterion is not satisfied is given in the last column of Table 5. Values outside parentheses in Table 5 refer to westward voyages from Split to Ancona, and values in parentheses are valid for the opposite direction. It can be seen the route from Ancona to Split has an appreciably larger frequency of exceeding operability criteria than the direction from Split to Ancona.

5.2. Probability Distributions of Extreme VWBM at Individual Locations

Short-term yearly extreme VWBMs are calculated in four points along the sailing route for all sea states and only for sea states satisfying operability criteria. The fitting of the Gumbel distribution to the histogram of VWBM is presented in Figures A1–A3 in Appendix A. It may be noticed that the Gumbel distribution in some of the cases does not follow VWBM histograms, so other distribution functions may also be considered for that purpose. However, as the Gumbel distribution is often used for modeling annual extreme values of waves and wave-induced responses (e.g., [18]) it is adopted in the present study as well.

Table 5. Frequencies (in %) of exceeding operability criteria limits for wave data at points 1, 2, 3, and 4 (Figure 2) for the route Split to Ancona (values for the opposite route in parentheses) (Adopted from [15]).

Point	Operability Criteria							
	Roll	Pitch	Vertical Acc. at FP	Slamming	Green Water	Propeller Emergence	WAMSI	Total
1	0.40 (0.17)	0 (0.04)	0.11 (2.12)	0 (0.11)	0 (0)	1.89 (1.14)	0.01 (0.14)	2.06 (2.62)
2	0.36 (0.16)	0 (0.07)	0.11 (2.76)	0 (0.17)	0 (0)	2.13 (1.34)	0.01 (0.15)	2.25 (3.24)
3	0.38 (0.22)	0 (0.07)	0.11 (2.89)	0 (0.18)	0 (0)	2.02 (1.38)	0.02 (0.18)	2.15 (3.40)
4	0.55 (0.36)	0 (0.08)	0.09 (3.44)	0.003 (0.23)	0 (0)	1.82 (1.56)	0.02 (0.23)	1.97 (4.03)

According to the Equation (7), VWBMs are the same in both sailing directions. Consequently, there are no differences in extreme values when operability criteria are not considered. However, the extreme VWBMs calculated only for those sea states satisfying operability criteria are different for two sailing directions due to differences in meeting the operability criteria.

For a 20-year return period, $F = 0.95$ is determined using Equations (10) and (11). The most probable long-term extreme value of VWBM is then obtained from the corresponding Gumbel probability distribution and presented in Table 6. It may be seen that in sea states meeting seakeeping operability criteria, extreme VWBMs are quite similar in different points along the route. Expectedly, extreme wave moments are considerably larger in sea states when operability criteria were not considered than moments when the operability criteria were considered.

Table 6. The most probable long-term extreme values of vertical wave bending moments for a 20-year return period at individual locations along the route (MNm).

Point (Figure 2)	Long-Term Extreme VWBM for 20-Year Return Period (MNm)		
	All Sea States	Sea States Satisfying Operability Criteria	
		Split–Ancona 275°	Ancona–Split 95°
1	124	76	68
2	139	76	73
3	152	78	73
4	150	77	68

5.3. System Probabilities

System probabilities of extreme VWBM for all sea states and for only those sea states satisfying operability criteria are shown in Figure 5. The curves obtained in Figure 5 are determined numerically by the procedure described in Section 4.

For the long return period of 20 years, bounds of extreme VWBM can be determined from diagrams in Figure 5. These values are presented in Table 7, where the left column represents values determined from fully correlated system probabilities and the right column represents statistically independent system probabilities.

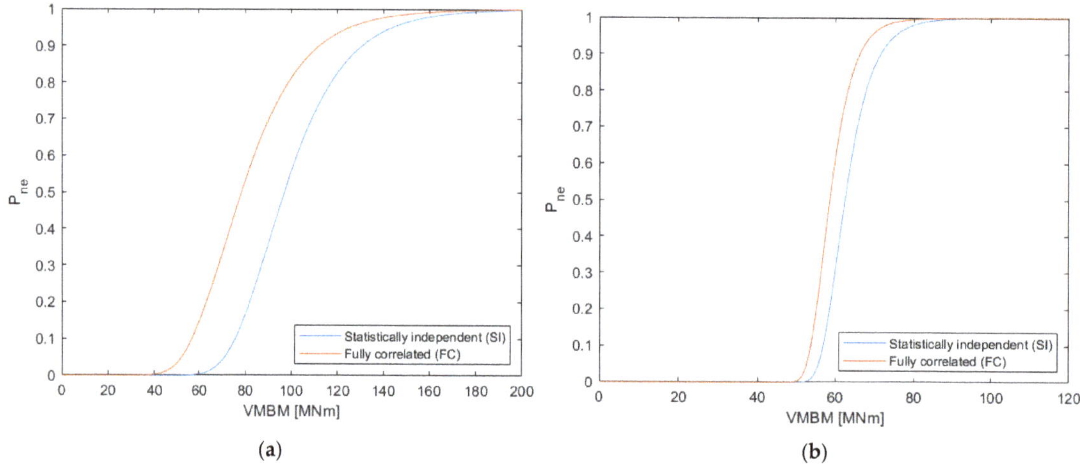

Figure 5. System probabilities of non-exceedance of extreme VWBM calculated for (**a**) all sea states and (**b**) sea states satisfying operability criteria.

Table 7. Bounds of extreme vertical bending moments.

	VWBM (MNm) Determined from $P_{ne,FC}$	VWBM (MNm) Determined from $P_{ne,SI}$
All sea states	125	143
Sea states satisfying operability criteria	69	75

It may be seen from Table 7 that the effect of possible operational restrictions is huge, as the extreme VWBMs are halved if operational restrictions are imposed. The effect of the assumption of the statistical independence is to increase VWBM by about 10% compared to the assumption of full correlation between different locations along the route. The assumption of full correlation seems to be more realistic, according to Table 3. Namely, extreme sea states and consequently extreme wave bending moments at different locations are likely to be achieved during the same voyage.

5.4. The Effect of the Speed Reduction in Severe Sea States

It is unlikely that navigation will be completely suspended during heavy seas. If waves increase during the voyage, and the possibility of exceeding the limits of operability criteria appears, the captain will either reduce the speed or change the direction of navigation.

Extreme VWBMs in Sections 5.2 and 5.3 are calculated for a case where the ship is sailing with a constant speed of 17 knots. To investigate the effect of speed reduction, only for those sea states where the operability criteria limits are exceeded, speed is reduced to 5 knots. Histograms of annual maximum VWBM calculated for courses 275° and 95° are shown in Figures A4 and A5 in Appendix B, together with the Gumbel distribution fitted. The most probable long-term extreme VWBMs for a return period of 20 years at each location are shown in Table 8. Figure 6 presents system probabilities of non-exceedance of the extreme VWBM, and Table 9 presents the bounds of the most probable long-term extreme VWBM for this case.

Table 8. The most probable long-term extreme values of vertical wave bending moments for a 20-year return period calculated for a speed of 17 knots on sea states satisfying operability criteria and a speed of 5 knots on sea states not satisfying operability criteria.

Point (Figure 2)	Long-Term Extreme VWBM for 20-Year Return Period (MNm)	
	Split–Ancona 275°	Ancona–Split 95°
1	104	103
2	116	117
3	126	126
4	124	125

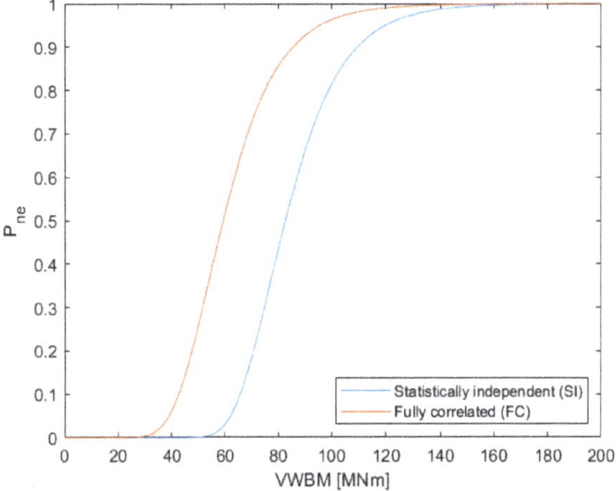

Figure 6. System probabilities of non-exceedance of extreme VWBM for reduced speed.

Table 9. Bounds of extreme vertical bending moments calculated for speed of 5 knots on sea states not satisfying operability criteria.

VWBM (MNm) Determined from $P_{ne,FC}$	VWBM (MNm) Determined from $P_{ne,SI}$
95	120

6. Discussion

The most probable extreme VWBM for the return period of 20 years along the shipping route is summarized in Figure 7. Results are shown for statistically independent and fully correlated extreme values along the shipping route and using different assumptions about the consequences of the operability analysis.

The first, albeit unrealistic option is to use full ship speed in all sea states recorded along the traveling route (in Figure 7a). If all sea states not satisfying operability criteria are avoided, the most probable extreme wave loads are reduced by a factor of 2 (in Figure 7b). If these sea states are not avoided, but ship speed is reduced to the minimum cruising speed, then extreme VWBM decreases by about 20% compared to the full speed (in Figure 7c). The effect of the correlation between short-term sea states along the route is to reduce extremes obtained by assuming statistical independence by 8–21%.

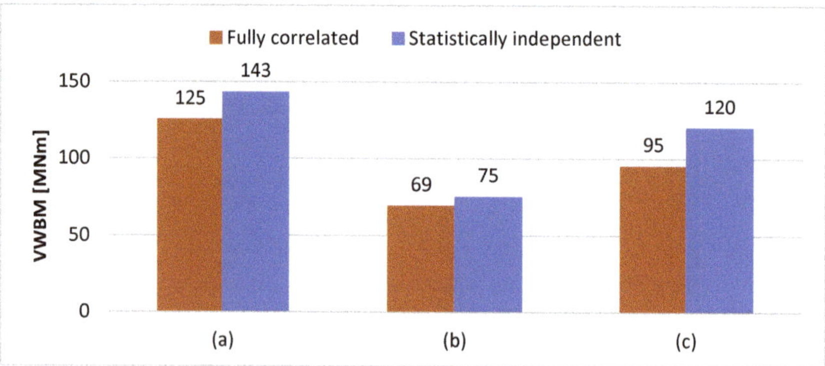

Figure 7. Comparison of extreme VWBMs determined from fully correlated and statistically independent system probabilities for (**a**) all sea states, (**b**) only sea states satisfying operability criteria with speed of 17 knots, and (**c**) a case when ship speed is reduced to 5 knots on sea states not satisfying operability criteria.

The results presented in Figure 7 are compared to the commonly used long-term procedure for the calculation of the extreme wave loads [1]. A combined wave scatter diagram is obtained for four considered locations. All heading angles in the interval $0–2\pi$ are assumed to be equally probable and the JONSWAP wave spectrum is used. Constant ship speed of 17 and 5 knots is used in two separate analyses. VWBM corresponding to 10^{-8} probability level is extracted from the long-term distribution and considered as the extreme value for the return period of 20 years. Software developed in [26] is used for a long-term analysis. The most probable extreme VWBM of 145 and 123 MNm is obtained for ship speed of 17 and 5 knots, respectively. These values are almost identical to the corresponding results presented in Figure 7 for the statistically independent case.

Validation of the method for the prediction of extreme wave loads could ideally be done by comparison with full-scale measurements. Such long-term measurement campaigns are, however, rare and difficult to organize [27]. An alternative approach would be to compare the results of the operability analysis with the practice of shipmasters of actual ships in service. This can be done by organizing an interview with shipmasters having experience with similar ships and in a similar wave environment to the one analyzed in the present study. An example of such an interview regarding analysis of a large container ship is presented in [28].

The rare examples of passenger ship accidents in the Adriatic Sea caused by large waves are a possible source of validation of the operability analysis. One such accident happened in November 2012, at 3 a.m., to the ferry sailing from Ancona to Split. Seventy vehicles collided with other vehicles and sides of the ferry, which caused great damage. The accident happened because the roll amplitudes were too large [28]. Significant wave height at the time of the accident was just above 5 m, while wave heading angle was about 140°. Comparing these values with the operability plot in Figure 4, one may conclude that ship was sailing at the operability limit for roll motion while limiting values for bow acceleration and pitch motion were largely exceeded. It is interesting to notice that the shipmaster had to change the course to following waves to avoid large rolling amplitudes [28].

The present analysis is performed for a ship sailing in the Adriatic Sea, as a specific sea environment. Similar analysis can be performed for other ship types and other shipping routes. When other ship types are analyzed, differences are expected in the first place regarding operability criteria limiting values, as the values adopted in the present study are valid for the passenger ships. Regarding different wave environments, the North Atlantic, which is considered as the design wave environment for ocean-going ships sailing without limitation, is of particular interest. Such analysis for the sailing route in the North Atlantic could be performed using, e.g., the ERA5 wave database [25].

It is to be mentioned that the Gumbel distribution is not always a perfect fit to the histogram of annual extreme VWBM, as may be seen in the appendices. Therefore, other distribution functions may also be considered for that purpose. The goodness-of-fit analysis is to be performed to find out the appropriate probability distribution.

To perform the extreme value analysis, it is necessary to have wave information for a period of at least 20 years [29], which is satisfied in the present study. For the analysis based on the long-term distribution, a minimum length of 20% of the desired return period is advised [30]. A convergence study to determine the sensitivity of predicted extreme values on the length of the wave data set is nevertheless recommended.

It should be noted that only linear wave loads are considered in this study. The most important effect of nonlinearity is the difference between sagging and hogging VWBM. The effect is particularly important for fine-form ships with low block coefficients, where sagging and hogging bending moments are larger and smaller than linear values, respectively. Simplified corrections for nonlinear VWBM that can be implemented in the present procedure to extend linear computation are presented in [31,32].

7. Conclusions

The method for calculating long-term extreme values of wave loads on a ship along a defined traveling route is described. The method is based on the hindcast wave database at discrete locations along the shipping route and ship operability analysis. Multiple operability criteria are considered, while transfer functions of wave-induced motions and loads are determined using closed-form expressions developed by Jensen et al. [12]. As the hindcast databases provide sea states in regular time intervals, the most probable extreme values of wave loads are determined for short-term sea states with a duration equal to the time resolution of the database (6 h). Annual maximum values of wave-induced loads are then determined for each location along the route, and Gumbel distribution is fitted. Gumbel distributions are then statistically combined along the route considering sea states at individual locations as independent or fully correlated members of a serial probabilistic system. The procedure enables a determination of the most probable values of wave loads for long return periods that may be used in ship structural design. The present method and the commonly used long-term distribution method result in almost the same long-term extreme values if the assumption of statistical independence among sea states at different locations is adopted. Furthermore, the presented method enables consideration of correlation among wave zones and quantification of the sensitivity of wave loads regarding operational criteria when direct calculation methods are used in ship structural design.

The following conclusions in terms of numerical results are obtained in the study:

- Operational criteria are not satisfied in 2–4% of sea states in four locations along the shipping route.
- If sea states where operability criteria are not satisfied are completely avoided, long-term extreme vertical wave bending moment is reduced by a factor of 2.
- If ship speed is reduced to the minimum cruising speed in sea states where operability criteria are not satisfied, instead of avoiding those sea states, long-term extreme vertical wave bending moment is reduced by about 20%.
- If the assumption of full statistical correlation among sea states along the shipping route is adopted, long-term extreme vertical wave bending moment is reduced by 8–21%. It should be mentioned that wave data in the database indicate that assumption of correlation is justified for this specific, relatively short shipping route.

Author Contributions: Conceptualization, J.P. and T.P.; methodology, J.P. and T.P.; software, A.M. and T.P.; validation, J.P., T.P. and M.Ć.; formal analysis, J.P. and T.P.; investigation, J.P. and T.P.; resources, J.P and M.K.; data curation, M.K.; writing—original draft preparation, T.P.; writing—review and editing, J.P.; visualization, J.P.; supervision, J.P.; project administration, J.P.; funding acquisition, J.P. All authors have read and agreed to the published version of the manuscript.

Funding: This work has been fully supported by Croatian Science Foundation under the project IP-2019-04-2085.

Institutional Review Board Statement: Not applicable.

Acknowledgments: This work has been fully supported by Croatian Science Foundation under the project MODUS (IP-2019-04-2085). The WorldWaves data used in the study are provided by Fugro OCEANOR AS.

Conflicts of Interest: The authors declare no conflict of interest. The funders had no role in the design of the study; in the collection, analyses, or interpretation of data; in the writing of the manuscript; or in the decision to publish the results.

Appendix A. Annual Maximum Vertical Wave Bending Moments

Appendix A.1. All Sea States

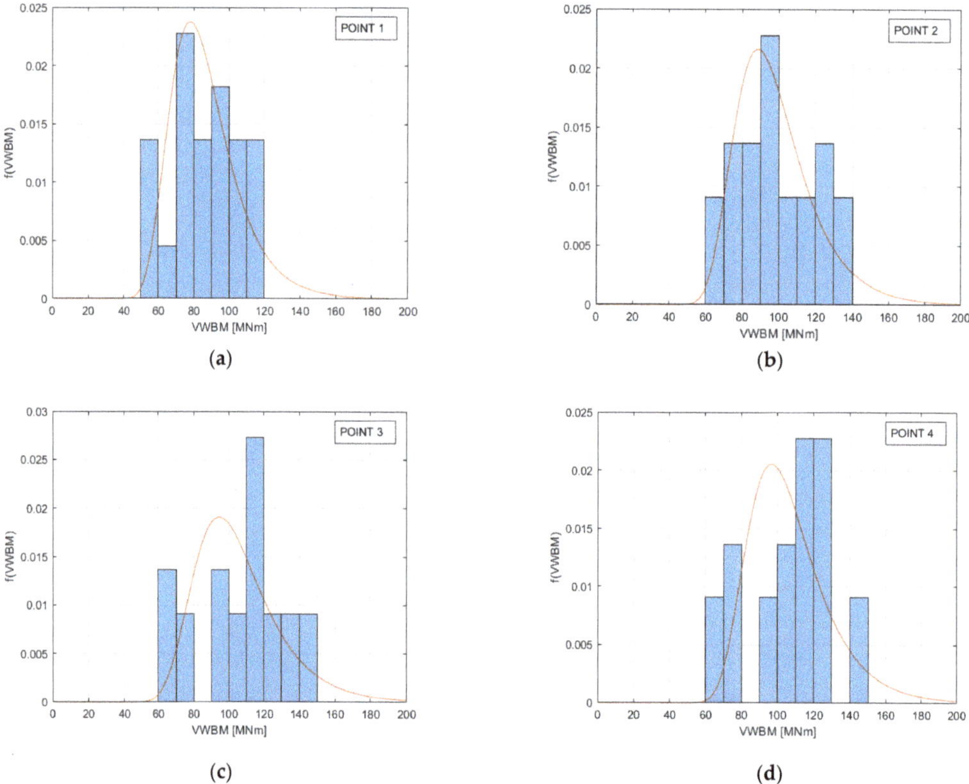

Figure A1. Annual maximum vertical wave bending moments calculated for all sea states: (**a**) at point 1; (**b**) at point 2; (**c**) at point 3; (**d**) at point 4.

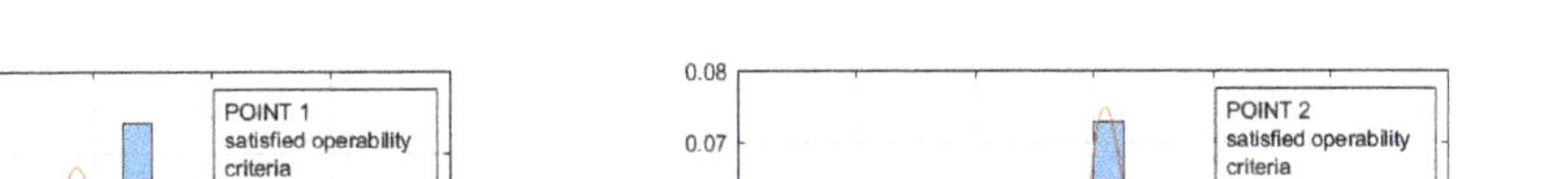

Appendix A.2. Sea States Satisfying Operability Criteria on the Sailing Route from Split to Ancona

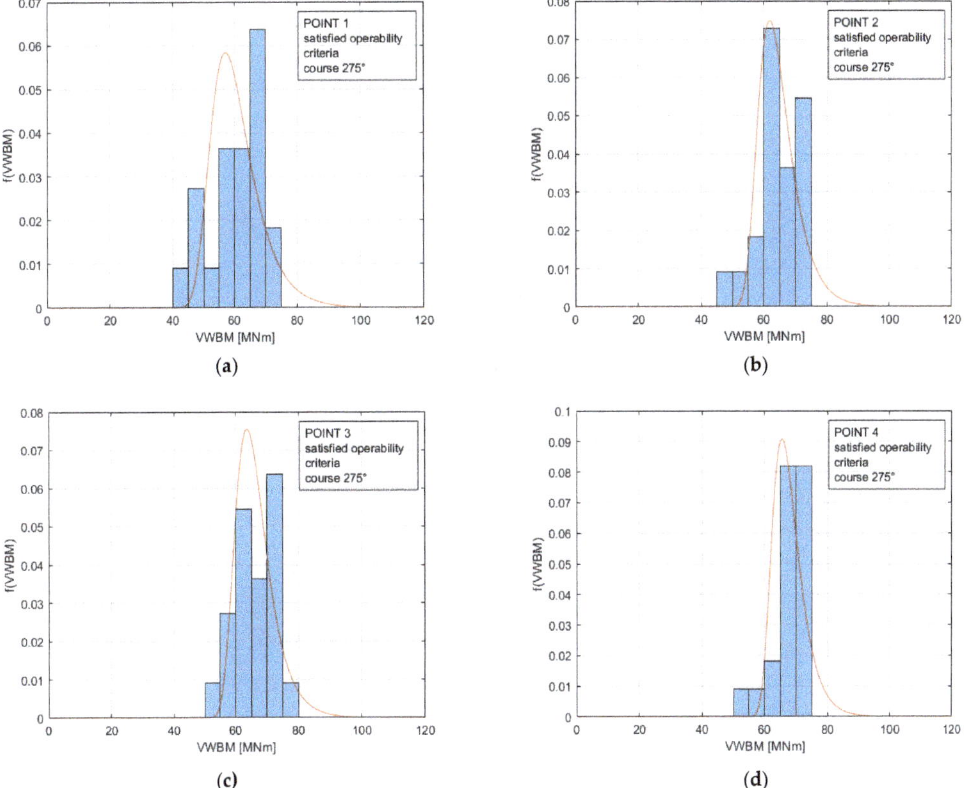

(a)

(b)

(c)

(d)

Figure A2. Annual maximum vertical wave bending moments calculated for the ship sailing from Split to Ancona for sea states satisfying operability criteria: (**a**) at point 1; (**b**) at point 2; (**c**) at point 3; (**d**) at point 4.

Appendix A.3. Sea States Satisfying Operability Criteria on the Sailing Route from Ancona to Split

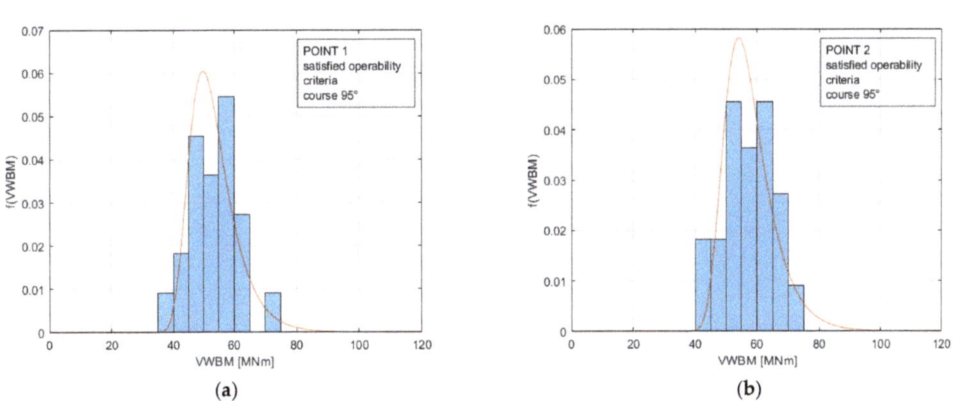

(a)

(b)

Figure A3. *Cont.*

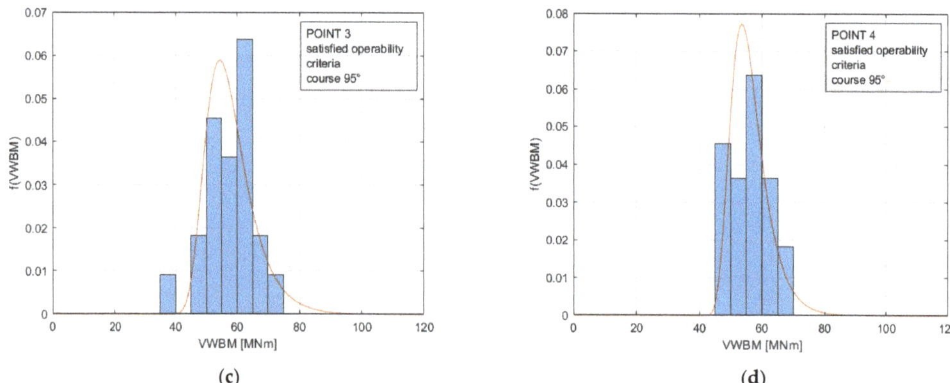

(c) (d)

Figure A3. Annual maximum vertical wave bending moments calculated for the ship sailing from Ancona to Split for sea states satisfying operability criteria: (**a**) at point 1; (**b**) at point 2; (**c**) at point 3; (**d**) at point 4.

Appendix B. Annual Maximum Vertical Wave Bending Moments for a Case of Reduced Speed on Sea States That Do Not Meet Operability Criteria

Appendix B.1. Voyage from Split to Ancona

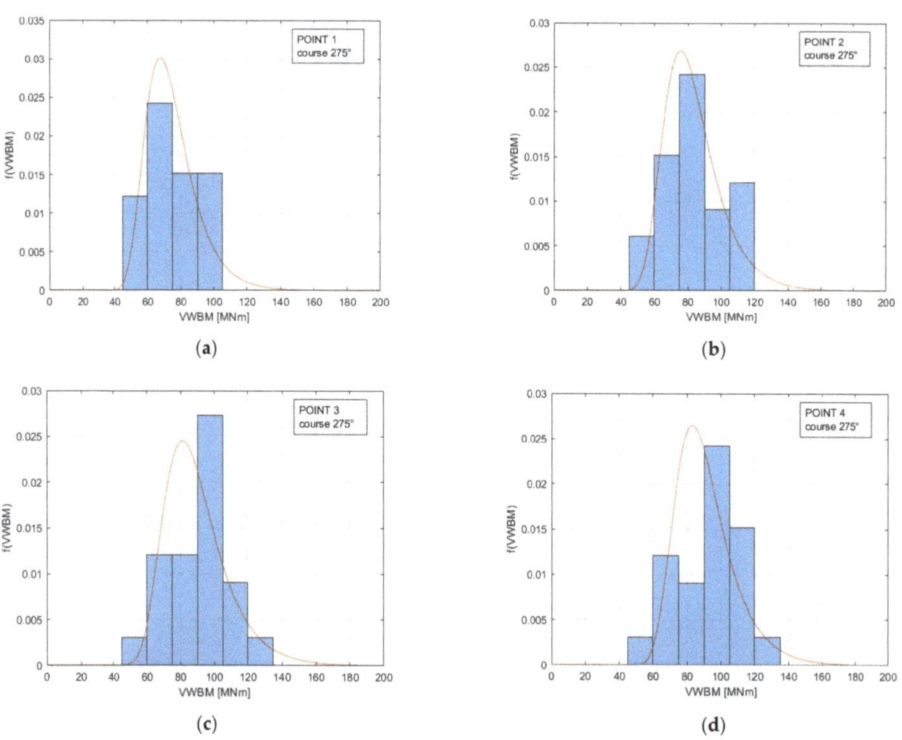

(a) (b)

(c) (d)

Figure A4. Annual maximum VWBM calculated for a speed of 17 knots on sea states satisfying operability criteria and a speed of 5 knots on sea states not satisfying operability criteria: (**a**) at point 1; (**b**) at point 2; (**c**) at point 3; (**d**) at point 4 on course 275°.

Appendix B.2. Voyage from Ancona to Split

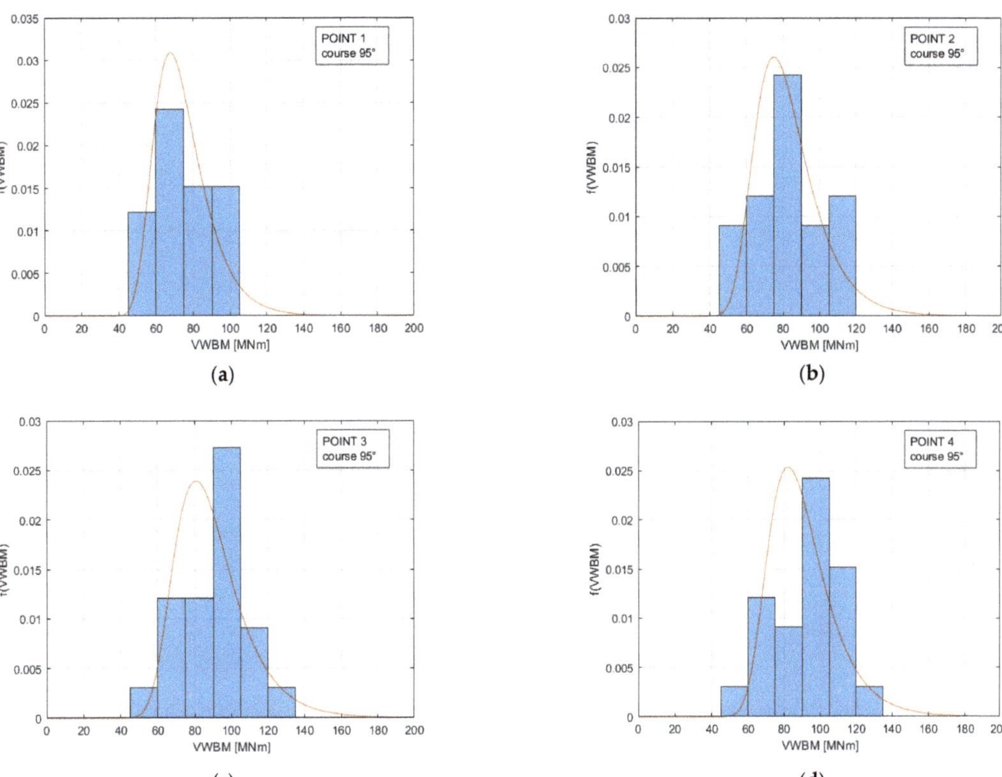

Figure A5. Annual maximum VWBM calculated for a speed of 17 knots on sea states satisfying operability criteria and a speed of 5 knots on sea states not satisfying operability criteria: (**a**) at point 1; (**b**) at point 2; (**c**) at point 3; (**d**) at point 4 on course 95°.

References

1. IACS Rec. *No. 34. Standard Wave Data*; IACS: London, UK, 2001.
2. Moan, T.; Shu, Z.; Drummen, I.; Amlashi, H. Comparative reliability analysis of ships—Considering different ship types and the effect of ship operations on loads. *Trans. Soc. Nav. Archit. Mar. Eng.* **2007**, *114*, 16–54.
3. Guedes Soares, C. On the Definition of Rule Requirements for Wave Induced Vertical Bending Moments. *Mar. Struct.* **1996**, *9*, 409–425. [CrossRef]
4. Prpić-Oršić, J.; Parunov, J.; Šikić, I. Operation of ULCS—Real Life. *Int. J. Nav. Archit. Ocean. Eng.* **2014**, *6*, 1014–1023. [CrossRef]
5. Hogben, N.; Dacunha, N.M.C.; Oliver, G.F. *Global Wave Statistics*; Feltham: British Maritime Technology Ltd.: London, UK, 1986.
6. Bitner-Gregersen, E.M.; Dong, S.; Fu, T.; Ma, N.; Maisondieu, C.; Miyake, R.; Rychlik, I. Sea state conditions for marine structures' analysis and model tests. *Ocean. Eng.* **2016**, *119*, 309–322. [CrossRef]
7. Derbanne, Q.; Shiguntov, V.; Storhaug, G.; Xie, G.; Zheng, G. Rule formulation of vertical hull girder wave loads based on direct computation. In Proceedings of the PRADS 2016, Copenhagen, Denmark, 4–8 September 2016.
8. de Hauteclocque, G.; Johnson, M.; Zhu, T.; Austefjord, H.; Bitner-Gregersen, E. Assesment of global wave datasets for long term response of ships. In Proceedings of the ASME 2020 39th International Conference on Ocean, Offshore and Arctic Engineering OMAE 2020, Fort Lauderdale, FL, USA, 28 June–3 July 2020.
9. Perrault, D.E. Probability of Sea Condition for Ship Strength, Stability, and Motion Studies. *J. Ship Res.* **2021**, *65*, 1–14. [CrossRef]
10. Schirmann, M.L.; Collette, M.D.; Gose, J.W. Significance of wave data source selection for vessel response prediction and fatigue damage estimation. *Ocean. Eng.* **2020**, *216*, 107610. [CrossRef]
11. Vettor, R.; Guedes Soares, C. Assessment of the Storm Avoidance Effect on the Wave Climate along the Main North Atlantic Routes. *J. Navig.* **2016**, *69*, 127–144. [CrossRef]

12. Jensen, J.J.; Mansour, A.E.; Olsen, A.S. Estimation of ship motions using closed-form expressions. *Ocean. Eng.* **2004**, *31*, 61–85. [CrossRef]
13. Ghaemi, M.H.; Olszewski, H. Total ship operability -review, concept and criteria. *Pol. Marit. Res.* **2017**, *24*, 74–81. [CrossRef]
14. Katalinić, M.; Parunov, J.; Mikulić, A. Toward operability analysis of a passenger ship in the Adriatic Sea based on the JONSWAP-Adriatic wave spectrum. In Proceedings of the XXIV Symposium Theory and Practice of Naval Architectures (SORTA 2020), Malinska, Croatia, 15–17 October 2020.
15. Petranović, T.; Katalinić, M.; Mikulić, A.; Parunov, J. Operability study of passenger ship in the Adriatic Sea using hindcast database. In Proceedings of the 8th International Conference on Marine Structures (MARSTRUCT 2021), Trondheim, Norway, 7–9 June 2021.
16. Katalinić, M.; Parunov, J. Comprehensive Wind and Wave Statistics and Extreme Values for Design and Analysis of Marine Structures in the Adriatic Sea. *J. Mar. Sci. Eng.* **2021**, *9*, 522. [CrossRef]
17. Katalinić, M.; Ćorak, M.; Parunov, J. Optimized Wave Spectrum Definition for the Adriatic Sea. *Naše More* **2020**, *67*, 19–23. [CrossRef]
18. DNV GL. *Recommended Practice DNVGL RP C-205: Environmental Conditions and Environmental Loads*; Edition August 2017; Det Norske Veritas Germanischer Lloyd: Oslo, Norway, 2017.
19. Lloyd, A.R.J.M. *Seakeeping: Ship Behaviour in Rough Water*; Gosport: Hampshire, UK, 1998.
20. Scamardella, A.; Piscopo, V. Passenger ship seakeeping optimization by the Overall Motion Sickness Incidence. *Ocean. Eng.* **2014**, *76*, 86–97. [CrossRef]
21. DNV GL. *Class Guideline DNVGL CG 0130: Wave Loads*; Det Norske Veritas Germanischer Lloyd: Oslo, Norway, 2018.
22. Mansour, A.; Liu, D. *The Principles of Naval Architecture Series: Strength of Ships And Ocean Structures*; The Society of Naval Architects and Marine Engineers: Jersey City, NJ, USA, 2008; pp. 4–56.
23. Katalinić, M.; Parunov, J. Uncertainties of Estimating Extreme Significant Wave Height for Engineering Applications Depending on the Approach and Fitting Technique—Adriatic Sea Case Study. *J. Mar. Sci. Eng.* **2020**, *8*, 259. [CrossRef]
24. Mansour, A.E.; Preston, D.B. Return periods and encounter probabilities. *Appl. Ocean. Res.* **1995**, *17*, 127–136. [CrossRef]
25. Mikulić, A.; Katalinić, M.; Ćorak, M.; Parunov, J. The effect of spatial correlation of sea states on extreme wave loads of ships. *Ships Offshore Struct.* **2021**, *16* (Suppl. S1), 22–32. [CrossRef]
26. Ćorak, M.; Parunov, J.; Guedes Soares, C. Long-term prediction of combined wave and whipping bending moments of container ships. *Ships Offshore Struct.* **2015**, *10*, 4–19. [CrossRef]
27. Jiao, J.; Ren, H.; Guedes Soares, C. A review of large-scale model at-sea measurements for ship hydrodynamics and structural loads. *Ocean. Eng.* **2021**, *227*, 108863. [CrossRef]
28. Mudronja, L.; Katalinić, M.; Vidan, P.; Parunov, J. Route planning based on ship roll in numerically modelled heavy seas. In Proceedings of the XXII Symposium Theory and Practice of Naval Architectures (SORTA 2016), Trogir, Croatia, 6–8 October 2016.
29. World Meteorological Organization. *Guide to Wave Analysis and Forecasting*; Secretariat of the World Meteorological Organization: Geneva, Switzerland, 1998.
30. Lavidas, G.; Venugopal, V. Wave energy resource evaluation and characterisation for the Libyan Sea. *Int. J. Mar. Energy* **2017**, *18*, 1–14. [CrossRef]
31. Mansour, A.E.; Wasson, J.-P. Charts for Estimating Nonlinear Hogging and Sagging Bending Moments. *J. Ship Res.* **1995**, *39*, 240–249. [CrossRef]
32. Mansour, A.E.; Jensen, J.J. Slightly Nonlinear Extreme Loads and Load Combinations. *J. Ship Res.* **1995**, *39*, 139–149. [CrossRef]

Journal of
*Marine Science
and Engineering*

Article

A Proposal of Mode Polynomials for Efficient Use of Component Mode Synthesis and Methodology to Simplify the Calculation of the Connecting Beams

Jeong Hee Park [1],* and Duck Young Yoon [2]

[1] Structural Design Department, Hyundai Samho Heavy Industries, Jeollanam-do 58462, Korea
[2] Department of Naval Architecture and Ocean Engineering, Chosun University, Gwangju 61452, Korea; dyyun@chosun.ac.kr
* Correspondence: parkjh@hshi.co.kr

Abstract: Analytical method using Rayleigh–Ritz method has not been widely used recently due to intensive use of finite element analysis (FEA). However as long as suitable mode functions together with component mode synthesis (CMS) can be provided, Rayleigh–Ritz method is still useful for the vibration analysis of many local structures in a ship such as tanks and supports for an equipment. In this study, polynomials which combines a simple and a fixed support have been proposed for the satisfaction of boundary conditions at a junction. Higher order polynomials have been generated using those suggested by Bhat. Since higher order polynomials used only satisfy geometrical boundary conditions, two ways are tried. One neglects moment continuity and the other satisfies moment continuity by sum of mode polynomials. Numerical analysis have been performed for typical shapes, which can generate easily more complicated structures. Comparison with FEA result shows good agreements enough to be used for practical purpose. Frequently dynamic behavior of one specific subcomponent is more concerned. In this case suitable way to estimate dynamic and static coupling of subcomponents connected to this specific subcomponent should be provided, which is not easy task. Elimination of generalized coordinates for subcomponents by mode by mode satisfaction of boundary conditions has been proposed. These results are still very useful for initial guidance.

Keywords: mode polynomials; CMS (component mode synthesis); FEA (finite element analysis); PMSC (proposal of methodology for a simplification of computation)

Citation: Park, J.H.; Yoon, D.Y. A Proposal of Mode Polynomials for Efficient Use of Component Mode Synthesis and Methodology to Simplify the Calculation of the Connecting Beams. *J. Mar. Sci. Eng.* 2021, 9, 20. https://dx.doi.org/10.3390/jmse9010020

Received: 20 November 2020
Accepted: 23 December 2020
Published: 26 December 2020

Publisher's Note: MDPI stays neutral with regard to jurisdictional claims in published maps and institutional affiliations.

1. Introduction

Each tank installed on the ship is arranged in the stern and engine room of the ship considering the cargo loading space, and there is a possibility that excessive vibration may occur due to the main excitation forces (main engine and propeller) that causes the ship vibration. If excessive vibration occurs after drying, it occurs significant restrictions and high cost on reinforcing work such as welding and special painting inside the tank. Therefore, anti-vibration measures are required at the design stage, along with a commercial program (MSC Patran/Nastran), in some cases, a calculation program [1,2] that can simply check the natural frequency has been developed and used.

In order to have an anti-vibration design of structures through calculation methods, it is necessary to analyze the normal mode for resonance avoidance with the main excitation force. For the normal mode analysis of structures, analytical methods such as Rayleigh–Ritz method are widely used together with finite element method. Although the finite element method is widely used in recent years, the analytical method which can be easily and simply reviewed at the initial design stage is still useful because the calculation time is longer than that of the analytical method.

Analytical methods require an approach that can yield more reliable results. In general, when the analytical method is applied, the Euler's beam function is used.

However, since the operation is very complicated, a study has been made on a polynomial having a beam property in order to simplify it. Park and Yang [3] performed the calculation of the natural frequency of the connected rectangular plate using a polynomial. Bhat [4] uses a polynomial as mode functions. Han [5,6] analyzed the complex vibration of the panel using the assumed mode method as an analytical method.

Kim [7] carried out the vibration analysis of the rectangular reinforced plate using polynomial with Timoshenko beam function property which can consider the rotational inertia and shear deformation effect of plate and stiffener. However, all calculations in the above mentioned studies are performed with given boundary conditions. The above mentioned studies were performed with differentiated boundary condition such as simple and fixed boundary conditions for a single structure.

However, the ship structure is not a single but a connection structure. Therefore, it is not appropriate to calculate the connection structure by simply assigning a simple or fixed boundary condition. CMS (component mode synthesis) method was applied to calculate the connection structure [8,9]. The first CMS method was presented by Hurty [10] in 1960. Alessandro Cammarata [11] introduced a wide range of CMS content, described an algorithm applied to flexible multi-objects, and a method of reducing degrees of freedom. In order to calculate the normal mode analysis of the connected structure using the CMS method, it is important to define the constraint at the connection part, and various studies on the constraint conditions at the junction were performed by Hurty et al. [12–18].

Carrera et al. [19] is developed theory that can be solved by converting a three-dimensional model for large deformation of a structure into one dimension was developed and applied to the calculation. Pagani et al. [20] is explained that the natural frequency and mode shape can be changed significantly when the metal structure is subjected to large displacement and rotation under geometrical nonlinear conditions.

Further, geometric nonlinear total Lagrange formula including cross-sectional deformation was developed to implement the vibration mode of the composite beam structure in the nonlinear region [21].

As mentioned above, many studies have been conducted, but it is still important to find a way to minimize the convergence of boundary conditions at the junction. This study proposed the following method to minimize the convergence of boundary conditions at the junction.

Firstly, we have proposed polynomials combining fixed and simple supports to satisfy boundary condition at junctions between each subsystem. We know that this approach has never been tried.

Secondly, although Bhat [4] proposed a fixed and simple support function, the calculation was performed by applying it to a simple plate. In addition, the function proposed by Bhat does not satisfy the natural condition in higher order terms of second or higher order.

In this study, in order to compensate for this problem, calculations were performed for the two cases mentioned below at the connection point and the results were compared in Section 4.1.

(1) Displacement, slope, and moment continuity (total sum of natural conditions is continuous);
(2) Displacement, and slope continuity (ignoring natural conditions).

For reference, the geometrical boundary condition mentioned in this manuscript refers to the boundary condition for displacement and slope, and the natural boundary condition refers to the boundary condition for moment. [4]

Third, in order to confirm the usefulness of the proposed method, a numerical analysis was performed on the representative shape of two and three components typical.

In particular, for the two component type, various verifications were performed in the entire length range $0 \ll x \ll 1$ according to the length ratio (L_A:L_B).

Fourth, frequently, only specific subcomponent is more concerned for vibration analysis. In this case, the suitable boundary conditions to consider the static and dynamic coupling from the other subcomponent through junctions should be provided. However, the suggestions for such boundary conditions are hardly found. In this study, in order to calculate the above case, A simplified method that can reduce the degree of freedom up to 50% by matching the subcomponent mode and the interest component mode as a constraint condition at the junctions is also proposed.

The purpose of the simplification method presented in this study is to show that it is possible to calculate a method that can reduce the degree of freedom by 50%, rather than a method for comparing numerical calculation results with the existing method. Although this method is somewhat excessive, as a result, it satisfies the finite element analysis (FEA) result and the analysis error of 15%, which is appropriate as an approximate numerical methodology, so it is considered to be efficient for approximate numerical calculation.

In addition, a three component structure was used for the calculation of structures in which symmetric and asymmetric modes occur repeatedly. A mode function having an appropriate boundary condition for a three component structure is proposed. The case of three component structures, fixed-fixed, fixed-simple, simple-fixed, fixed-free, simple-simple, and displacement functions were used.

A method of simplifying and calculating it for asymmetric and symmetric modes was proposed.

2. Definition of Assumed Mode Functions

The component mode synthesis is suitable for the vibration analysis of many local structures in ships such as tanks and supports for equipment, in which structures are divided into smaller subcomponents.

In the component mode synthesis, each mode function does not need to satisfy the junction conditions as long as their combined sum allows these junction conditions to be satisfied.

Nevertheless efficient mode functions to improve convergence are still very important for the practical use. Polynomials are frequently considered as mode functions [4]. Junction conditions among subcomponent are neither fixed nor simple support. It is a reasonable guess for mode functions to be represented by combined sum of functions for fixed and simple support. To represent the basic ideas of the method of modal synthesis, an example shown in Figure 1 is used. Vibration only in the plane of paper is considered.

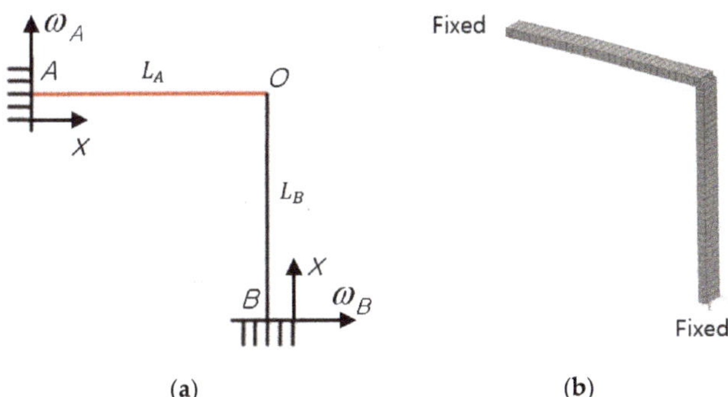

Figure 1. Structure model of two section type connected beam. (a) Simplify model; (b) finite element analysis (FEA) Model.

2.1. Two Components Type Connected Structure

The beam is separated into two sections OA and OB, whose coordinates are shown as w_A; x and w_B; x.

The properties of structures used in the numerical analysis are shown in Table 1.

Table 1. Properties and cross section of model.

Category	W_A	W_B	Cross Section	
Density [kg/m^3]	7850	7850		
Length [m]	5	5/4/2.5		0.2m
Area [m^2]	0.1	0.1		
Young's modulus [N/mm^2]	2.10×10^5	2.10×10^5	0.5m	
2nd moment of area [mm^4]	3.33×10^8	3.33×10^8		

Deflections can be shown as below.

$$W_A(x,t) = \sum_{i=1}^{m} w_{Ai}(x) \cdot p_{Ai}(t) \tag{1}$$

$$W_B(x,t) = \sum_{i=1}^{n} w_{Bi}(x) \cdot p_{Bi}(t) \tag{2}$$

For the simple explanation of component mode synthesis, Euler beam are assumed and $(EI)_A = (EI)_B = EI$.

Where E and I are Young's modulus and 2nd moment of area, and L_A and L_B are length of beams.

The deflection of subcomponent OA and OB using fixed and simple boundary condition can be expressed as below.

Where x_A and x_B are non-dimensional such that $\zeta = \frac{x_A}{L_A}, \xi = \frac{x_B}{L_B}$:

$$W_A(\zeta,t) = \sum_{i=1}^{m} (\psi_i(\zeta)p_{Ai}(t) + \phi_i(\zeta)q_{Ai}(t)) \tag{3}$$

$$W_B(\xi,t) = \sum_{j=1}^{n} (\psi_j(\xi)p_{Bj}(t) + \phi_j(\xi)q_{Bj}(t)) \tag{4}$$

and $p_{Ai}(t), q_{Ai}(t), p_{Bj}(t), q_{Bj}(t)$ are the general coordinate system in the mode function of beam.

The polynomials mode function for ψ_i, ϕ_i are suggested such that ψ_i for fixed-fixed boundary condition and ϕ_i for fixed-simple boundary conditions.

ψ_1 can be derived by assuming fourth order polynomial and boundary conditions.

Looking at the process of deriving a fixed-fixed function:

$$\psi_1(\zeta) = a_0 + a_1 \times \zeta + a_2 \times \zeta^2 + a_3 \times \zeta^3 + a_4 \times \zeta^4 \tag{5}$$

Applying geometric boundary condition $\psi_1(0) = 0$, $\psi_1{}'(0) = 0$, $\psi_1(1) = 0$ and $\psi_1{}'(1) = 0$ to Equation (5) yields a fixed-fixed fundamental mode function such as Equation (6).

$$\psi_1(\zeta) = \left(\zeta^4 - 2\zeta^3 + \zeta^2\right) A_1 \tag{6}$$

ϕ_1 can be derived by assuming 3rd order (only used geometric boundary condition) and fourth order (combine used geometric and natural boundary condition) polynomial and boundary conditions.

First, the fundamental mode function of a third polynomial is

$$\phi_1(\zeta) = a_0 + a_1 \times \zeta + a_2 \times \zeta^2 + a_3 \times \zeta^3 \tag{7}$$

Applying $\phi_1(0) = 0$, $\phi_1{}'(0) = 0$ and $\phi_1(1) = 0$ to Equation (7) yields a fixed-simple fundamental mode function such as Equation (8):

$$\phi_1(\zeta) = \left(\zeta^3 - \zeta^2\right) B_1 \tag{8}$$

and the fundamental mode function of a fixed-simple 4th polynomial is

$$\phi_1(\zeta) = a_0 + a_1 \times \zeta + a_2 \times \zeta^2 + a_3 \times \zeta^3 + a_4 \times \zeta^4 \tag{9}$$

Applying $\phi_1(0) = 0$, $\phi_1{}'(0) = 0$, $\phi_1(1) = 0$ and $\phi_1''(1) = 0$ to Equation (9) yields a fixed-simple fundamental mode function such as Equation (10):

$$\phi_1(\zeta) = \zeta(\zeta - 1)\left(2\zeta^2 - 3\zeta\right)B_1 \tag{10}$$

The coefficients A_1 and B_1 are implemented using the orthogonal formula of the beam function:

$$\int_0^1 \psi_i \psi_j d\zeta = \delta_{ij} \tag{11}$$

$$\int_0^1 \phi_i \phi_j d\zeta = \delta_{ij} \tag{12}$$

where i, j is the vibration order, and δ_{ij} is Kronecker delta.

$$A_1 = \frac{\int_0^1 \psi_1^2 d\zeta}{\sqrt{\int_0^1 \left(\zeta^4 - 2\zeta^3 + \zeta^2\right)^2 d\zeta}} \tag{13}$$

$$B_1 = \frac{\int_0^1 \phi_1^2 d\zeta}{\sqrt{\int_0^1 \left(\zeta^4 - 2.5\zeta^3 + 1.5\zeta^2\right)^2 d\zeta}} \tag{14}$$

and the mode function of the second mode or more can be implemented from the following Equations (15) to (16). The expansion to higher mode is the same regardless of the third or fourth fundamental mode function:

$$\psi_k = A_k\left[\psi_{k-1} \times \zeta - \sum_{i=1}^{k-1} a_{ki} \times \psi_{k-1}\right] \tag{15}$$

$$\phi_k = B_k\left[\phi_{k-1} \times \zeta - \sum_{i=1}^{k-1} b_{ki} \times \phi_{k-1}\right] \tag{16}$$

The coefficients a_{ki}, b_{ki} can be obtained from the orthogonal relation of the Equations (15) to (16).

The mass and stiffness matrix was obtained for each of the defined mode functions, and slope was given as constraint at the connection point to implement the natural frequency and mode of the connected beams.

The mass matrix and stiffness matrix using suggested polynomials are shown in Equations (17) and (18):

$$[M_A] = mL_A \begin{vmatrix} \psi_1\psi_1 & \cdots & \psi_1\psi_m & \psi_1\phi_1 & \cdots & \psi_1\phi_m \\ \vdots & \ddots & \vdots & \vdots & & \vdots \\ \psi_m\psi_1 & \cdots & \psi_m\psi_m & \psi_m\phi_1 & \cdots & \psi_m\phi_m \\ \phi_1\psi_1 & \cdots & \phi_1\psi_m & \phi_1\phi_1 & \cdots & \phi_1\phi_m \\ \vdots & \ddots & \vdots & \vdots & & \vdots \\ \phi_m\psi_1 & \cdots & \phi_m\psi_m & \phi_m\phi_1 & \cdots & \phi_m\phi_m \end{vmatrix} \tag{17}$$

$$[K_A] = \frac{8EI}{L_A^3} \begin{vmatrix} \psi_1''\psi_1'' & \cdots & \psi_1''\psi_m'' & \psi_1''\phi_1'' & \cdots & \psi_1''\phi_m'' \\ \vdots & \ddots & \vdots & \vdots & & \vdots \\ \psi_m''\psi_1'' & \cdots & \psi_m''\psi_m'' & \psi_m''\phi_1'' & \cdots & \psi_m''\phi_m'' \\ \phi_1''\psi_1'' & \cdots & \phi_1''\psi_m'' & \phi_1''\phi_1'' & \cdots & \phi_1''\phi_m'' \\ \vdots & \ddots & \vdots & \vdots & & \vdots \\ \phi_m''\psi_1'' & \cdots & \phi_m''\psi_m'' & \phi_m''\varphi_1'' & \cdots & \phi_m''\phi_m'' \end{vmatrix} \tag{18}$$

where m is mass per unit length, we can express the mass matrix as follows by using orthogonality:

$$[M_A] = mL_A \begin{vmatrix} 1 & \cdots & 0 & \psi_1\phi_1 & \cdots & \psi_1\phi_m \\ \vdots & \ddots & \vdots & \vdots & & \vdots \\ 0 & \cdots & 1 & \psi_m\phi_1 & \cdots & \psi_m\phi_m \\ \phi_1\psi_1 & \cdots & \phi_1\psi_m & 1 & \cdots & 0 \\ \vdots & \ddots & \vdots & \vdots & \ddots & \vdots \\ \phi_m\psi_1 & \cdots & \phi_m\psi_m & 0 & \cdots & 1 \end{vmatrix} \tag{19}$$

and M_B, K_B can be expressed in a similar.

Therefore, the mass and stiffness matrix of the subcomponents in Figure 1 can be expressed as in Equations (20) and (21):

$$[M] = \begin{vmatrix} M_A & 0 \\ 0 & M_B \end{vmatrix} \tag{20}$$

$$[K] = \begin{vmatrix} K_A & 0 \\ 0 & K_B \end{vmatrix} \tag{21}$$

Note that no coupling between the displacement of OA and that of OB.

Note that the number of generalized coordinates shall be 2 $(m + n)$, in order to simplify the calculation process, only slope was assigned as a constraint condition at the connection point.

The coordinate system reflecting the constraints is expressed in Equation (22); where α is the ratio of length for subcomponents ($\alpha = L_B / L_A$):

$$\begin{Bmatrix} p_{A1} \\ \vdots \\ p_{Am} \\ q_{A1} \\ \vdots \\ q_{Am} \\ p_{B1} \\ \vdots \\ p_{Bn} \\ q_{B1} \\ \vdots \\ q_{Bn} \end{Bmatrix} = \begin{vmatrix} 1 & \cdots & 0 & 0 & \cdots & 0 & 0 & \cdots & 0 & 0 & \cdots & 0 \\ \vdots & \ddots & \vdots & \vdots & \ddots & \vdots & \vdots & \ddots & \vdots & \vdots & \ddots & \vdots \\ 0 & \cdots & 1 & 0 & \cdots & 0 & 0 & \cdots & 0 & 0 & \cdots & 0 \\ 0 & \cdots & 0 & 1 & \cdots & 0 & 0 & \cdots & 0 & 0 & \cdots & 0 \\ \vdots & \ddots & \vdots & \vdots & \ddots & \vdots & \vdots & \ddots & \vdots & \vdots & \ddots & \vdots \\ 0 & \cdots & 0 & 0 & \cdots & 1 & 0 & \cdots & 0 & 0 & \cdots & 0 \\ 0 & \cdots & 0 & 0 & \cdots & 0 & 1 & \cdots & 0 & 0 & \cdots & 0 \\ \vdots & \ddots & \vdots & \vdots & \ddots & \vdots & \vdots & \ddots & \vdots & \vdots & \ddots & \vdots \\ 0 & \cdots & 0 & 0 & \cdots & 0 & 0 & \cdots & 1 & 0 & \cdots & 0 \\ 0 & \cdots & 0 & 0 & \cdots & 0 & 0 & \cdots & 0 & 1 & \cdots & 0 \\ \vdots & \ddots & \vdots & \vdots & \ddots & \vdots & \vdots & \ddots & \vdots & \vdots & \ddots & \vdots \\ 0 & \cdots & 0 & \alpha\frac{\phi_1'}{\phi_n'} & \cdots & \alpha\frac{\phi_m'}{\phi_n'} & 0 & \cdots & 0 & -\frac{\phi_1'}{\phi_n'} & \cdots & -\frac{\phi_{n-1}'}{\phi_n'} \end{vmatrix} \begin{Bmatrix} p_{A1} \\ \vdots \\ p_{Am} \\ q_{A1} \\ \vdots \\ q_{Am} \\ p_{B1} \\ \vdots \\ p_{Bn-1} \\ q_{B1} \\ \vdots \\ q_{Bn-1} \end{Bmatrix} \tag{22}$$

the $[M]$, $[K]$, and $[C]$ matrix obtained in this way were substituted into the Lagrange equation of motion to calculate the natural frequencies, and the results are mentioned in Section 4.

2.2. Three Components Type Connected Structure

The beam is separated into three sections AB, BD, and CD, whose coordinates are shown as $w_A; x, w_B; x$, and $w_C; x$.

The properties of structures used in the numerical analysis are shown in Table 2.

Table 2. Properties & cross section of model.

Category	W_A	W_B	W_C	Cross Section
Density [kg/m^3]	7850	7850	7850	
Length [m]	5	5/2.5	5	
Area [m^2]	0.1	0.1	0.1	0.2m
Young's modulus [N/mm^2]	2.10×10^5	2.10×10^5	2.10×10^5	0.5m
2nd moment of area [mm^4]	3.33×10^8	3.33×10^8	3.33×10^8	

Deflections can be shown as below

$$W_A(x,t) = \sum_{i=1}^{m} w_{Ai}(x) \cdot p_{Ai}(t) \tag{23}$$

$$W_B(x,t) = \sum_{i=1}^{n} w_{Bi}(x) \cdot p_{Bi}(t) \tag{24}$$

$$W_C(x,t) = \sum_{i=1}^{k} w_{Ci}(x) \cdot p_{Ci}(t) \tag{25}$$

The deflection of subcomponent *AB*, *CD* and *BD* using fixed and simple boundary condition can be expressed as below:

$$W_A(\zeta,t) = \sum_{i=1}^{m} (\psi_i(\zeta)p_{Ai}(t) + \phi_i(\zeta)q_{Ai}(t) + \gamma_i(\zeta)r_{Ai}(t)) \tag{26}$$

$$W_B(\xi,t) = \sum_{j=1}^{n} (\phi_j(\xi)q_{Bj}(t) + v_j(\xi)s_{Bj}(t) + \lambda_j(\xi)o_{Bj}(t)) \tag{27}$$

$$W_C(\eta,t) = \sum_{k=1}^{s} (\psi_k(\eta)p_{Ck}(t) + \phi_k(\eta)q_{Ck}(t) + \gamma_k(\eta)r_{Ck}(t)) \tag{28}$$

$$U_{B0}(\xi,t) = u_{b0}(t) \tag{29}$$

$$U_{B1}(\xi,t) = u_{b1}(t) \tag{30}$$

where x_A, x_B, and x_c are non-dimensional such that $\zeta = \frac{x_A}{L_A}, \xi = \frac{x_B}{L_B}, \eta = \frac{x_C}{L_C}$. $p_{Ai}(t), q_{Ai}(t), r_{Ai}(t), q_{Bj}(t), s_{Bj}(t), o_{Bj}(t), p_{Ck}(t), q_{Ck}(t), r_{Ck}(t), u_{b0}(t), u_{b1}(t)$ are the general coordinate system in the mode function of beams, and U_{B0} and U_{B1} are displacement functions of both sides in horizontal beam. L_A, L_B and L_C are length of beams, separately.

Polynomial mode functions ψ_i and ϕ_i are mentioned in Section 2.1. Where γ_i, v_j, and λ_j are functions for fixed-free, simple-fixed, and simple-simple boundary conditions, respectively.

The fundamental mode function reflecting each boundary condition is shown in Tables 3–5.

Table 3. Fixed-free fundamental mode function (γ_1).

Boundary Condition	Mathematical Expression	Fundamental Mode Function
GBC (2) + NBC (2)	$\gamma_1(0) = 0,\ \gamma_1'(0) = 0$ $\gamma_1''(1) = 0,\ \gamma_1'''(1) = 0$	$\zeta^4 - 4\zeta^3 + 6\zeta^2$

GBC (2): Geometric Boundary Condition (number of boundary condition). NBC (2): Natural Boundary Condition (number of boundary condition).

Table 4. Simple-fixed fundamental mode function (v_1).

Boundary Condition	Mathematical Expression	Fundamental Mode Function
GBC (3) + NBC (1)	$v_1''(0) = 0,\ v_1(0) = 0$ $v_1(1) = 0,\ v_1'(1) = 0$	$\zeta^4 - 1.5\zeta^3 + 0.5\zeta$

Table 5. Simple-simple fundamental mode function (λ_1).

Boundary Condition	Mathematical Expression	Fundamental Mode Function
GBC (2) + NBC (2)	$\lambda_1(0) = 0,\ \lambda_1(1) = 0$ $\lambda_1''(0) = 0,\ \lambda_1''(1) = 0$	$\zeta^4 - 2\zeta^3 + \zeta$

The expansion to the higher-order term for the fundamental mode function for each boundary condition defined in Tables 3–5 is the same as the method mentioned in Equations (11) to (16). It shows the mass and stiffness matrix for the vertical member among the three component structures with the expanded mode function, as shown in Equations (31) and (32):

$$[M_A] = mL_A
\begin{vmatrix}
\psi_1\psi_1 & \cdots & \psi_1\psi_m & \psi_1\phi_1 & \cdots & \psi_1\phi_m & \psi_1\gamma_1 & \cdots & \psi_1\gamma_m \\
\vdots & \ddots & \vdots & \vdots & \ddots & \vdots & \vdots & \ddots & \vdots \\
\psi_m\psi_1 & \cdots & \psi_m\psi_m & \psi_m\phi_1 & \cdots & \psi_m\phi_m & \psi_m\gamma_1 & \cdots & \psi_m\gamma_m \\
\phi_1\psi_1 & \cdots & \phi_1\psi_m & \phi_1\phi_1 & \cdots & \phi_1\phi_m & \phi_1\gamma_1 & \cdots & \phi_1\gamma_m \\
\vdots & \ddots & \vdots & \vdots & \ddots & \vdots & \vdots & \ddots & \vdots \\
\phi_m\psi_1 & \cdots & \phi_m\psi_m & \phi_m\phi_1 & \cdots & \phi_m\phi_m & \phi_m\gamma_1 & \cdots & \phi_m\gamma_m \\
\gamma_1\psi_1 & \cdots & \gamma_1\psi_m & \gamma_1\phi_1 & \cdots & \gamma_1\phi_m & \gamma_1\gamma_1 & \cdots & \gamma_1\gamma_m \\
\vdots & \ddots & \vdots & \vdots & \ddots & \vdots & \vdots & \ddots & \vdots \\
\gamma_m\psi_1 & \cdots & \gamma_m\psi_m & \gamma_m\phi_1 & \cdots & \gamma_m\phi_m & \gamma_m\gamma_1 & \cdots & \gamma_m\gamma_m
\end{vmatrix}
\tag{31}$$

$$[K_A] = \frac{8EI}{L_A^3}
\begin{vmatrix}
\psi_1''\psi_1'' & \cdots & \psi_1''\psi_m'' & \psi_1''\phi_1'' & \cdots & \psi_1''\phi_m'' & \psi_1''\gamma_1'' & \cdots & \psi_1''\gamma_m'' \\
\vdots & \ddots & \vdots & \vdots & \ddots & \vdots & \vdots & \ddots & \vdots \\
\psi_m''\psi_1'' & \cdots & \psi_m''\psi_m'' & \psi_m''\phi_1'' & \cdots & \psi_m''\phi_m'' & \psi_m''\gamma_1'' & \cdots & \psi_m''\gamma_m'' \\
\phi_1''\psi_1'' & \cdots & \phi_1''\psi_m'' & \phi_1''\phi_1'' & \cdots & \phi_1''\phi_m'' & \phi_1''\gamma_1'' & \cdots & \phi_1''\gamma_m'' \\
\vdots & \ddots & \vdots & \vdots & \ddots & \vdots & \vdots & \ddots & \vdots \\
\phi_m''\psi_1'' & \cdots & \phi_m''\psi_m'' & \phi_m''\phi_1'' & \cdots & \phi_m''\phi_m'' & \phi_m''\gamma_1'' & \cdots & \phi_m''\gamma_m'' \\
\gamma_1''\psi_1'' & \cdots & \gamma_1''\psi_m'' & \gamma_1''\phi_1'' & \cdots & \gamma_1''\phi_m'' & \gamma_1''\gamma_1'' & \cdots & \gamma_1''\gamma_m'' \\
\vdots & \ddots & \vdots & \vdots & \ddots & \vdots & \vdots & \ddots & \vdots \\
\gamma_m''\psi_1'' & \cdots & \gamma_m''\psi_m'' & \gamma_m''\phi_1'' & \cdots & \gamma_m''\phi_m'' & \gamma_m''\gamma_1'' & \cdots & \gamma_m''\gamma_m''
\end{vmatrix}
\tag{32}$$

where we can express the mass matrix as Equation (33) by using orthogonality:

$$[M_A] = mL_A \begin{vmatrix} 1 & \cdots & 0 & \psi_1\phi_1 & \cdots & \psi_1\phi_m & \psi_1\gamma_1 & \cdots & \psi_1\gamma_m \\ \vdots & \ddots & \vdots & \vdots & \ddots & \vdots & \vdots & \ddots & \vdots \\ 0 & \cdots & 1 & \psi_m\phi_1 & \cdots & \psi_m\phi_m & \psi_m\gamma_1 & \cdots & \psi_m\gamma_m \\ \phi_1\psi_1 & \cdots & \phi_1\psi_m & 1 & \cdots & 0 & \phi_1\gamma_1 & \cdots & \phi_1\gamma_m \\ \vdots & \ddots & \vdots & \vdots & \ddots & \vdots & \vdots & \ddots & \vdots \\ \phi_m\psi_1 & \cdots & \phi_m\psi_m & 0 & \cdots & 1 & \phi_m\gamma_1 & \cdots & \phi_m\gamma_m \\ \gamma_1\psi_1 & \cdots & \gamma_1\psi_m & \gamma_1\phi_1 & \cdots & \gamma_1\phi_m & 1 & \cdots & 0 \\ \vdots & \ddots & \vdots & \vdots & \ddots & \vdots & \vdots & \ddots & \vdots \\ \gamma_m\psi_1 & \cdots & \gamma_m\psi_m & \gamma_m\phi_1 & \cdots & \gamma_m\phi_m & 0 & \cdots & 1 \end{vmatrix} \tag{33}$$

and M_B, M_C, and K_B, K_C can be expressed similarly to Equations (31) and (32).

Therefore, the mass and stiffness matrix of the subcomponents in Figure 2 can be expressed as in Equations (34) and (35).

$$[M] = \begin{vmatrix} M_A & 0 & 0 \\ 0 & M_B & 0 \\ 0 & 0 & M_C \end{vmatrix} \tag{34}$$

$$[K] = \begin{vmatrix} K_A & 0 & 0 \\ 0 & K_B & 0 \\ 0 & 0 & K_C \end{vmatrix} \tag{35}$$

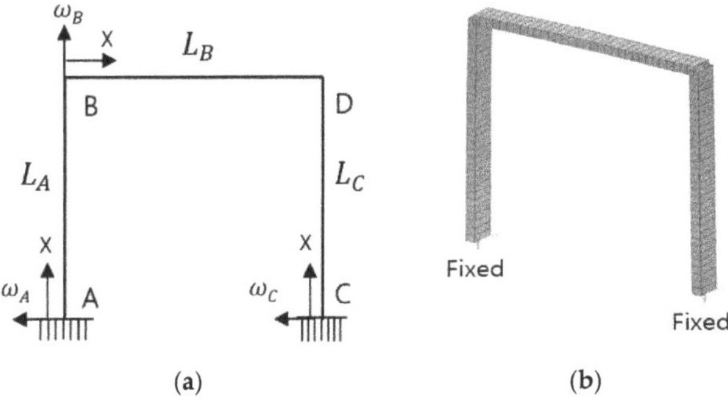

Figure 2. Structure model of three components type connected beam, (a) simplified model (b) FEA model.

Note that the number of generalized coordinates shall be 3 ($m + n + k$), since the number of constraints at junction B, D are 2, separately.

Displacement continuity:

$$w_A(1) - u_B(0) = 0 \tag{36}$$

$$w_C(1) - u_B(1) = 0 \tag{37}$$

Slope continuity:

$$w_A{}'(1) - w_B{}'(0) = 0 \tag{38}$$

$$w_A{}'(1) - w_B{}'(1) = 0 \tag{39}$$

The coordinate system reflecting the constraints is expressed in Equation (40):

$$
\begin{Bmatrix}
p_{A1} \\ \vdots \\ p_{Am} \\ q_{A1} \\ \vdots \\ q_{Am} \\ r_{A1} \\ \vdots \\ r_{Am} \\ q_{B1} \\ \vdots \\ q_{Bm} \\ s_{B1} \\ \vdots \\ s_{Bn} \\ O_{B1} \\ \vdots \\ O_{Bn} \\ p_{C1} \\ \vdots \\ p_{Ck} \\ q_{C1} \\ \vdots \\ q_{Ck} \\ r_{C1} \\ \vdots \\ r_{Ck} \\ u_{b0} \\ u_{b1}
\end{Bmatrix}
= [C]
\begin{Bmatrix}
p_{A1} \\ \vdots \\ p_{Am} \\ q_{A1} \\ \vdots \\ q_{Am} \\ r_{A1} \\ \vdots \\ r_{Am} \\ q_{B1} \\ \vdots \\ q_{Bn-1} \\ s_{B1} \\ \vdots \\ s_{Bn-1} \\ O_{B1} \\ \vdots \\ O_{Bn} \\ p_{C1} \\ \vdots \\ p_{Ck} \\ q_{C1} \\ \vdots \\ q_{Ck} \\ r_{C1} \\ \vdots \\ r_{Ck}
\end{Bmatrix}
\tag{40}
$$

where the [C] matrix represents a constraint and is implemented in the same way as Equation (18).

The [M], [K], and [C] matrices implemented in this way were substituted into the Lagrange equation of motion to calculate the natural frequencies, and the results are mentioned in Section 4.

3. Proposal of Methodology for Simplification of Computation

3.1. Two Components Type Connected Structure

Frequently, we may more concern about the vibration of one subcomponent. Suppose we concern the vibration of part OA in Figure 1.

The suitable boundary conditions at junction O can be obtained by removal of generalized coordinates of subcomponent OB through satisfaction of constraints at junction O as shown in Equations (41) and (42):

$$
q_{Bj} = \alpha q_{Ai} \tag{41}
$$

$$
p_{Bj} = -\alpha^2 p_{Ai} \text{ for } i = 1 \text{ to } m \ (m = n) \tag{42}
$$

where α is the ratio of length for subcomponents ($\alpha = L_B/L_A$):

$$W_A(\zeta, t) = \sum_{i=1}^{m} (\psi_i(\zeta) p_{Ai}(t) + \phi_i(\zeta) q_{Ai}(t)) \tag{43}$$

$$W_B(\xi, t) = \sum_{j=1}^{n} \left(\psi_j(\xi) \times -\alpha^2 p_{Ai}(t) + \phi_j(\xi) \times \alpha q_{Ai}(t) \right) \tag{44}$$

Although the assumption of simplification by the constraint at the junction O is excessive, the natural frequency and mode shape in the range $\alpha = 0$ to 1 are similar to the FEA results. Therefore, we may expect that this will give reasonable result for all cases ($0 \leq \alpha \leq 1$).

$$[M] = mL_A \begin{vmatrix} \psi_1\psi_1(1+\alpha^5) & \cdots & \psi_1\psi_n(1+\alpha^5) & \psi_1\phi_1(1-\alpha^4) & \cdots & \psi_1\phi_n(1-\alpha^4) \\ \vdots & \ddots & \vdots & \vdots & & \vdots \\ \psi_m\psi_1(1+\alpha^5) & \cdots & \psi_m\psi_n(1+\alpha^5) & \psi_m\phi_1(1-\alpha^4) & \cdots & \psi_m\phi_n(1-\alpha^4) \\ \phi_1\psi_1(1-\alpha^4) & \cdots & \phi_1\psi_n(1-\alpha^4) & \phi_1\phi_1(1+\alpha^3) & \cdots & \phi_1\phi_n(1+\alpha^3) \\ \vdots & \ddots & \vdots & \vdots & & \vdots \\ \phi_m\psi_1(1-\alpha^4) & \cdots & \phi_m\psi_n(1-\alpha^4) & \phi_m\phi_1(1+\alpha^3) & \cdots & \phi_m\phi_n(1+\alpha^3) \end{vmatrix} \tag{45}$$

$$[K] = \frac{16EI}{L_A^3} \begin{vmatrix} \psi_1''\psi_1''(1+\alpha) & \cdots & \psi_1''\psi_n''(1+\alpha) & 0 & \cdots & 0 \\ \vdots & \ddots & \vdots & \vdots & & \vdots \\ \psi_m''\psi_1''(1+\alpha) & \cdots & \psi_m''\psi_n''(1+\alpha) & 0 & \cdots & 0 \\ 0 & \cdots & 0 & \phi_1''\phi_1''\left(1+\frac{1}{\alpha}\right) & \cdots & \phi_1''\phi_n''\left(1+\frac{1}{\alpha}\right) \\ \vdots & \ddots & \vdots & \vdots & & \vdots \\ 0 & \cdots & 0 & \phi_m''\phi_1''\left(1+\frac{1}{\alpha}\right) & \cdots & \phi_m''\phi_n''\left(1+\frac{1}{\alpha}\right) \end{vmatrix} \tag{46}$$

The mass and stiffness matrix can be created from Equations (43) and (44). Then, the degree of freedom can reduce 50% compared to the previous calculation method, and mass and stiffness matrix are shown in Equations (45) and (46).

The $[M]$ and $[K]$ matrices were substituted into the Lagrange equation of motion to calculate the natural frequencies, and the results are mentioned in Section 4.

3.2. Three Components Type Connected Structure

In the case of a structures in Figure 2, these structures have the symmetric and asymmetric modes as like Figure 3. In order to reflect the behavior of the structure and simplify the calculation, the mode function was applied separately according to the symmetric and asymmetric modes.

Firstly, in the case of asymmetry, the structure has a mode shape similar to the simple support condition affected by the slope at the middle point of the horizontal member, and in the case of symmetry, it has a mode similar to the fixed support with the slope close to 0 at the middle point.

Of course, it is not strictly a fixed condition because deflection occurs at the middle point, but it is suitable as a simplification method as it satisfies the allowable range of analysis when calculated considering the fixed boundary condition.

This can be seen visually in the FEA results.

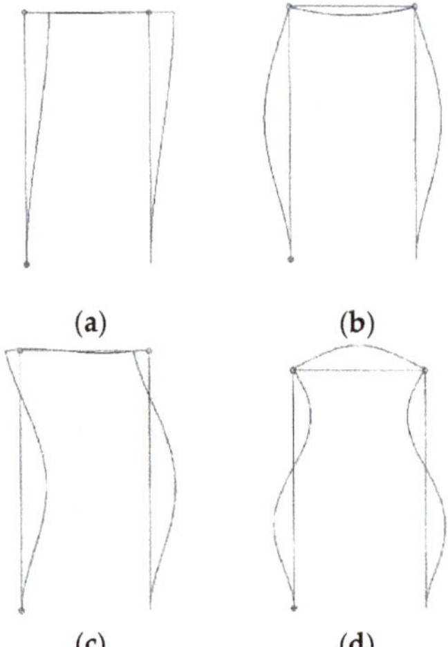

Figure 3. FEA Result of n-type structure: (**a**) 1st mode—asymmetry; (**b**) 2nd mode—symmetry; (**c**) 3rd mode—asymmetry; (**d**) 4th mode—symmetry.

In the case of three components type structure, Figures 4 and 5 show the shape of the symmetric and asymmetric modes in the intermediate position. It can be concluded that this can be implemented by a combination of mode functions suitable for conditions in symmetrical and asymmetry.

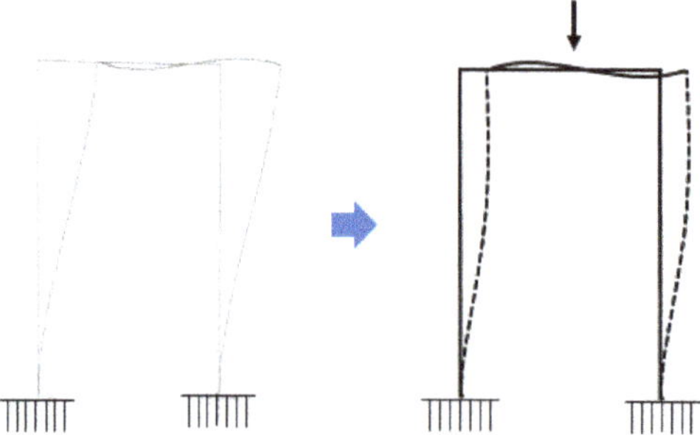

Figure 4. Idealization of approximate analytical approach for asymmetric mode.

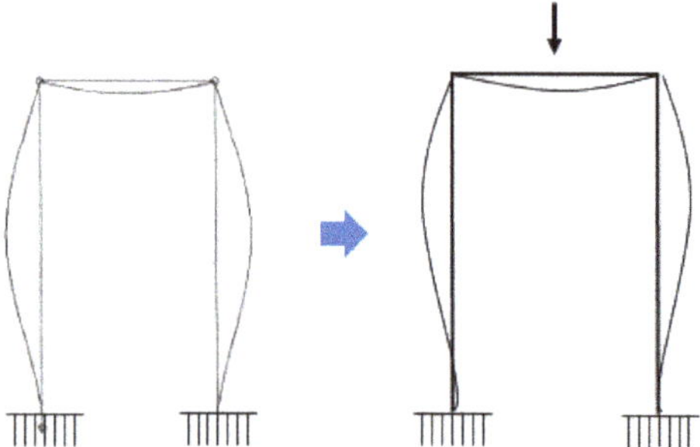

Figure 5. Idealization of approximate analytical approach for symmetric mode.

In the case of asymmetry as shown in Figure 4, the mode function of a simple boundary condition of 1/2 L (L: the length of horizontal beam) length is used for the horizontal member to the mode function of the existing vertical member, and in the case of Figure 5, the mode function of the fixed boundary condition is used to compare the results. The results performed in Section 4.2 were shown valid results within 15%.

In Table 6, the mode shapes for length ratios of 5 m:2.5 m:5 m are shown, the mode of 5 m:5 m:5 m is similar to the previous case.

Table 6. The mode shape for each length ratio (L_A:L_B:L_C = 5 m:2.5 m:5 m).

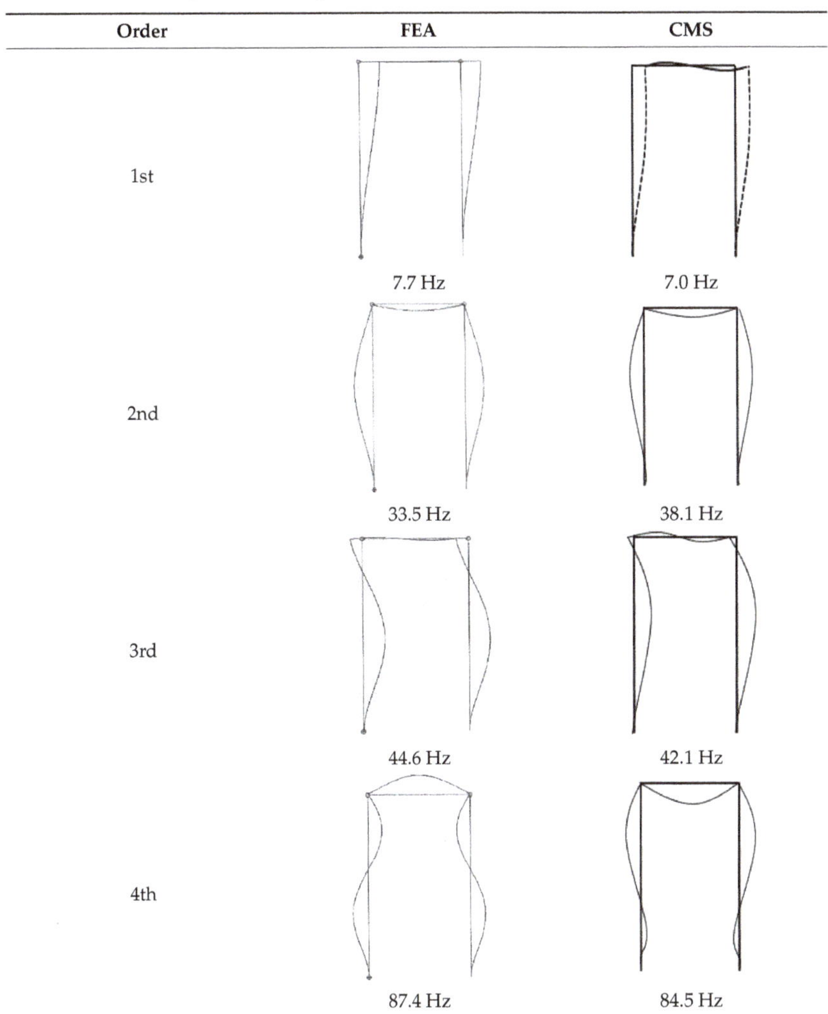

Order	FEA	CMS
1st	7.7 Hz	7.0 Hz
2nd	33.5 Hz	38.1 Hz
3rd	44.6 Hz	42.1 Hz
4th	87.4 Hz	84.5 Hz

4. Comparison of Numerical Results and Mode Shape

4.1. Comparison of Numerical Calculation Results by Fundamental Mode Function

The difference was confirmed by applying the fundamental mode function for each boundary condition defined above to the two components type connected structure.

The application cases are described in Table 7, and the results are shown in Table 8.

Table 7. The case of study for two components type.

Case	W_A	W_B	Junction Constraint
Case 1	F-F (4) + F-S (3)	F-F (4) + F-S (3)	
Case 2		F-F (4) + F-S (4)	Slope and Moment
Case 3	F-F (4) + F-S (4)	F-F (4) + F-S (3)	
Case 4		F-F (4) + F-S (4)	
Case 5	F-F (4) + F-S (3)	F-F (4) + F-S (3)	
Case 6		F-F (4) + F-S (4)	Slope Only
Case 7	F-F (4) + F-S (4)	F-F (4) + F-S (3)	
Case 8		F-F (4) + F-S (4)	

Table 8. The result of numerical calculation for fundamental mode of two components type.

L_A:L_B	Order	FEA	Case 1	Case 2	Case 3	Case 4	Case 5	Case 6	Case 7	Case 8
	1	29.1	29.4	29.4	29.4	29.4	29.4	29.4	29.4	29.4
1:1	2	41.9	42.6	42.6	42.6	42.6	42.6	42.6	42.6	42.6
	3	92.9	95.3	95.3	95.3	95.3	95.3	95.3	95.3	95.3
	1	35.2	39.2	39.2	39.2	39.2	39.1	39.1	39.1	39.1
1:0.6	2	86.5	85.8	85.8	85.8	85.8	85.6	85.6	85.6	85.6
	3	112.7	119.6	119.6	119.6	119.6	119.5	119.5	119.5	119.5
	1	36.8	42.6	42.6	42.6	42.6	42.5	42.5	42.5	42.5
1:0.4	2	101.8	116.1	116.1	116.1	116.1	116.0	116.0	116.0	116.0
	3	180.6	187.3	187.3	187.3	187.3	187.3	187.3	187.3	187.3

From the result of above study, it can be seen that the contribution of the geometric boundary condition to the natural frequency is dominant than the natural boundary condition. In addition, it was confirmed that the influence was insignificant even when moment continuity was considered as a constraint condition at the junction.

4.2. Two Components Type Connected Structure

The properties of structures used in the numerical analysis are shown in Table 1. Table 9 shows the results of component mode synthesis using the proposed polynomial function and the numerical results using the proposed simplification method.

Table 9. Comparison of numerical result (Unit:Hz).

Ratio (L_A:L_B)		1st	2nd	3rd	Ratio (L_A:L_B)		1st	2nd	3rd
	FEA	29.1	41.9	92.9		FEA	36.0	97.8	136.8
1:1	CMS	29.1	42.5	95.2	1:0.5	CMS	36.3	100.7	146.8
	PMSC	29.3	42.5	98.8		PMSC	36.4	101.6	147.5
	FEA	33.2	56.5	102.6		FEA	36.8	101.8	180.6
1:0.8	CMS	33.3	57.6	106.2	1:0.4	CMS	37.1	105.4	197.3
	PMSC	33.5	57.7	108.9		PMSC	37.2	106.3	200.9
	FEA	35.2	86.5	112.7		FEA	38.4	105.9	205.6
1:0.6	CMS	35.4	88.7	118.8	1:0.2	CMS	39.2	110.9	224.8
	PMSC	35.6	89.2	120.1		PMSC	39.3	111.5	233.8

CMS: Component mode synthesis. PMSC: Proposal of methodology for simplification of computation.

In Tables 10 and 11, the mode shapes for length ratios of 1:1, 1:0.5 are shown, respectively.

Table 10. The mode shape for each length ratio (L_A:L_B = 1:1).

Order	FEA	CMS	PMSC
1st	29.1 Hz	29.1 Hz	29.3 Hz
2nd	41.9 Hz	42.5 Hz	42.5 Hz
3rd	92.9 Hz	95.2 Hz	98.8 Hz

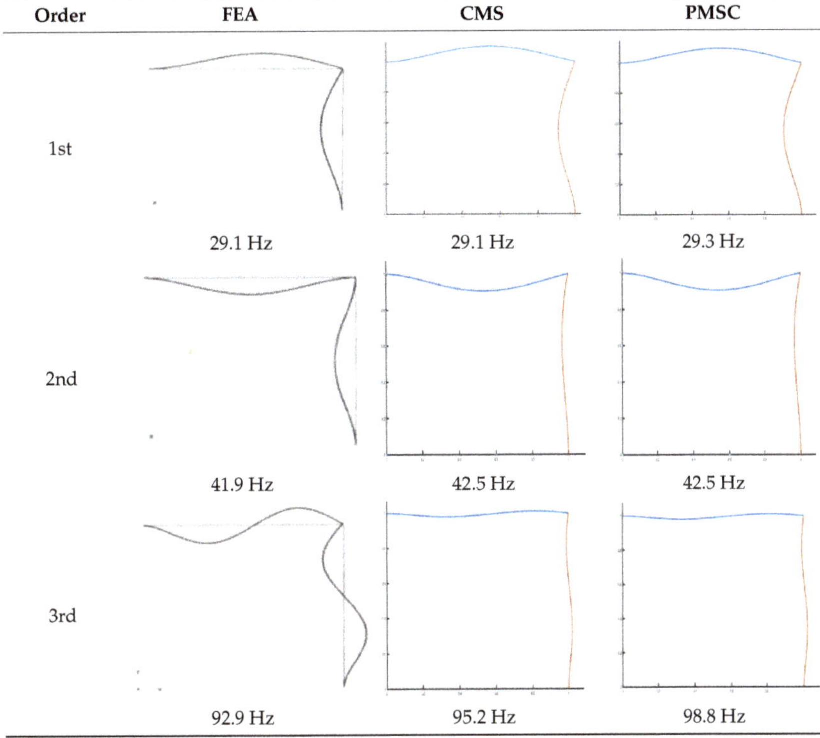

Table 11. The mode shape for each length ratio (L_A:L_B = 1:0.5).

Order	FEA	CMS	PMSC
1st	36.0 Hz	36.3 Hz	36.4 Hz
2nd	97.8 Hz	100.7 Hz	101.6 Hz
3rd	136.8 Hz	146.8 Hz	147.5 Hz

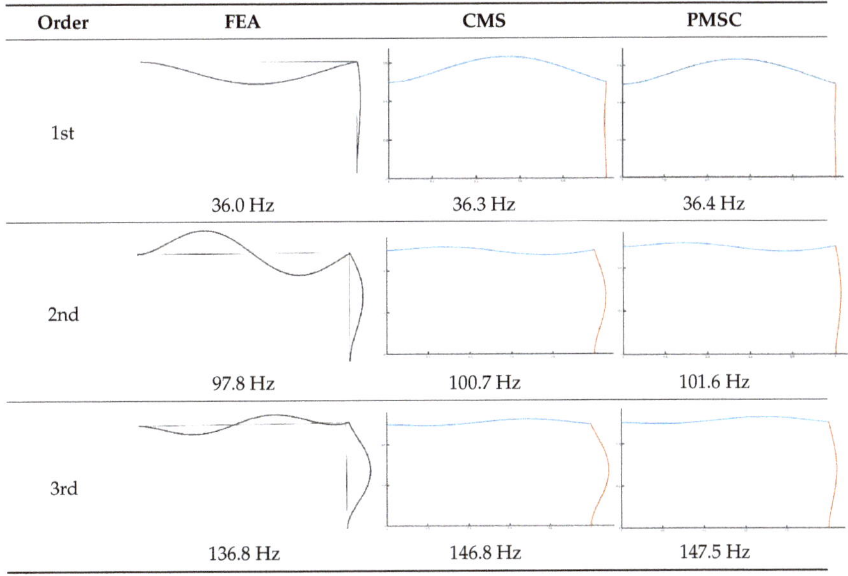

4.3. Three Components Type Connected Structure

The properties of structures used in the numerical analysis are shown in Table 2. Table 12 shows the results of component mode synthesis using the proposed polynomial function and the numerical results using the proposed simplification method, and the mode shape is shown in Section 3.2.

Table 12. Comparison of numerical result (Unit:Hz).

Ratio (L_A:L_B:L_C)	Calculation Method	1st	2nd	3rd	4th
5 m:5 m:5 m	FEA	6.1	23.9	38.9	41.9
	CMS	6.7	20.8	37.4	42.0
	PMSC	6.8	21.2	37.7	42.4
5 m:2.5 m:5 m	FEA	7.7	33.5	44.6	87.4
	CMS	7.0	38.1	42.1	84.5
	PMSC	7.3	41.2	44.3	86.2

5. Conclusions

We proposed the use of polynomials that can satisfy the boundary conditions at the junction between subcomponents, and a method that can be calculated by dramatically reducing the infinitely increasing degree of freedom.

To do this, numerical results of a structural subcomponent OA considering dynamic and static coupling of subcomponent OB are also given, and these numerical results are also compared with result of FEA. Numerical results prove the following:

1. In the two and three components, which are the typical shapes of the structure, it is proposed that the effective boundary conditions of the junction can be implemented by a combination of the appropriate boundary conditions of fixed, simple, and free.
2. Proposed polynomial mode functions allows wide and mode effective application of component mode synthesis for the vibration analysis of many ship local structures.
3. Numerical results based on the suggested method to reflect the dynamic and static coupling of connected subcomponent are proved to show very good agreement.
4. The proposed polynomial function is efficient enough to be compared with the FEA results in terms of natural frequency and mode.
5. The application of method suggested can be easily expanded for the analysis of Timoshenko beam or other more complicated structures such as reinforced plates.

Author Contributions: Conceptualization, J.H.P., D.Y.Y.; methodology, J.H.P.; D.Y.Y.; software, J.H.P.; validation, J.H.P., D.Y.Y.; formal analysis, J.H.P., D.Y.Y.; investigation, J.H.P.; writing—original draft preparation, J.H.P., D.Y.Y.; writing—review and editing, J.H.P., D.Y.Y.; funding acquisition, J.H.P. All authors have read and agreed to the published version of the manuscript.

Funding: This research received no external funding.

References

1. Cho, D.-S.; Kim, J.-H.; Choi, T.-M.; Kim, K.-S.; Choi, S.-W.; Jung, T.-S.; Lee, D.-K.; Seok, H.-I. Development of a Framework for Improving Efficiency of Ship Vibration Analysis. *KSNVE* **2011**, *21*, 761–767. [CrossRef]
2. Han, S.W.; Kweon, H. Development of vibration analysis program for local panel with additive. In Proceedings of the Conference of KSNVE, Jeju, Korea, 22–24 June 2000; pp. 311–321.
3. Park, J.-H.; Yang, J.-H. Normal Mode Analysis for Connected Plate Structure Using Efficient Mode Polynomials with Component Mode Synthesis. *Appl. Sci.* **2020**, *10*, 7717. [CrossRef]
4. Bhat, R. Natural frequencies of rectangular plates using characteristic orthogonal polynomials in rayleigh-ritz method. *J. Sound Vib.* **1985**, *102*, 493–499. [CrossRef]
5. Han, S.Y. Vibration Analysis of Local Panel for the Hull Structure (Written in Korean). *Bull. Soc. Nav. Archit. Korea* **1994**, *31*, 33–35.
6. Han, S.Y.; Huh, Y.C. Vibration Analysis of Quadrangular plate having attachments by the Assumed Mode Method (Written in Korean). *Trans. Soc. Nav. Archit. Korea* **1995**, *32*, 116–125.

7. Kim, B.H.; Kim, J.H.; Cho, D.S. Free Vibration Analysis of Stiffened Plates Using Polynomials Having the Property of Timoshenko Beam Functions. In *Proceedings of the Korean Society for Noise and Vibration Engineering Conference*; The Korean Society for Noise and Vibration Engineering: Seoul, Korea, 2004; pp. 623–628.
8. Thomson, W.T. *Theory of Vibration with Applications*, 4th ed.; Prentice Hall: Upper Saddle River, NJ, USA, 1993; pp. 360–365.
9. Thorby, D. *Structural Dynamics and Vibration in Practice*, 1st ed.; Elsevier: London, UK, 2008; pp. 194–213.
10. Hurty, W.C. Vibrations of structural systems by component mode synthesis. *J. Eng. Mech. Div.* **1960**, *86*, 51–69.
11. Cammarata, A.; Pappalardo, C. On the use of component mode synthesis methods for the model reduction of flexible multibody systems within the floating frame of reference formulation. *Mech. Syst. Signal Process.* **2020**, *142*, 106745. [CrossRef]
12. Hurty, W.C. Dynamic analysis of structural systems using component modes. *AIAA J.* **1965**, *3*, 678–685. [CrossRef]
13. Hintz, R.M. Analytical Methods in Component Modal Synthesis. *AIAA J.* **1975**, *13*, 1007–1016. [CrossRef]
14. Bourquin, F. Component mode synthesis and eigenvalues of second order operators: Discretization and algorithm. *ESAIM Math. Model. Numer. Anal.* **1992**, *26*, 385–423. [CrossRef]
15. Bamford, R.M. *A Model Combination Program for Dynamic Analysis of Structures*; Tech Memo 33-290; Jet Propulsion Laboratory: Pasadena, CA, USA, 1967.
16. Hou, S. Review of modal synthesis techniques and a new approach. *Shock Vibr. Bull.* **1969**, *40*, 25–39.
17. Rubin, S. Improved Component-Mode Representation for Structural Dynamic Analysis. *AIAA J.* **1975**, *13*, 995–1006. [CrossRef]
18. Benfield, W.A.; Hruda, R.F. Vibration Analysis of Structures by Component Mode Substitution. *AIAA J.* **1971**, *9*, 1255–1261. [CrossRef]
19. Carrera, E.; Pagani, A.; Giusa, D.; Augello, R. Nonlinear analysis of thin-walled beams with highly deformable sections. *Int. J. Non-linear Mech.* **2021**, *128*, 103613. [CrossRef]
20. Pagani, A.; Augello, R.; Carrera, E. Frequency and mode change in the large deflection and post-buckling of compact and thin-walled beams. *J. Sound Vib.* **2018**, *432*, 88–104. [CrossRef]
21. Carrera, E.; Pagani, A.; Augello, R. Effect of large displacements on the linearized vibration of composite beams. *Int. J. Non-linear Mech.* **2020**, *120*, 103390. [CrossRef]

Journal of
Marine Science and Engineering

Review

Collapse Strength of Intact Ship Structures

Mesut Tekgoz and Yordan Garbatov *

Centre for Marine Technology and Engineering (CENTEC), Instituto Superior Técnico, Universidade de Lisboa, 1049-001 Lisboa, Portugal; mesut.tekgoz@centec.tecnico.ulisboa.pt
* Correspondence: yordan.garbatov@tecnico.ulisboa.pt

Abstract: Ship structures are subjected to complex sea loading conditions, leading to a sophisticated structural design to withstand and avoid structural failure. Structural capacity assessment, particularly of the longitudinal strength, is crucial to ensure the safety of ships, crews, the marine environment, and the cargoes carried. This work aims to overview the ultimate strength assessment of intact ship structures in recent decades. Particular attention is paid to the ultimate strength of plates, stiffened panels, box girders, and entire ship hull structures. A discussion about numerical and experimental analyses is also provided. Finally, some conclusions and suggestions about potential future work are noted.

Keywords: intact plates; intact stiffened panels; box girders; ship structures; ultimate strength

Citation: Tekgoz, M.; Garbatov, Y. Collapse Strength of Intact Ship Structures. *J. Mar. Sci. Eng.* **2021**, *9*, 1079. https://doi.org/10.3390/jmse9101079

Academic Editor: Cristiano Fragassa

Received: 10 September 2021
Accepted: 28 September 2021
Published: 1 October 2021

Publisher's Note: MDPI stays neutral with regard to jurisdictional claims in published maps and institutional affiliations.

1. Introduction

The hull girder capacity assessment concept has been transformed drastically as more insights have been collected about how ships respond to an externally applied complex load. Initially, the breaking strength was the criterion to assess a ship's structural capacity. Still, it has evolved to include buckling strength. It later has moved into more sophisticated concepts involving the geometrical and material non-linearity of the structures, developing the ultimate strength or ultimate limit state criteria.

In the past, the criteria and procedures for the structural design of ships and offshore platforms were primarily based on the allowable stresses and simplified buckling checks for structural components. However, it is now well recognized that the ultimate limit state-based approaches are better suited for structural design and strength assessment than the traditional working stress-based approaches. The latter are typically formulated as a fraction of the material yield strength [1].

When it comes to the simple definition of the ultimate state, it is considered the stage where the ship structure cannot bear any further load increases and moves into its post-collapse or unstable phase. All the components that constitute the ship structure have their contributions to the ultimate strength. Each structural component, namely, stiffened plates or unstiffened ones, has its ultimate limit stages that add to the complexity of the ultimate limit state definition of a complex structure such as ships, which are composed of many components.

Many parameters influence the ultimate load carrying capacity of ship structures and range from the uncertainty of the materials that the ships are built of to the externally induced factors; namely, manufacturing defects that specifically involve welding-induced initial deformations and stresses, corrosion, fatigue-related cracks, the load that is acting on the ship's structural components and their interactions, which are the bending moments and shear forces from a local and also global point of view, not to mention human error and poorly maintained on-board equipment, which lead to ship collisions and groundings, etc.

This review aims to provide a general overview of the work that has been performed over recent years, including on the strength of intact unstiffened plates or stiffened ones, box girders and finally, hull girders built of steel.

2. Plates

Plates are the main structural components in ship structures. They are subjected to a variety of external loads.

The load type that leads the plate to buckle, the compressive load, has been a study of interest because it involves unstable behavior and influences the safety of the hull girder [2]. Figure 1 shows the fundamental behavior of thin and thick plates under uniaxial compressive load. The buckling strength of the thick perfect plate is decisive in evaluating the collapse behavior that occurs in the plastic regime. However, for an ideal thin plate where the buckling occurs in the elastic regime, it does not indicate its true collapse strength since, after the buckling, the thin plate can carry a further load until the external load reaches the stage where the resistance of the plate supports no more load increases and the plate collapses and moves into its unstable stage.

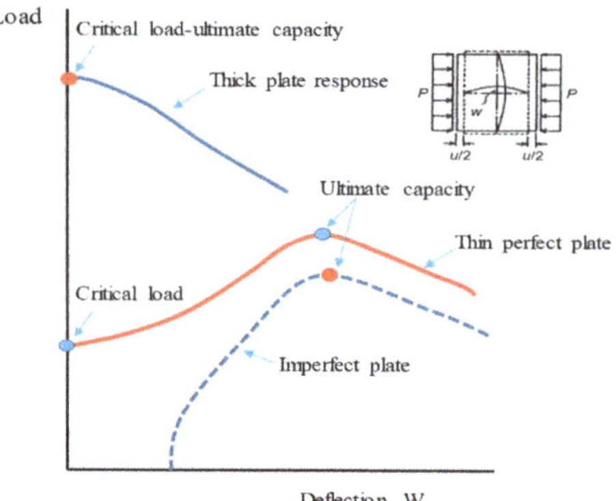

Figure 1. Fundamental behavior of plates under uniaxial compressive load.

However, in the case of typical ship plating, which is considered a thin or moderately thick plate, plate buckling generally does not occur in the critical sense due to the initial plate imperfections, and the deformations arise as soon as the load is applied. The plate carries the load until it collapses with a combination of buckling and plasticity. One of the primary failure modes of stiffened panels is the buckling and plastic collapse of the plates surrounded by the support members. Thus an evaluation of the plate buckling and plastic collapse behavior of the plates is essential in identifying the failure of ship structures [3,4].

2.1. Plate Collapse Assessment

Several design formulas have been presented under uniaxial compressive load over the years to predict the ultimate load carrying capacity of the intact plate components. The models proposed in [5,6] have been commonly used in marine structures. Frankland's model is less conservative than Faulkner's, which may be attributed to the assumed initial imperfections, boundary conditions or welding-induced stresses, an option for design approaches that prefer conservatism. These models predict a lower strength than the full plastic load, and this can be associated with the idea that only part of the plate is effective due to the buckling phenomenon, leading to a typical expression of plate effectiveness or effective width; that is, to reduce the strength of a plate by equating it to the strength of another plate that has an effective width and collapses at the nominal yield stress [7], which is a function of plate slenderness, given the fact that plate slenderness is a critical indicator of plate strength.

As suggested by Guedes Soares [8], it is possible to conclude that the most straightforward design method should include only plate slenderness because plate strength can change by as much as 60% over the useful range of the slenderness. Suppose an improvement of the accuracy is desired: in that case, explicit attention must be given to the variables that can produce changes of 20% in the plate strength, the initial plate imperfections, welding-induced residual stresses, and plate boundary conditions. It was indicated that the plate aspect ratio is only relevant at low plate slenderness from 0.6 to 1.0 and leads to about 5% of plate strength. However, this is more appropriate when the initial imperfection shape is concerned. The type of loading is to be explicitly accounted for.

The plate strength approach, using the effective width accounting for the welding-induced residual stresses, was explicitly proposed by Faulkner [6]. That formulation has been extended to expressly account for the effect of initial imperfections [8]. However, considering that the main impact on the collapse strength depends on the plate slenderness and that the initial imperfections are random, only properly treated by probabilistic methods [9], alternative formulations to [6] have been proposed by Guedes Soares [2,10], which treat the initial imperfections in a probabilistic way and derive design equations only dependent on the slenderness instead of including the effect of residual stresses explicitly as in the Faulkner's formulation, which incidentally covers both simply supported and clamped conditions. Carlsen [11], who also used the concept of the effective width given by a modified Faulkner's formula, proposed other simplified formulations adopted by DnV at the time.

Instead of using the type of effective width formulation, Dwight and Little [12], and Little [13], preferred to use the effective yield stress associated with a Perry-Robertson formulation, which is a function of the breadth to the thickness that governs the plate strength. They also considered different classes of plate curves depending on the level of residual stresses.

The plate buckling strength under a uniaxial load is influenced by several external parameters, including manufacturing defects, initial imperfections, welding-induced stresses, and plate edge boundary conditions. It is essential to point out that some of the formulas presented were built to assume that the plate edges are simply supported. This can be justified, as indicated by Fujikubo and Yao [14]. In most cases, the increase in buckling strength due to the stiffness provided by the stiffeners around the plate is only slightly more substantial than the decrease due to residual welding stresses. The resulting elastic buckling strength is close to a simply supported plate with no residual stress for both longitudinal and transverse thrust cases.

The importance of the formulations of plate strength results from the fact that ship-stiffened panels are commonly designed so that the plate fails first, and only at a later stage does the total failure of the combination of plate and stiffener occur. As pointed out by Faulkner et al. [15], the inability of the structure to carry an additional load will be limited either by the panel collapse or by grillage instability in which the plate elements and stiffeners show unstable behavior together. Therefore, stiffened panel collapse is represented by the failure of the plate and stiffener together. The contribution of the plate after its failure to the stiffener must be quantified. An approximate incremental method was proposed to estimate the load-shortening curves of a stiffener assembly with associated plates, including the post-collapse behavior accounting for the initial imperfection and welding-induced residual stresses [7].

Several studies have attempted to estimate the plate's average stress–strain relationship analytically by combining the larger elastic behavior of the plate with the rigid plastic method in the post-collapse region to determine the full load-shortening curves of the plates [16–19]. However, the analytical solutions are still to be improved because the ship plate collapse occurs in the plate elastic-plastic stage. It is also difficult to evaluate plate collapse involving complex failure modes under combined loading conditions.

It is challenging to formulate the non-linear governing equations representing both the geometrical and material non-linearity for plating, although it is not impossible. A significant source of difficulty is that an analytical treatment of plasticity with increased applied loads is quite cumbersome. Even if such treatment were possible, it would not be easy to solve the resulting non-linear equations analytically [20].

This approach was also adopted in a recent study performed in [21]. The solution of combining the large elastic displacement with small strain analysis using the concept of the principle of the virtual work is given for a uni-modal plate with initial imperfection under uni-axial thrust if the initial imperfection shape does not alter. However, the magnitude of the initial imperfection changes during external load exposure and the material non-linearity is not considered given the small strain assumption. They also implemented a simple solution to the post-collapse behavior of the plates using the concept of rigid plastic material behavior with the Von misses yield definition (see Figure 2).

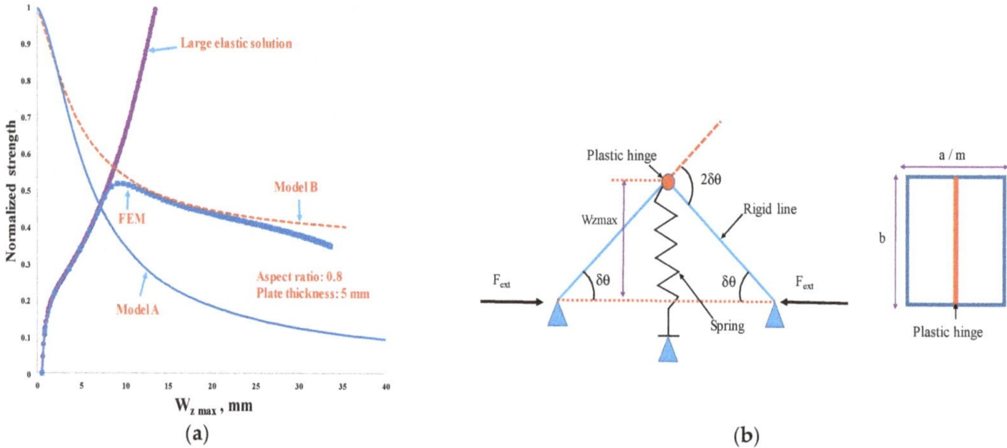

Figure 2. (a) Normalized strength and mid-plate vertical displacement, W_z, relationship; (b) simple rigid plastic solution.

2.2. Impact of Initial Imperfection

When a plate with initial defects and local bending stresses due to the initial defects and welding-induced stresses is subjected to thrust, it reduces the plate rigidity and ultimate load capacity [22]. It has generally been found that initial imperfections tend to decrease the rigidity and ultimate strength of plates. The weld-induced residual stresses have been shown to reduce the plates' stiffness and strength [23]. Therefore, numerous studies have investigated the influence of the initial imperfections, welding-induced residual stresses, different loading, and boundary conditions on the plates' structural response.

Kmiecik [24] investigated the influence of the general initial imperfections formulated by Fourier components, which were pioneering. The initial theoretical imperfections had been assumed as the buckling mode in most cases. He concluded that the initial imperfection buckling mode component has a significant reducing effect, whereas the rest have a stiffening influence. The impact of different modes of imperfections was also studied in [25,26] using the finite element method. In this period, several authors dealt with the behavior of plates using numerical methods, such as in [27–29], among others.

Carlsen and Czujko [30] discussed the specification of tolerances for post-welding distortions of plates. They demonstrated that only the buckling mode component of the Fourier series expansion of the distortion shape significantly influences the plate strength. In contrast, other Fourier components only have a marginal strengthening effect, consolidating the findings in [24].

Dow and Smith [31] studied the effects of the localized imperfections on the compressive strength of long rectangular plates, where they investigated several initial imperfection modes, including localized distortions. They showed that the initial imperfection leads to ultimate load capacity and stiffness reductions. They concluded that localized initial distortions are associated with less pre-collapse loss of stiffness and more rapid post-collapse unloading than equivalent periodic distortions. The magnitude of the localized distortions governs the plate collapse. It was also concluded that the Fourier distortion components corresponding to the initial elastic buckling mode will give a non-conservative representation of the initial shape if a localized initial distortion exists.

Ueda and Yao [32] investigated the influence of the complex initial defect modes on the elasto-plastic large deflection behavior and ultimate strength of rectangular plates under compression. They showed the importance of the initial imperfection. The magnitude and the components of the initial imperfection also significantly influence the plates' ultimate load carrying capacity, which consolidates the study performed by Kmiecik [24].

Guedes Soares and Kmiecik [33] investigated the compressive strength of unstiffened plates under uniaxial compression, accounting for the initial distortions using the finite element solution. They concluded that the uncertainty in plate strength depends strongly on the plate slenderness. Its most significant values are in the intermediate slenderness range and decrease with stocky plates, which agrees with the earlier formulation presented in [34].

Sadovský et al. [35,36] studied the influence of initial defects on the collapse strength of thin rectangular plates in longitudinal compression by normalizing the initial defects by the energy measure instead of the amplitude. They concluded that comparing the effects of the theoretical and measured initial defects upon the collapse strength and the buckling mode proved to yield a very conservative lower bound when adopting the amplitude to thickness ratio to measure the imperfections. A reasonably close lower bound was obtained using the buckling mode to normalize the initial defects by the energy measure.

Sadovský et al. [37] proposed an approach to the numerical study of the influence of the initial defects considered a random field on the resistance of in-plane loaded plates. The study demonstrated that a more realistic treatment of the effects of imperfections on the plate strength might lead to significantly less conservative design loads. Given that the initial imperfections are random and their impact is difficult to predict from a limited description of their shape, Sadovský and Guedes Soares [38] proposed the use of a neural network model trained with results of finite element calculations to predict the capacity of plates with imperfections and managed to obtain good results. An enhanced method in predicting the structural behavior of thick steel plates by using a random field approach has been proposed in [39,40].

Cubells et al. [41] used the photogrammetric technique to measure plates' real initial imperfection to assess the plates' structural capacity. They demonstrated the difference in the structural response using the initial imperfect geometry as calculated by this technique and the one modelled by the Fourier model. They concluded that the asymmetries of the initial imperfection are significant and influence the maximum capacity of the structures. The photogrammetry technique was used earlier to measure the deformation of three-dimensional structures [42,43] and in close range to determine the detailed distortions induced by welding on plates.

2.3. Impact of Welding Induced Stress

The importance of the residual stresses in plate strength has motivated the interest in calculating residual welding stresses [44–46] and the associated temperature distributions [47,48]. In this field, significant developments have been made by using finite element analyses to assess the temperature fields and the residual stresses. Many studies have adopted the Goldak et al. [49] model for modelling the transient welding process and calibrating the model's parameters to the welding conditions [50]. Others compared predictions with experimental results [51–57]. While the validation of the permanent defects induced by welding is efficiently conducted, the validation of the residual stresses is much more complicated and uncertain. Thus much fewer studies exist with this type of validation [58].

More complex cases have also been studied, including multi-pass welding [59] and the effect of the sequence in multiple welding [60]. Proper sequences are chosen to reduce the weld induced distortions, but restraining the plate during welding is another approach adopted and studied [61–65].

The initial imperfection and welding-induced residual stresses on the square plate load capacity were studied in [22]. They concluded that the influence of the welding-induced residual stresses on the plate buckling is significant. When the plate is thin, the welding-induced residual stresses have little effect on the ultimate strength. In contrast, they significantly influence the ultimate capacity of thick plates if local bending is present.

Ueda and Yao [66] presented the fundamental behavior of plates and stiffened plates subjected to welding imperfections. They concluded that welding-induced stresses reduce the plate in-plane rigidity and ultimate load carrying capacity. The reduction is more significant when the plate slenderness is about 1.8. Only one component of the initial imperfection is amplified when the load is above the buckling load for thin and long plates. They become stable as the load increases, and only the deflection of this component influences the ultimate strength. As for long thick plates, the most considerable curvature of the initial imperfection is a leading cause that initiates the spread of the plastic zone in the plate, resulting in the load carrying capacity reduction. The maximum curvature influences the ultimate load-carrying capacity reduction more than the maximum magnitude of the initial imperfection.

Paik and Sohn [67] studied the influence of welding-induced residual stresses on high-tensile plates. They concluded that longitudinal welding-induced stresses significantly impact thick plates in terms of their load carrying capacity. Transverse welding-induced stresses have a negligible effect on the ultimate load carrying capacity if the longitudinal compression is predominant. However, they have a significant influence on the post-collapse regime of the plates.

The shakeout effect on welding-induced residual stress for a plate under uniaxial cyclic load was investigated in [68], where it was demonstrated that it might lead to a 63.1% and 27.2% welding-induced residual stress reduction in the tensile and compressive stresses, respectively, when the applied cyclic load is 75% of the material yield strength, or cyclic load 2. When the applied cyclic load is 50% of yield stress, cyclic load 1 reduces the tensile stress by 40% and the compressive stress by 22.2%, demonstrating that the ships may gradually shake out the longitudinal welding-induced stresses.

Tekgoz et al. [69] studied the influence of welding-induced stresses with a moving heat source on the load carrying capacity of the plates, accounting for several plate thicknesses. They demonstrated that tension stresses are developed around the weld beads when the plates are welded in the middle. The angular plate distortions may increase the plate load carrying capacity in plates of a certain slenderness.

Gordo [70] proposed a methodology to evaluate the structural performance of rectangular plates accounting for the welding-induced stresses. Several strength influencing parameters were considered, and it was concluded that the welding-induced stresses influenced the plate pre-collapse significantly and the ultimate load capacity.

2.4. Load and Boundary Condition Impact

Guedes Soares and Kmiecik [71] investigated the influence of boundary conditions, accounting for initial imperfections, in square plate strength. They concluded that for stocky plates, the boundary conditions have an impact almost equivalent to those in perfect plates. However, when the initial imperfection is significant, their strength is insensitive to the boundary conditions. As for thin plates, it was concluded that the influence of initial imperfections is small, but the boundary conditions have significant effects.

A study about the influence of initial imperfections on a restrained plate was presented in [72]. It was demonstrated that restrained plates might have a larger ultimate capacity than the yield stress in stocky plates with small initial imperfection amplitude. Guedes Soares [8,10] had already considered this in a study where the design equations predicted an ultimate capacity larger than 1.0 in some conditions. It was concluded that high-order modes lead to minimum ultimate strength compared to the critical mode. A stress-induced direction perpendicular to the loading can reach very high values, generating non-negligible forces in the support structures.

The study of transverse strength is often related to analyzing the strength of plates in biaxial loading [73–75], and the compressive strength of rectangular plates under biaxial load and lateral pressure was studied in [76]. A method based on the formula used by the American Bureau Veritas or Det Norske Veritas for the plates having plate slenderness of more than 1.3 was suggested. The proposed method was extended to account for the initial defects and lateral pressure.

An investigation related to the transverse strength of the rectangular plates under transverse compression based on the test data collected and numerical results was performed in [77]. Several formulations have been compared and based on the results, and they proposed an alternative formula for the transverse strength of the plates. Alternative formulations have been discussed in [15,78,79], while numerical results are available in [75] and experimental results in [80].

Numerous authors have also studied the combined compressive loading and lateral pressure effects on the load capacity of plates [17,20,81–84].

A benchmark study was performed in [85,86] using several methods to estimate the limit state of unstiffened plates under bi-axial compressive load lateral pressure. It concluded that plate ultimate strength behavior is significantly affected by plate initial deflection shape, boundary conditions, and loading conditions. It was concluded that DNV-PULS [87] and ALPS/ULSAP [88] methods are beneficial for the ULS assessment of unstiffened plates in terms of the computational effort and the resulting accuracy, compared to more refined non-linear finite element solutions [89].

3. Stiffened Panels

In plate panels, longitudinal stiffeners provide the necessary support to the plates, ensuring that they retain the required strength [15]. A stiffened panel is an assembly of plate elements and support members, namely longitudinal stiffeners, and the interaction between the plate elements and support members regarding their geometrical and material properties. Other factors such as loading condition and initial imperfections play an essential role in the ultimate strength, buckling, and plastic collapse patterns of stiffened panels [90].

3.1. Failure Modes of Stiffened Panels

Failure of panels is usually classified as plate-induced failure, column-like failure, tripping of stiffeners and overall grillage failure [23,91]. However, a panel will be subjected to all failure modes, and it will finish up collapsing in the mode corresponding to its lowest strength [92]. It is common to calculate the strength of the weakest collapse pattern, representing the stiffened panel in question.

Plate-induced failure assumes that the stiffeners will carry the loads up to the material yield stress; therefore, they are fully effective throughout the external load exposure. In contrast, the plate itself induces failure, which may provide, in a sense, a higher bound of the strength. The effective plate width approach represents the reduction of plate stiffness. When the maximum stress acting on the plate reaches the material yield stress, overall collapse occurs. The response of a short stiffened panel with a length approximately equal to the width of the plate between stiffeners is dominated by plate failure [23]. However, as pointed out in [15], the inability of the structure to carry an additional load will be limited either by panel collapse or by grillage instability in which the plate elements and stiffeners weaken together. Therefore, stiffened panel collapse is represented by the failure of the plate and stiffener together.

Additionally, ship panels are longer in the critical locations where the ship sustains higher load levels, in which plate-induced failure possibility is low, and grillage failure is normally avoided by ensuring that the transverse frames are of adequate size. They are stiff enough to stimulate inter-frame panel collapse. Several design formulas have been proposed to predict the load-shortening response of stiffened panels accounting for different collapse modes.

3.2. Collapse Strength Assessment

Design formulas for two primary failure modes—strut or column failure of the stiffened plate and tripping of the stiffeners for stiffened panels under compressive load—were proposed in [15]. The beam-column failure of stiffened plates is based on the formulation suggested by the Johnson-Ostenfeld beam-column approach, which accounts for the inelastic weakening effect of column buckling of the plates and struts that have very high elastic buckling stress where the buckling occurs in the elastic–plastic regime. The plate and stiffener are subjected to the same maximum stress load, and the plate effective width approach to account for the weakening effect when plates are in the post-buckling stage is also accounted for.

Carlsten [93] proposed another method based on the Perry-Robertson method, which assumes that when the maximum compressive stress at the outermost fiber of the column cross-section reaches the material yield stress, the collapse is considered to take place.

Guedes Soares and Gordo [92] demonstrated the performance of three methods for beam-column failure as proposed in [15,93,94], where stiffened panels are under in-plane uniaxial compressive load based on the comparison between the numerical and experimental results. They demonstrated that the effective width prediction for plate-induced failure proposed by the Faulkner method falls between the ABS, which has an optimistic prediction, and that of Carlsen, which has a conservative prediction as a function of plate slenderness.

Paik and Kim [95] pointed out that the stiffener-induced failure mode based on the Perry-Robertson formula generally provides overly pessimistic results. In contrast, the plate-induced failure mode based on the Perry-Robertson formula reasonably predicts the ultimate panel strength in a specific range of stiffener dimensions, following the beam-column-type collapse mode.

Paik and Thayamballi [96] proposed an empirical formula as a function of the column slenderness and the attached plate slenderness coefficients based on extensive experimental data to estimate the ultimate load carrying capacity of stiffened panels, which implicitly accounts for stiffener web buckling or tripping as well as the beam-column-type collapse. Several empirical formulas to predict the load carrying capacity of the stiffened panels were reported in [97–100].

Ozdemir et al. [101] proposed a new method to predict the ultimate load capacity of the stiffened panels using the large elastic and initial yield solution. The proposed model was shown to be good enough to predict the ultimate load capacity of the stiffened panels compared with finite element solutions.

However, the calculation of the prediction of the collapse strength of stiffened panels is not sufficient to adequately represent the collapse behavior of hull girders. Therefore, other authors proposed several methods to predict the load-shortening curves of a stiffened panel, accounting for the collapse mode of stiffened panels.

An approximate incremental method to estimate the load-shortening curves of stiffened panels, including post-collapse behavior accounting for the initial imperfection and welding-induced residual stresses based on the proposed model, has been reported in [7,15]. The estimated load-shortening curves were compared to those predicted by the finite element solution, demonstrating a good agreement.

In the study of Yao and Nikolov [91] and its follow-up, Yao and Nikolov [102] proposed a method to calculate the average stress-strain behavior of stiffened panels to estimate a hull girder's ultimate load capacity. The method involves formulating the coupled flexural-torsional behavior of angle bar stiffeners, including stiffener tripping, which is one of the most critical stiffened panel failures since it comes with a very steep reduction of post-collapse strength and largely influences both the ultimate and post-collapse capacity of hull girders.

The International Association of Classification of Societies, IACS [103,104] has provided several collapse modes for evaluating the average stress-strain relationship of stiffened panels in a progressive collapse analysis based on the work in [7].

Li et al. [105] proposed an adaptable algorithm to predict stiffened panels' complete load-shortening curve under uniaxial compressive load. It has been shown that the proposed model is practical and efficient and can be further incorporated in the definition of hull girder strength assessments.

3.3. Impact of Governing Factors

Numerous studies have been focused on the impact of factors governing the load carrying capacity of stiffened panels, namely the external loading conditions, initial imperfections, welding-induced residual stress, etc.

3.3.1. Impact of Load and Boundary Conditions

Byklum and Amdahl [106] presented a computational model to predict the local buckling and post-buckling of stiffened panels, accounting for biaxial in-plane compression or shear and lateral pressure. The method predictions have been compared with the finite element solution and found to be in good agreement.

A series of elastic-plastic analyses for continuous stiffened plates subjected to combined transverse thrust and lateral pressure were performed in [107]. They proposed a formula based on FEM to estimate the ultimate strength of continuous stiffened panels under combined transverse and lateral pressure. It was demonstrated that the proposed formula is in good agreement with the FEM.

A benchmark study of several methods used to estimate the limit state of stiffened plates under bi-axial compressive load and lateral pressure was performed in [86]. It was concluded that DNV-PULS (2006) and ALPS/ULSAP (2006) methods are beneficial for the ULS assessment of stiffened plates in terms of the computational effort and the resulting accuracy.

Xu et al. [108] investigated the influence of model geometry and boundary conditions on the ultimate strength of stiffened panels under uniaxial compressive loading. They concluded that the boundary condition influences the ultimate strength and the collapsed shape of the stiffened panels. The fundamental advantage of periodic symmetric boundary conditions is that they can deal with either the symmetric or asymmetric collapse deformation mode. However, symmetric boundary conditions can only represent symmetric deformation. The influence of the boundary conditions on the load capacity of the stiffened panel using the numerical solution and experimental tests have also been reported in [109].

Tanaka et al. [110] performed a series of collapse analyses on 720 stiffened panels to evaluate the ultimate strength of the stiffened panels under longitudinal thrust by varying the numbers, types, and sizes of stiffeners and aspect ratio and slenderness ratio of local panels partitioned by stiffeners. They presented several conclusions. When smaller stiffeners are attached to thicker panels, overall buckling dominates the collapse behavior. The ultimate strength widely differs depending on the number of stiffeners and the aspect ratio of the whole stiffened panel. When larger stiffeners are provided, the collapse of local panels partitioned by the stiffeners dominates the collapse behavior. The ultimate strength of a stiffened panel under longitudinal thrust evaluated by CSR-B is in general lower compared to that assessed by FEM analysis.

Xu et al. [111] studied the ultimate load capacity of continuous stiffened panels under a combined longitudinal compressive load and lateral pressure using the analytical and finite element solutions. A two-span model with periodical boundary conditions was implemented. They demonstrated that with the increase of lateral pressure, the ultimate strength decreased with plate-induced failure, but it increased in the case of the stiffener-induced failure. The in-plane constraints on the longitudinal edges have a significant effect that enhances the load carrying capacity of the stiffened panels.

3.3.2. Impact of Material Properties and Welding Induced Effects

A study of the ultimate capacity of a stiffened panel accounting for different material and geometrical parameters was presented in [112]. Namely, material yield and ultimate tensile stress, Young's modulus, and plate initial imperfection and plate column slenderness were defined using the Monte Carlo simulation. The influence of these parameters was demonstrated using the ANOVA (analysis of variance) methodology to investigate the impact of these parameters on the load carrying capacity. The ultimate load capacity was estimated using the finite element solution.

Compressive tests on the stiffened panels with intermediate slenderness were performed in [113]. They studied the performance of conventional and U-type stiffened panels with hybrid material properties. It was demonstrated that hybrid panels have better performance than full S690 (material with a high tensile yield stress of 690 MPa) panels. The S690 material increases the average ultimate strength in the order of two times or higher when compared to mild steel when used in the plating. Experimental tests were also performed on short and long continuous panels in [114] and on long and narrow stiffened panels in [115,116].

Tekgoz et al. [117] investigated the influence of welding-induced stresses on the load capacity of stiffened panels. Two methods, namely, those testing the idealized welding stress field and modified material stress-strain curves, were implemented to study the impact of residual stresses. They concluded that the modified stress-strain curve approach is a fast approach showing a good agreement with the idealized welding stress field approach in terms of load capacity assessment of stiffened panels.

Tekgoz et al. [118] studied the impact of the welding sequences on the load carrying capacity of a stiffened panel, where the welding is simulated with a moving heat source. It was demonstrated that the welding sequence is the parameter that most affects the lateral deformation of the stiffener, which leads to more load carrying capacity reduction. The welding sequence profoundly influences the plate edge buckling pattern.

Gannon et al. [119] studied the influence of welding-induced residual stress release by the linear elastic shakedown phenomenon and its impact on stiffened panels' ultimate load carrying capacity. They demonstrated that the ultimate capacity might increase as much as 7%. They concluded that the ultimate load capacity increase depends on the failure location and the level of the welding-induced residual stresses.

Li et al. [120] investigated the influence of welding-induced residual stresses on the load capacity of stiffened panels using the finite element solution and the one provided by the standard structural rules. It was concluded that residual stress with average severity would reduce the ultimate load capacity significantly. The modified material's stress-strain curves were adopted to incorporate the influence of welding-induced stresses. The modified stress-strain curve approach did not correlate well with the one predicted by the finite element solution, but they concluded that it may still be a practical approach.

4. Box Structures

Box structures are used for different problems because they may represent ship structures on more minor scales. Therefore, numerous box girder case studies have been performed to understand ships' overall behavior using different structural and material configurations.

One of the first box girder studies oriented to represent hull girders was performed in [121], where the author performed collapse tests on seven box girders that were subjected to pure bending load. Ostapenko [122] proposed a method to determine the ultimate strength of the longitudinal hull girder under bending, shear, and torque. The method's validity has been verified by performing three box girder collapse tests under the specified loadings.

Nishihara [123] proposed a simple method to estimate the ultimate hull girder capacity under a pure bending load. He performed several box girder tests to represent different ship structures, namely, bulk carrier tanker and container ships, to validate the proposed method.

A four-point bending test on a box girder to represent the mid-ship region of conventional ships or FPSOs was presented in [124]. The authors presented two methods to evaluate the level of the residual stresses indirectly from the experimental data. The initial cycles to release welding-induced stresses led to loading memory effects on the structure, influencing the moment-curvature relationship.

The experimental behavior of a mild steel box girder under a pure bending moment was investigated in [125]. Several insights were provided in this experiment, including that the column slenderness controls the type of collapse of the structure. In essence, high column slenderness leads to more collapse, followed by a massive discharge of load during failure of the structure. The proposed approximate method based on the progressive collapse of stiffened plate elements gives a reasonable estimation of the ultimate load supported by the structure. It allows reproducing the effect of residual stresses on the box behavior.

Tekgoz et al. [126] studied the effect of the initial imperfection and transverse net section shapes on the ultimate strength of ship-shaped structures subjected to asymmetrical bending loading. They demonstrated that a triangular net section performs better than a square one when subjected to a hogging bending load. The net sectional sensitivity analysis demonstrated that a minimum ultimate bending moment is achieved at a different heeling angle to that of the other net sectional shapes. They also proposed a method to predict the continuous neutral axis rotation during an incremental load, which agrees with the one predicted by the finite element solution.

5. Ship Hull Structures

Ship structural collapse, or longitudinal strength, assessment has been a study of interest for a long time to ensure that ships withstand external loads without failure and provide safety for crews and traded cargoes throughout their lifetimes.

The prevention of hull collapse is the most critical task in ship structure design and safety assessment. When the strength of ship structures is assessed, it has been common to consider three strengths: longitudinal, transverse, and local strength. Among these, longitudinal strength, which is the hull girder strength against longitudinal bending, is the most fundamental and important strength to ensure the safety of ships [127]. Thus, an accurate and efficient method for computing the ultimate hull girder strength is always required in robust ship structural design [90].

Several examples of ship failure have led to catastrophic damage to the environment and loss of human life and the traded cargo. One of the most well-known is the structural collapse of the Titanic and subsequently the loss of thousands of human lives in 1912. A single-hull oil tanker, the Energy Concentration, broke in two during the unloading of its cargo in 1980. The single-hulled oil tanker Erika broke in two in rough weather in 1999; it was thought the sea weather was so severe that she could not withstand the external load. Reasons for these incidents include poor execution of cargo loading/unloading operations, age-related degradations, and unexpected events.

5.1. Ultimate Strength Assessment

The first known shipload capacity assessment dates from the 1850s when Isambard Kingdom Brunel designed the Great Eastern, built of iron. He applied the beam theory to estimate the bending stress acting on the ship where the failure concept was breakage of the iron plates in the deck and bottom under tension load. In 1874, John calculated the acting bending stresses assuming the wavelength equals the ship's length using the beam theory. However, it was observed that the beam theory underestimated the actual strain/displacement during a real ship collapse, that of the Wolf, which was attributed to the weakening effect of local buckling and the impact of shear lag [127]. These effects are neglected in the beam theory since it assumes a uniform stress distribution across the compressed flange.

Several actual or small-scaled ships were subsequently built to perform collapse tests to understand the ships' vertical bending collapse behavior [128–134]. These collapse experiments had similar findings in that they found the compressed flange of the ships led to buckling/plastic collapse, which consolidates the point that a full plastic bending moment is not attained due to the buckling phenomenon. It is an essential issue in the structural strength assessment of hull girders. The torsional strength of hull girders has also been experimentally documented in [135,136].

Yao [137] presented a review of hull girder collapse strength in which several advanced and straightforward methods were investigated in terms of their hull collapse prediction accuracy, namely simple techniques such as initial yielding, assumed stress distribution, and advanced methods such as progressive collapse methods with idealized or computed stress-strain curves, ISUM, and the finite element solution, FEM. It was concluded that only the advanced methods could capture the progressive collapse behavior of hull girders.

The initial yield assumption to estimate hull collapse strength is not an accurate indicator of the actual hull capacity given the fact that in real ship collapse, it involves geometrical non-linearity, being the panel buckling, and the material non-linearity, being the material plasticity factors that need considering of the evaluation of the hull collapse. It requires a progressive approach from the initial to the final collapse stage, which includes interacting with the hull girder's elements. Each element's strength contributes to the overall hull strength. Therefore, ship structural capacity assessment has moved into a more sophisticated phase, which involves the geometrical and material non-linearity of the structures; that is, the ultimate strength or ultimate limit state criteria.

Caldwell [138] pioneered the ship longitudinal ultimate capacity assessment that accounts for the weakening effect of the buckling phenomenon in the compressed flange of the hull girder by reducing the material yield stress. He assumed a bending stress distribution over the ship cross-section at the ultimate stage; then, the fully plastic vertical bending moment was calculated. The elements under tension are yielded, and those under compression reach their ultimate individual capacity, accounting for the buckling phenomenon. However, in this method, the transition from the linear stage to the fully plastic stage is neglected without considering the time lag between the components in terms of their failure. This assumption is not possible because of the weakening effect of the buckling phenomenon. However, the stress distribution presumed by Caldwell does not represent the ultimate limit states of modern ship structures, resulting in overestimated calculations of the ultimate hull girder strength [90].

Ueda and Rased [139] proposed an idealized unit method (ISUM) to estimate the ultimate transverse capacity of ship structures. The new ISUM elements were developed and reported in [140,141].

Smith [142] proposed a progressive collapse method, also widely used as the Common Structural Rules (IACS, 2015), to estimate the longitudinal strength under pure bending load. The ship cross-section is divided into components defined as plates with associated stiffeners, hard corners, or plate elements. The average stress and average strain relationships of the elements are firstly calculated, accounting for the plasticity and buckling. The average stress-strain curves later are used in the progressive collapse analysis, assuming that the load is path-dependent and incrementally increased in the form of curvature. More details on the progressive collapse procedure based on Smith´s method have been reported in [143].

In this method, two main assumptions are made. Firstly, the ship cross-section remains on a plane throughout the external load exposure. Each element is independent and progressively loses its strength and stiffness during the incremental permissible curvature, which includes the time lag of the component failure enabling the collapse to be progressive.

The first assumption rules out the possible influence of the vertically acting shear force that may induce cross-sectional warping deformations. The linear strain distribution over the hull girder cross-section needs to be updated at each curvature increment, accounting for the additional warping strains. It also rules out the influence of the shear on the components buckling and material yield strength definition, meaning the shear's impact on the individual component buckling strength needs to be accounted for with an updated material yield definition. It is a fact that probable ship failure occurs where the maximum bending moment takes place, and shear forces are negligible. However, the importance of the time-dependent load combination, shear, and bending forces should not be neglected.

The second assumption neglects the possible influence of the elements on each other as far as their stress-strain response is concerned, and the non-uniform external bending load that the ship side shell elements are subjected to during the vertical bending moment, which leads to both the compression and in-plane bending load that influences the side shell buckling strength. The Smith method prediction is profoundly affected by the accuracy of the individual component stress-strain curves.

For example, Rigo et al. [144] performed a sensitivity analysis of an ultimate hull girder bending moment using the Smith method. They studied the influence of three influencing factors in the progressive collapse analysis, PCA, namely decomposition of the hull cross-section into the elements, the meshed model, evaluation of the average stress-strain relationship, and execution of the progressive collapse analysis. They demonstrated that the assessment of the ultimate strength of individual elements is the most influencing factor and the average stress-strain relationship of the individual elements is not negligible. Several conclusions were presented; for example, a PCA model is better than another if the mesh model of the structure is less simplified.

The International Association of Classification of Societies, IACS [103,104] has initiated the calculation of the ultimate load carrying capacity of hull girders. Several collapse modes for evaluating the average stress-strain relationship of stiffened panels have been presented to be used in the progressive collapse analysis, as initially proposed in [142]. The progressive collapse method has also been documented in several benchmark studies [1,145,146].

Paik and Mansour [147] proposed a simple analytical formula by assuming a credible bending stress distribution over the hull cross-section at the collapse stage to predict the ultimate load capacity of hull girders under a vertical bending moment involving single and double hull ships. However, the method assumes a uniform stress distribution along the flange under compression. The experimental ultimate load capacities of several box girders and small-scale ships were presented. The ultimate experimental capacities were compared with the proposed method and several existing analytical and empirical formulations [148–152]. There are significant differences observed in the ultimate moment results from these formulas used in the comparison. They concluded that the proposed method is in good agreement with the experimental and numerical predictions and may be helpful in preliminary design estimates of the ultimate strength of ships under a vertical bending moment.

5.2. Buckling and Lateral Pressure Impact on Collapse Strength

Several authors have implemented Smith's method over the years. Yao and Nikolov [91,102] implemented the Smith approach to predict the load carrying capacity of several box girders and ship structures. They proposed an analytical solution to calculate the plate and stiffened panel stress-strain curves to be used in the progressive collapse analysis. It was shown that the cross-section could not carry the full plastic bending moment due to the local buckling in the compression side of bending. More decrease was observed in ultimate strength under sagging conditions than hogging conditions. It was demonstrated that the tripping of the stiffened panels leads to a 5% ultimate capacity decrease. They concluded that the presented method deals with only one cross-section. It can be improved by incorporating the loads acting in transverse directions and the vertical and horizontal shear forces.

Gordo and Guedes Soares [153] proposed an approximate method using the Smith method to evaluate the hull girder collapse strength under pure bending load. They presented several box girder collapse tests and 1/3-scaled frigate ship collapse results and compared the accuracy of the approximate method.

Yao et al. [154] investigated the influence of lateral load pressure on the ultimate load carrying capacity of a ship's hull girder under a longitudinal vertical bending moment. The proposed method to estimate the average stress-strain relationships of the plate and stiffened panels to be used in the progressive collapse analysis was compared with the one predicted by the finite element solution accounting for the lateral pressure where some discrepancy was observed, citing the assumption of the plastic formations. It was demonstrated that in the studied ship case, MV Energy Concentration, the lateral pressure effect on the load carrying capacity was not significant.

Garbatov et al. [155] presented an approach to verify the hull girder ultimate bending moment capacity of a real structure according to the Class Society Rules based on experimental results scaled by the dimensional theory. For this purpose, three experimentally tested box girders were used to verify the applicability of the presented approach. They demonstrated that the possible deviation of the correct estimation from the predicted one, for the analyzed real and model structural configurations, in the worst case, may be underestimated by 8.3% in the present example. They concluded that the proposed method might validate the ultimate global strength of real ship hull structures, in the phase of the new structural design or during the service life, and calibrate the newly developed codes.

The finite element solution is commonly used for the ultimate collapse analysis of marine structures, initially performed in [156]. Several studies have been conducted to predict a hull girder's load carrying capacity, accounting for pure bending or combined loading conditions and the lateral pressure using the finite element solution, as documented in [157–166].

6. Conclusions

A review of advances in the ultimate strength assessment of intact ship hull structures has been provided, showing that the structural behavior of steel plates, stiffened panels, ship-shaped box girders, and ship hulls had been extensively studied under different loading and environmental conditions.

It is concluded that there is still room for further studies on the impact of manufacturing and environmental conditions during ship service on the collapse modes of plates and stiffened panels.

Possible analytical solutions to estimate the average stress–strain relationship of plates and stiffened panels under monotonic and cyclic loads and the influence of the acting shear on the collapse strength of hull girders are also worth investigating.

Additional attention should be paid to the ultimate strength assessment of ship structures in service, which are typically subjected to very severe environmental conditions; these add to initial manufactory imperfections, and more imperfections are induced related to corrosion degradation, crack growth, collision, etc. [167].

Author Contributions: Conceptualization, M.T. and Y.G.; methodology, M.T. and Y.G.; investigation, M.T. and Y.G.; writing—original draft preparation, M.T. and Y.G.; writing—review and editing, M.T. and Y.G. All authors have read and agreed to the published version of the manuscript.

Funding: This research received no external funding.

Institutional Review Board Statement: Not applicable.

Informed Consent Statement: Not applicable.

Acknowledgments: This work was performed within the Strategic Research Plan of the Centre for Marine Technology and Ocean Engineering (CENTEC), which is financed by the Portuguese Foundation for Science and Technology (Fundação para a Ciência e Tecnologia-FCT) under contract UIDB-UIDP/00134/2020.

Conflicts of Interest: The authors declare no conflict of interest.

References

1. Paik, J.K.; Amlashi, H.; Boon, B.; Branner, K.; Caridis, P.; Das, P.; Fujikubo, M.; Huang, C.H.; Josefson, L.; Kaeding, P.; et al. ISSC Committee III.1 Ultimate Strength. In Proceedings of the 18th International Ship and Offshore Structures Congress, Rostock, Germany, 9–13 September 2012; pp. 285–364.
2. Guedes Soares, C. Uncertainty Modelling in Plate Buckling. *Struct. Saf.* **1988**, *5*, 17–34. [CrossRef]
3. Paik, J.K.; Thayamballi, A.K. *Ultimate Limit State Design of Steel-Plated Structures*; Wiley: Chichester, UK, 2003.
4. Paik, J.K.; Thayamballi, A.K. *Ship-Shaped Offshore Installations: Design, Building and Operation*; Cambridge University Press: Cambridge, UK, 2007.
5. Frankland, J.M. *The Strength of Ship Plating under Edge Compression*; U.S. Experimental Model Basin Progress Report 469; U.S. Experimental Model Basin: Washington, DC, USA, 1940.
6. Faulkner, D. A review of effective plating for use in the analysis of stiffened plating in bending and compression. *J. Ship Res.* **1975**, *19*, 1–17. [CrossRef]
7. Gordo, J.M.; Guedes Soares, C. Approximate load shortening curves for stiffened plates under uniaxial compression. *Integr. Offshore Struct.* **1993**, *5*, 189–211.
8. Guedes Soares, C. Design equation for the compressive strength of unstiffened plate elements with initial imperfections. *J. Constr. Steel Res.* **1988**, *9*, 287–310. [CrossRef]
9. Faulkner, D.; Guedes Soares, C.; Warwick, D.M. *Requirements for Structural Design and Assessment*; Elsevier: Amsterdam, The Netherlands, 1987.
10. Guedes Soares, C. Design equation for ship plate elements under uniaxial compression. *J. Constr. Steel Res.* **1992**, *22*, 99–114. [CrossRef]
11. Carlsen, C.A. Simplified collapse analysis of stiffened plates. *Nor. Marit. Res.* **1977**, *5*, 135–177.
12. Dwight, J.B.; Little, G.H. Stiffened steel compression flanges-a simpler approach. *Struct. Eng.* **1976**, *54*, 501–509.
13. Little, G.H. Stiffened steel compression panels—A theoretical failure analysis. *Struct. Eng.* **1976**, *54*, 489–500.
14. Fujikubo, M.; Yao, T. Elastic local buckling strength of stiffened plate considering plate/stiffener interaction and welding residual stress. *Mar. Struct.* **1999**, *12*, 543–564. [CrossRef]
15. Faulkner, D.; Adamchak, J.C.; Synder, G.J.; Vetter, M.F. Synthesis of welded grillages to withstand compression and normal loads. *Comput. Struct.* **1973**, *3*, 221–246. [CrossRef]
16. Cui, W.C.; Mansour, A.E. Generalization of a simplified method for predicting ultimate compressive strength of ship panels. *Int. Shipbuild. Prog.* **1999**, *447*, 291–303.
17. Cui, W.C.; Wang, Y.J.; Pedersen, P.T. Strength of ship plates under combined loading. *Mar. Struct.* **2002**, *15*, 75–97. [CrossRef]
18. Fujita, Y.; Nomoto, T.; Niho, O. Ultimate strength of stiffened plates subjected to compression. *J. Soc. Nav. Archit. Jpn.* **1977**, *141*, 190–197. [CrossRef]
19. Paik, J.K.; Pedersen, P.T. A simplified method for predicting ultimate compressive strength of ship panels. *Int. Shipbuild. Prog.* **1996**, *43*, 139–157.
20. Paik, J.K.; Thayamballi, A.K.; Lee, S.K.; Kang, S.J. A semi-analytical method for the elastic-plastic large deflection analysis of welded steel or aluminum plating under combined in-plane and lateral pressure loads. *Thin Wall Struct.* **2001**, *39*, 125–152. [CrossRef]
21. Tekgoz, M.; Garbatov, Y. Analysis of post-collapse behaviour of rectangular plate employing roof mode plastic solutions. In *Developments in Maritime Technology and Engineering*; Guedes Soares, C., Santos, T.A., Eds.; Taylor and Frances: London, UK, 2021; Volume 1, pp. 599–608.
22. Ueda, Y.; Yasukawa, W.; Yao, T.; Ikegami, H.; Ohminami, R. Effect of welding residual stresses and initial deflections on rigidity and strength of square plates subjected to compression. *Trans. JWRI* **1977**, *6*, 33–38.
23. Guedes Soares, C.; Soreide, T.H. Behavior and design of stiffened plates under predominantly compressive loads. *Int. Shipbuild. Prog.* **1983**, *30*, 13–27. [CrossRef]
24. Kmiecik, M. *Behaviour of Axially Loaded Simply Supported Long Rectangular Plates Having Initial Deformations*; Report No. 84; Ship Research Institute: Trondheim, Norway, 1971.
25. Crisfield, M.A. Full-range analysis of steel plates and stiffened plating under uniaxial compression. *Proc. Inst. Civ. Eng.* **1975**, *59*, 595–624. [CrossRef]
26. Frieze, P.A.; Dowling, P.J.; Hobbs, R.E. Ultimate load behaviour of plates in compression. In *Steel Plated Structures*; Dowling, P.J., Harding, J.E., Frieze, P.A., Eds.; Crosby Lockwod Staples: London, UK, 1977; pp. 26–50.
27. Bradfield, C.D.; Stonor, R.W.P. Simple collapse analysis of plates in compression. *J. Struct. Eng.* **1984**, *110*, 2976–2993. [CrossRef]
28. Harding, J.E.; Hobbs, R.E.; Neal, B.G. The elasto-plastic analysis of imperfect square plates under in-plane loading. *Proc. Inst. Civ. Eng.* **1977**, *63*, 137–158.
29. Matthies, H.; Payer, H.G. Elastic-plastic post-buckling behaviour of ship plating. *Comput. Struct.* **1981**, *13*, 745–750. [CrossRef]
30. Carlsen, C.A.; Czujko, J. The specification of post-welding distortion tolerances for stiffened plates in compression. *Struct. Eng.* **1978**, *56*, 133–141.

31. Dow, R.S.; Smith, C.S. Effects of localized imperfections on compressive strength of long rectangular plates. *J. Constr. Steel Res.* **1984**, *4*, 51–76. [CrossRef]
32. Ueda, Y.; Yao, T. The influence of complex initial deflection modes on the behaviour and ultimate strength of rectangular plates in compression. *J. Constr. Steel Res.* **1985**, *5*, 265–302. [CrossRef]
33. Guedes Soares, C.; Kmiecik, M. Simulation of the ultimate compressive strength of unstiffened rectangular plates. *Mar. Struct.* **1993**, *6*, 553–569. [CrossRef]
34. Guedes Soares, C. A code requirement for the compressive strength of plate elements. *Mar. Struct.* **1988**, *1*, 71–80. [CrossRef]
35. Sadovský, Z.; Teixeira, A.P.; Guedes Soares, C. Degradation of the compression strength of rectangular plates due to initial deflection. *Thin Wall Struct.* **2005**, *43*, 65–82. [CrossRef]
36. Sadovský, Z.; Teixeira, A.P.; Guedes Soares, C. Degradation of the compression strength of square plates due to initial deflection. *J. Constr. Steel Res.* **2006**, *62*, 369–377. [CrossRef]
37. Sadovský, Z.; Guedes Soares, C.; Teixeira, A.P. Random field of initial deflections and strength of thin rectangular plates. *Reliab. Eng. Syst. Saf.* **2007**, *92*, 1659–1670. [CrossRef]
38. Sadovský, Z.; Guedes Soares, C. Artificial neural network model of the strength of thin rectangular plates with weld induced initial imperfections. *Reliab. Eng. Syst. Saf.* **2011**, *96*, 713–717. [CrossRef]
39. Woloszyk, K.; Garbatov, Y. Random field modelling of mechanical behaviour of corroded thin steel plate specimens. *Eng. Struct.* **2020**, *212*, 110544. [CrossRef]
40. Woloszyk, K.; Garbatov, Y. An enhanced method in predicting tensile behaviour of corroded thick steel plate specimens by using random field approach. *Ocean Eng.* **2020**, *213*, 107803. [CrossRef]
41. Cubells, A.; Garbatov, Y.; Guedes Soares, C. Photogrammetry measurements of initial imperfections for the ultimate strength assessment of plates. *Int. J. Marit. Eng.* **2014**, *156*, A291–A302.
42. Chen, B.Q.; Garbatov, Y.; Guedes Soares, C. Displacement measurement of box girder based on photogrammetry. In Proceedings of the 11th International Symposium on Practical Design of Ships and Other Floating Structures, Rio de Janeiro, Brazil, 19–24 September 2010. PRADS2010-20083.
43. Chen, B.Q.; Garbatov, Y.; Guedes Soares, C. Measurement of weld-induced deformations in three-dimensional structures based on photogrammetry technique. *J. Ship Prod. Des.* **2011**, *27*, 51–62.
44. Kamtekar, A.G. The calculation of welding residual stresses in thin steel plates. *Int. J. Mech. Sci.* **1978**, *20*, 207–227. [CrossRef]
45. Porter Goff, R.F.D. A simplified analysis of the residual longitudinal stresses and strains due to gas-cutting and welding of thin steel plate. *Int. J. Mech. Sci.* **1979**, *21*, 287–300. [CrossRef]
46. Ueda, Y.; Yamakawa, T. Analysis of thermal elastic-plastic stress and strain during welding by finite element method. *Join. Weld. Res. Inst.* **1971**, *2*, 90–100.
47. Little, G.H.; Kamtekar, A.G. The effect of thermal properties and weld efficiency on transient temperatures during welding. *Comput. Struct.* **1998**, *68*, 157–165. [CrossRef]
48. Papazoglou, V.J.; Masubuchi, K. Numerical analysis of thermal stresses during welding including phase transformation effects. *J. Press. Vessel. Technol.* **1982**, *104*, 198–203. [CrossRef]
49. Goldak, J.; Chakravarti, A.; Bibby, M. A new finite element model for welding heat sources. *Metall. Trans. B* **1984**, *15*, 299–305. [CrossRef]
50. Fu, G.; Gu, J.; Lourenco, M.I.; Duan, M.; Estefen, S.F. Parameter determination of double-ellipsoidal heat source model and its application in the multi-pass welding process. *Ships Offshore Struct.* **2014**, *10*, 204–217. [CrossRef]
51. Biswas, P.; Mahpatra, M.; Mandal, N. Numerical and experimental study on prediction of thermal history and residual deformation of double-sided fillet welding. *J. Eng. Manuf.* **2009**, *223*, 1–10. [CrossRef]
52. Biswas, P.; Mandal, N.R.; Sha, O.P. Three-dimensional finite element prediction of transient thermal history and residual deformation due to line heating. *J. Eng. Marit. Environ.* **2007**, *221*, 17–30. [CrossRef]
53. Chang, P.H.; Teng, T.L. Numerical and experimental investigations on the residual stresses of the butt-welded joints. *Comput. Mater. Sci.* **2004**, *29*, 511–522. [CrossRef]
54. Chen, B.Q.; Hashemzadeh, M.; Garbatov, Y.; Guedes Soares, C. Numerical and parametric modelling and analysis of weld-induced residual stresses. *Int. J. Mech. Mater. Des.* **2014**, *11*, 439–453. [CrossRef]
55. Gery, D.; Long, H.; Maropoulos, P. Effects of welding speed, energy input and heat source distribution on temperature variations in butt joint welding. *J. Mater. Process. Technol.* **2005**, *167*, 393–401. [CrossRef]
56. Hashemzadeh, M.; Garbatov, Y.; Guedes Soares, C. Welding-induced residual stresses and distortions of butt-welded corroded and intact plates. *Mar. Struct.* **2021**, *79*, 103041. [CrossRef]
57. Hashemzadeh, M.; Garbatov, Y.; Guedes Soares, C.; O'Connor, A. Friction stir welding induced residual stresses in thick steel plates from experimental and numerical analysis. *Ships Offshore Struct.* **2021**, 1–9. [CrossRef]
58. Estefen, S.F.; Gurova, T.; Werneck, D.; Leontiev, A. Welding stress relaxation effect in butt-jointed steel plates. *Mar. Struct.* **2012**, *29*, 211–225. [CrossRef]
59. Fu, G.; Gurova, T.; Lourenco, M.I.; Estefen, S.F. Numerical and experimental studies of residual stresses in multipass welding of high strength shipbuilding steel. *J. Ship Res.* **2015**, *59*, 133–144. [CrossRef]
60. Biswas, P.; Mandal, N.R. Effect of welding tacks and sequences on residual stress in stiffened panel fabrication. *J. Ocean. Ship Technol.* **2011**, *2*, 99–113.

61. Adak, M.; Guedes Soares, C. Effects of different restraints on the weld-induced residual deformations and stresses in a steel plate. *Int. J. Adv. Manuf. Technol.* **2013**, *71*, 699–710. [CrossRef]
62. Adak, M.; Mandal, N.R. Numerical and experimental study of mitigation of welding distortion. *Appl. Math. Model.* **2010**, *34*, 146–158. [CrossRef]
63. Choobi, M.S.; Haghpanahi, M.; Sedighi, M. Investigation of the effect of clamping on residual stresses and distortions in butt-welded plates. *Mech. Eng.* **2010**, *17*, 387–394.
64. Hashemzadeh, M.; Garbatov, Y.; Soares, C. Reduction in weld induced distortions of butt-welded plates subjected to preventive measures. In *Analysis and Design of Marine Structures V*; Guedes Soares, C., Shenoi, A., Eds.; Taylor & Francis Group: London, UK, 2015; pp. 581–588.
65. Mahapatra, M.M.; Datta, G.L.; Pradhan, B.; Mandal, N.R. Modelling the effects of constraints and single-axis welding process parameters on angular distortions in one-sided fillet welds. *Proc. Inst. Mech. Eng. Part B J. Eng. Manuf.* **2007**, *221*, 397–407. [CrossRef]
66. Ueda, Y.; Yao, T. Fundamental Behavior of Plates and Stiffened Plates with Welding Imperfections. *Trans. JWRI* **1991**, *20*, 141–155.
67. Paik, J.K.; Sohn, J.M. Effects of welding residual stresses on high tensile steel plate ultimate strength: Non-linear finite element method investigations. In Proceedings of the 28th International Conference on Offshore Mechanics and Arctic Engineering, Honolulu, HI, USA, 31 May–5 June 2009. OMAE2009-79297.
68. Tekgoz, M.; Garbatov, Y.; Guedes Soares, C. Ultimate strength of a plate accounting for shakedown effect and corrosion degradation. In *Developments in Maritime Transportation and Exploitation of Sea Resources*; CRC Press: Boca Raton, FL, USA, 2014; pp. 395–403.
69. Tekgoz, M.; Garbatov, Y.; Guedes Soares, C. Ultimate strength assessment of welded stiffened plates. *Eng. Struct.* **2015**, *84*, 325–339. [CrossRef]
70. Gordo, J.M. Effect of Residual Stresses on the Elastoplastic Behaviour of Welded Steel Plates. *J. Mar. Sci. Eng.* **2020**, *8*, 702. [CrossRef]
71. Guedes Soares, C.; Kmiecik, M. Influence of the boundary conditions on the collapse strength of square plates with initial imperfections. In *Marine Technology and Transportation*; Graczyk, T., Jastrzebski, T., Brebbia, C.A., Burns, R.S., Eds.; Computational Mechanics Publications: Southampton, UK, 1995; pp. 227–235.
72. Gordo, J.M. Effect of initial imperfections on the strength of restrained plates. *J. Offshore Mech. Arct. Eng.* **2015**, *137*, 051401. [CrossRef]
73. Paik, J.K.; Thayamballi, A.K.; Lee, M.J. Effect of initial deflection shape on the ultimate strength behaviour of welded steel plates under biaxial compressive loads. *J. Ship Res.* **2004**, *48*, 45–60. [CrossRef]
74. Steen, E.; Byklum, E.; Hellesland, J. Elastic post-buckling stiffness of biaxially compressed rectangular plates. *Eng. Struct.* **2008**, *30*, 2631–2643. [CrossRef]
75. Valsgaard, S. Numerical design prediction of the capacity of plates in-plane compression. *Comput. Struct.* **1980**, *12*, 729–739. [CrossRef]
76. Guedes Soares, C.; Gordo, J.M. Compressive strength of rectangular plates under biaxial load and lateral pressure. *Thin Wall Struct.* **1996**, *24*, 231–259. [CrossRef]
77. Guedes Soares, C.; Gordo, J.M. Compressive strength of rectangular plates under transverse load and lateral pressure. *J. Constr. Steel Res.* **1996**, *36*, 215–234. [CrossRef]
78. Becker, H. Instability strength of polyaxially loaded plates and relation to design. In Proceedings of the Steel Plated Structures: An international symposium, London, UK, July 1977; pp. 51–88.
79. Becker, H.; Colao, A. *Compressive Strength of Ship Hull Girders*; Ship Structures Committee Report SSC-267; U.S. Coast Guard: Washington, DC, USA, 1977.
80. Bradfield, C.D.; Stonor, R.W.P.; Moxham, K.E. Tests of long plates under biaxial compression. *J. Construct. Steel Res.* **1993**, *24*, 25–56. [CrossRef]
81. Fujikubo, M.; Yao, T.; Khedmati, M.R.; Harada, M.; Yanagihara, D. Estimation of ultimate strength of continuous stiffened panel under combined transverse thrust and lateral pressure—Part 1: Continuous plate. *Mar. Struct.* **2005**, *18*, 383–410. [CrossRef]
82. Fujita, Y.; Nomoto, T.; Niho, O.; Yoshie, A. Ultimate strength of rectangular plates subjected to combined loading (2nd Report). *J. Soc. Nav. Archit. Jpn.* **1979**, *146*, 289–298. [CrossRef]
83. Smith, C.S.; Anderson, N.; Chapman, J.C.; Davidson, P.C.; Dowling, P.J. Strength of Stiffened Plating under Combined Compression and Lateral Pressure. *Trans. RINA* **1991**, *134*, 131–147.
84. Yao, T.; Fujikubo, M.; Mizulani, K. Collapse behaviour of rectangular plates subjected to combined thrust and lateral pressure. *Trans. West-Jpn. Soc. Nav. Archit.* **1996**, *92*, 249–262.
85. Paik, J.K.; Kim, B.J.; Seo, J.K. Methods for ultimate limit state assessment of ships and ship-shaped offshore structures: Part I—Unstiffened plates. *Ocean Eng.* **2008**, *35*, 261–270. [CrossRef]
86. Paik, J.K.; Kim, B.J.; Seo, J.K. Methods for ultimate limit state assessment of ships and ship-shaped offshore structures: Part II—Stiffened plates. *Ocean Eng.* **2008**, *35*, 271–280. [CrossRef]
87. DNV-PULS. *User's Manual (Version 2.05)*; Technical Report No. 2004-0406; Det Norske Veritas: Oslo, Norway, 2006.
88. ALPS/ULSAP. *A Computer Program for Ultimate Limit State Assessment for Stiffened Panels*; Proteus Engineering: Stevensville, MD, USA, 2006.

89. ANSYS. *User's Manual*; Swanson Analysis Systems Inc.: Houston, TX, USA, 2006.
90. Hughes, F.O.; Paik, J.K. *Ship Structural Analysis and Design*; The Society of Naval Architects and Marine Engineers: Alexandria, VA, USA, 1989.
91. Yao, T.; Nikolov, P.I. Progressive collapse analysis of a ship´s hull under longitudinal bending. *Soc. Nav. Archit. Jpn.* **1991**, *170*, 449–461. [CrossRef]
92. Guedes Soares, C.; Gordo, J.M. Design methods for stiffened plates under predominantly uniaxial compression. *Mar. Struct.* **1997**, *10*, 465–497. [CrossRef]
93. Carlsen, C.A. A parametric study of collapse of stiffened plates in compression. *Struct. Eng.—Part B* **1980**, *58*, 33–40.
94. American Bureau of Shipping. *Rules Restatement Report*; American Bureau of Shipping: Spring, TX, USA, 1991.
95. Paik, J.K.; Kim, B.J. Ultimate strength formulations for stiffened panels under combined axial load, in-plane bending and lateral pressure: A benchmark study. *Thin Wall Struct.* **2002**, *40*, 45–83. [CrossRef]
96. Paik, J.K.; Thayamballi, A.K. An empirical formulation for predicting the ultimate compressive strength of stiffened panels. In Proceedings of the 7th International Offshore and Polar Engineering Conference, Honolulu, HI, USA, 25–30 May 1997; pp. 328–338.
97. Kim, D.K.; Lim, H.L.; Kim, M.S.; Hwang, O.J.; Park, K.S. An empirical formulation for predicting the ultimate strength of stiffened panels subjected to longitudinal compression. *Ocean Eng.* **2017**, *140*, 270–280. [CrossRef]
98. Kim, D.K.; Lim, H.L.; Yu, S.Y. A technical review on ultimate strength prediction of stiffened panels in axial compression. *Ocean Eng.* **2018**, *170*, 392–406. [CrossRef]
99. Kim, D.K.; Lim, H.L.; Yu, S.Y. Ultimate strength prediction of T-bar stiffened panel under longitudinal compression by data processing: A refined empirical formulation. *Ocean Eng.* **2019**, *192*, 106522. [CrossRef]
100. Kim, D.K.; Young, S.Y.; Lim, H.L.; Cho, N.K. Ultimate compression strength of stiffened panel: An empirical formulation for flat-bar type. *J. Mar. Sci. Eng.* **2020**, *8*, 605. [CrossRef]
101. Ozdemir, M.; Ergin, A.; Yanagihara, D.; Tanaka, S.; Yao, T. A new method to estimate ultimate strength of stiffened panels under longitudinal thrust based on analytical formulas. *Mar. Struct.* **2018**, *59*, 510–535. [CrossRef]
102. Yao, T.; Nikolov, P.I. Progressive collapse analysis of a ship´s hull under longitudinal bending (2nd Report). *Soc. Nav. Archit. Jpn.* **1992**, *172*, 437–446. [CrossRef]
103. IACS. *Common Structural Rules for Bulk Carriers with a Length of 90 Meters and Above*; DNV: Høvik, Norway, 2006.
104. IACS. *Common Structural Rules for Double Hull Oil Tankers with a Length of 150 Meters and Above*; DNV: Høvik, Norway, 2006.
105. Li, S.; Kim, D.K.; Benson, S. An adaptable algorithm to predict the load-shortening curves of stiffened panels in compression. *Ships Offshore Struct.* **2021**, *16*, 122–139. [CrossRef]
106. Byklum, E.; Amdahl, J. A simplified method for elastic large deflection analysis of plates and stiffened panels due to local buckling. *Thin Wall Struct.* **2002**, *40*, 925–953. [CrossRef]
107. Fujikubo, M.; Harada, M.; Yao, T.; Reza Khedmati, M.; Yanagihara, D. Estimation of ultimate strength of continuous stiffened panel under combined transverse thrust and lateral pressure Part 2: Continuous stiffened panel. *Mar. Struct.* **2005**, *18*, 411–427. [CrossRef]
108. Xu, M.C.; Fujikubo, M.; Guedes Soares, C. Influence of model geometry and boundary conditions on the ultimate strength of stiffened panels under uniaxial compressive loading. *J. Offshore Mech. Arct.* **2013**, *135*, 041603. [CrossRef]
109. Woloszyk, K.; Garbatov, Y.; Kowalski, J.; Samson, L. Numerical and experimental study on effect of boundary conditions during testing of stiffened plates subjected to compressive loads. *Eng. Struct.* **2021**, *235*, 112027. [CrossRef]
110. Tanaka, S.; Yanagihara, D.; Yasuoka, A.; Harada, M.; Okazawa, S.; Fujikubo, M.; Yao, T. Evaluation of ultimate strength of stiffened panels under longitudinal thrust. *Mar. Struct.* **2014**, *36*, 21–50. [CrossRef]
111. Xu, M.C.; Song, Z.J.; Pan, J.; Guedes Soares, C. Ultimate strength assessment of continuous stiffened panels under combined longitudinal compressive load and lateral pressure. *Ocean Eng.* **2017**, *139*, 39–53. [CrossRef]
112. Garbatov, Y.; Tekgoz, M.; Guedes Soares, C. Uncertainty assessment of the ultimate strength of a stiffened panel. In *Advances in Marine Structures*; Taylor & Francis Group: London, UK, 2011; pp. 659–668.
113. Gordo, J.M.; Guedes Soares, C. Compressive tests on stiffened panels of intermediate slenderness. *Thin Wall Struct.* **2011**, *49*, 782–794. [CrossRef]
114. Gordo, J.M.; Guedes Soares, C. Compressive tests on short continuous panels. *Mar. Struct.* **2008**, *21*, 113–126. [CrossRef]
115. Xu, M.C.; Guedes Soares, C. Comparison of numerical results with experiments on the ultimate strength of long stiffened panels. In Proceedings of the 30th International Conference on Offshore Mechanics and Arctic Engineering, Rotterdam, The Netherlands, 19–24 June 2011; pp. 915–922.
116. Xu, M.; Guedes Soares, C. Assessment of the ultimate strength of narrow stiffened panel test specimens. *Thin Wall Struct.* **2012**, *55*, 11–21. [CrossRef]
117. Tekgoz, M.; Garbatov, Y.; Guedes Soares, C. Finite element modelling of the ultimate strength of stiffened plates with residual stresses. In *Analysis and Design of Marine Structures*; CRC Press: Boca Raton, FL, USA, 2013; pp. 309–317.
118. Tekgoz, M.; Garbatov, Y.; Guedes Soares, C. Ultimate strength assessment of a stiffened plate accounting for welding sequences. In Proceedings of the PRADS2013, Changwon City, Korea, 20–25 October 2013; pp. 1089–1095.
119. Gannon, L.G.; Pegg, N.G.; Smith, M.J.; Liu, Y. Effect of residual stress shakedown on stiffened plate strength and behaviour. *Ships Offshore Struct.* **2013**, *8*, 638–652. [CrossRef]

120. Li, S.; Kim, D.K.; Benson, S. The influence of residual stress on the ultimate strength of longitudinally compressed stiffened panels. *Ocean Eng.* **2021**, *231*, 108839. [CrossRef]
121. Reckling, K.A. Behaviour of box girders under bending and shear. In Proceedings of the 7th International Ship and Offshore Structures Congress, Paris, France, 1979; pp. 46–49.
122. Ostapenko, A. Strength of ship hull girders under moment, shear, and torque. In Proceedings of the SSC-SNAME Symposium on Extreme Loads Response, Arlington, VA, USA, 19–20 October 1981; pp. 149–166.
123. Nishihara, S. Ultimate longitudinal strength of midships cross-section. *Nav. Archit. Ocean Eng.* **1984**, *22*, 200–214.
124. Gordo, J.M.; Guedes Soares, C. Experimental evaluation of the ultimate bending moment of a box girder. *Mar. Syst. Ocean. Technol.* **2004**, *1*, 33–46. [CrossRef]
125. Gordo, J.M.; Guedes Soares, C. Experimental evaluation of the behaviour of a mild steel box girder under bending moment. *Ships Offshore Struct.* **2008**, *3*, 347–358. [CrossRef]
126. Tekgoz, M.; Garbatov, Y.; Guedes Soares, C. Strength analysis of ship-shaped structures subjected to asymmetrical bending moment. In *Analysis and Design of Marine Structures*; Guedes Soares, C., Shenoi, R.A., Eds.; Taylor & Francis Group: Abingdon, UK, 2015; pp. 415–423.
127. Yao, T.; Fujikobo, M. *Buckling and Ultimate Strength of Ship and Ship-Like Floating Structures*; Butterworth-Heinemann: Oxford, UK, 2016.
128. Dow, R.S. Testing, and analysis of a 1/3-scale welded steel frigate model. In *Advances in Marine Structures 2*; Spon Press: London, UK, 1991; pp. 749–773.
129. Kell, C. Investigation of structural characteristics of destroyers "Preston" and "Bruce" part I—Description. *Trans. SNAME* **1931**, *39*, 35–64.
130. Kell, C. Investigation of structural characteristics of destroyers "Preston" and "Bruce" part I—Analysis of data and results. *Trans. SNAME* **1940**, *48*, 125–172.
131. Lang, D.; Warren, W. Structural strength investigation of destroyer Abuera. *Trans. Inst. Nav. Arch.* **1952**, *94*, 243–286.
132. Mansour, A.E.; Yang, J.M.; Thayamballi, A.K. An experimental investigation of ship hull ultimate strength. *Trans. SNAME* **1990**, *98*, 411–439.
133. Sugimura, T.; Nozaki, M.; Suzuki, T. Destructive experiment of ship hull model under longitudinal bending. *J. Soc. Nav. Arch. Jpn.* **1966**, *199*, 209–220. (In Japanese) [CrossRef]
134. Yao, T.; Fujikubo, M.; Yanagihara, D.; Fujii, I.; Matrui, R.; Furui, N.; Kuwamura, Y. Buckling collapse strength of chip carrier under longitudinal bending (1st Report)-collapse test on 1/10-scale hull girder model under pure bending. *J. Soc. Nav. Archit. Jpn.* **2002**, *191*, 265–274. [CrossRef]
135. Sun, H.H.; Guedes Soares, C. An experimental study of ultimate torsional strength of a ship-type hull girder with a large deck opening. *Mar. Struct.* **2003**, *16*, 51–67. [CrossRef]
136. Tanaka, Y.; Ando, T.; Anai, Y.; Yao, T.; Fujikubo, M.; Iijima, K. Longitudinal strength of container ships under combined torsional and bending moments. In Proceedings of the 19th International Offshore and Polar Engineering Conference, Osaka, Japan, 21–26 June 2009; pp. 748–755.
137. Yao, T. Hull girder strength. *Mar. Struct.* **2003**, *16*, 1–13. [CrossRef]
138. Caldwell, J.B. Ultimate longitudinal strength. *Trans. RINA* **1965**, *107*, 411–430.
139. Ueda, Y.; Rased, S.M.H. An ultimate transverse strength analysis of ship structures. *J. Soc. Nav. Archit. Jpn.* **1974**, *136*, 309–324. [CrossRef]
140. Fujikubo, M.; Kaeding, P. ISUM rectangular plate element with new lateral shape function (2nd report): Stiffened plates under biaxial thrust. *J. Soc. Nav. Archit. Jpn.* **2000**, *188*, 479–487. [CrossRef]
141. Fujikubo, M.; Kaeding, P.; Yao, T. ISUM rectangular plate element with new lateral shape function (1st report): Longitudinal and transverse thrust. *J. Soc. Nav. Archit. Jpn.* **2000**, *187*, 209–219. [CrossRef]
142. Smith, C. Influence of local compressive failure on ultimate longitudinal strength of a ship hull. In Proceedings of the International Symposium on Practical Design in Shipbuilding, Tokyo, Japan, 18–20 October 1977; pp. 73–79.
143. Sun, H.H.; Wang, X. Procedure for calculating hull girder ultimate strength of ship structures. *Mar. Syst. Ocean. Technol.* **2005**, *1*, 127–136. [CrossRef]
144. Rigo, P.; Toderan, C.; Yao, T. Sensitivity analysis on ultimate hull bending moment. *Pract. Des. Ships Other Float. Struct.* **2001**, 987–995.
145. Yao, T.; Brunner, E.; Cho, S.R.; Choo, Y.S.; Czujko, J.; Estefen, S.F.; Gordo, J.M.; Hess, P.E.; Naar, H.; Pu, Y.; et al. Committee III.1 Ultimate Strength. In Proceedings of the 16th International Ship and Offshore Structures Congress, Southampton, UK, 20–25 August 2006; pp. 369–458.
146. Yoshikawa, T.; Bayatfar, A.; Kim, B.J.; Chen, C.P.; Wang, D.; Boulares, J.; Gordo, J.M.; Josefson, L.; Smith, M.; Kaeding, P.; et al. Committee III. 1 Ultimate Strength. In Proceedings of the 19th International Ship and Offshore Structures Congress, Cascais, Portugal, 7–10 September 2015.
147. Paik, J.K.; Mansour, A.E. A simple formulation for predicting the ultimate strength of ships. *J. Mar. Sci. Technol.* **1995**, *1*, 52–62. [CrossRef]
148. Frieze, P.A.; Lin, Y.T. Ship longitudinal strength modelling for reliability analysis. In Proceedings of the Marine Structural Inspection, Maintenance and Monitoring Symposium, Arlington, VA, USA, 18–19 March 1991; pp. III.C.1–III.C.20.

149. Mansour, A.; Faulkner, D. On applying the statistical approach to extreme sea loads and ship hull strength. *Trans. RINA* **1973**, *115*, 277–314.
150. Valsgaard, S.; Steen, E. Ultimate hull girder strength margins in present class requirements. In Proceedings of the Marine Structural Inspection, Maintenance and Monitoring Symposium, Arlington, VA, USA, 18–19 March 1991.
151. Vasta, J. Lessons learned from full-scale structural tests. *Trans. SNAME* **1958**, *66*, 165–243.
152. Viner, A.C. Development of ship strength formulations. In Proceedings of the International Conference on Advances in Marine Structures, Dunfermline, UK, 20–23 May 1986; pp. 152–173.
153. Gordo, J.M.; Guedes Soares, C. Approximate method to evaluate the hull girder collapse strength. *Mar. Struct.* **1996**, *9*, 449–470. [CrossRef]
154. Yao, T.; Fujikubo, M.; Khedmati, M.R. Progressive collapse analysis of a ship´s hull girder under longitudinal bending considering local pressure loads. *J. Soc. Nav. Archit. Jpn.* **2000**, *188*, 507–515. [CrossRef]
155. Garbatov, Y.; Saad-Eldeen, S.; Guedes Soares, C. Hull girder ultimate strength assessment based on experimental results and the dimensional theory. *Eng. Struct.* **2015**, *100*, 742–750. [CrossRef]
156. Chen, K.Y.; Kutt, L.M.; Piaszczyk, C.M.; Bieniek, M.P. Ultimate strength of ship structures. *Trans. SNAME* **1983**, *91*, 149–168.
157. Amlashi, H.K.K.; Moan, T. Ultimate strength analysis of a bulk carrier hull girder under alternate hold loading condition—A case study Part 1: Non-linear finite element modelling and ultimate hull girder capacity. *Mar. Struct.* **2008**, *21*, 327–352. [CrossRef]
158. Benson, S.; Downes, J.; Dow, R.S. An automated finite element methodology for hull girder progressive collapse analysis. In Proceedings of the 13th International Marine Design Conference, Glasgow, UK, 11–14 June 2012.
159. Benson, S.; Downes, J.; Dow, R.S. Compartment level progressive collapse analysis of lightweight ship structures. *Mar. Struct.* **2013**, *31*, 44–62. [CrossRef]
160. Benson, S.; Downes, J.; Dow, R.S. Overall buckling of lightweight stiffened panels using an adapted orthotropic plate method. *Eng. Struct.* **2015**, *85*, 107–117. [CrossRef]
161. Kim, D.K.; Park, D.H.; Kim, H.B.; Kim, B.J.; Seo, J.K.; Paik, J.K. Lateral pressure effects on the progressive hull collapse behaviour of a Suezmax-class tanker under vertical bending moments. *Ocean Eng.* **2013**, *63*, 112–121. [CrossRef]
162. Mansour, A.E.; Lin, Y.H.; Paik, J.K. Ultimate strength of ships under combined vertical and horizontal moments. In Proceedings of the Sixth International Symposium on PRADS, Seoul, Korea, 17–22 September 1995; pp. 844–851.
163. Paik, J.K.; Kim, B.J.; Seo, J.K. Methods for ultimate limit state assessment of ships and ship-shaped offshore structures: Part III—Hull girders. *Ocean Eng.* **2008**, *35*, 281–286. [CrossRef]
164. Paik, J.K.; Thayamballi, A.K.; Che, J.S. Ultimate strength of ship hulls under combined vertical bending, horizontal bending, and shear forces. *Trans. SNAME* **1996**, *104*, 31–59.
165. Tekgoz, M.; Garbatov, Y.; Guedes Soares, C. Strength assessment of an intact and damaged container ship subjected to asymmetrical bending loadings. *Mar. Struct.* **2018**, *58*, 172–198. [CrossRef]
166. Xu, M.; Garbatov, Y.; Guedes Soares, C. Ultimate strength assessment of a tanker hull based on experimentally developed master curves. *J. Mar. Sci. Appl.* **2013**, *12*, 127–139. [CrossRef]
167. Tekgoz, M.; Garbatov, Y.; Guedes Soares, C. Review of ultimate strength assessment of ageing and damaged ship structures. *J. Mar. Sci. Appl.* **2021**, *19*, 512–533. [CrossRef]

Journal of
*Marine Science
and Engineering*

Article

Strength Assessment of Rectangular Plates Subjected to Extreme Cyclic Load Reversals

Mesut Tekgoz and Yordan Garbatov *

Centre for Marine Technology and Engineering (CENTEC), Instituto Superior Técnico, Universidade de Lisboa, P-1049-001 Lisbon, Portugal; mesut.tekgoz@centec.tecnico.ulisboa.pt
* Correspondence: yordan.garbatov@tecnico.ulisboa.pt

Received: 21 December 2019; Accepted: 19 January 2020; Published: 21 January 2020

Abstract: The objective of this study is to investigate the strength of the rectangular plates subjected to cyclic load reversals with varying strain ranges. The finite element solution is implemented to estimate the load-carrying capacity. The influence of the initial imperfections, plate thicknesses and aspect ratio parameters have been accounted for. The cyclic response is predicted by using the material model assumed to follow the combined non-linear isotropic and kinematic strain hardening rules with Von Misses yield criterion accounting for the Bauschinger effect. It has been shown that the type of plastic formation during the cyclic load has a significant influence on the structural capacity and stiffness reduction. The initial imperfection has a significant impact on the ultimate load capacity reduction where the uni-modal initial imperfection type leads to a more stable load transition and plastic formation, reducing the structural capacity during the cyclic load exposure.

Keywords: ultimate strength; cyclic load; Bauschinger effect

1. Introduction

The cyclic load phenomenon is a common load type that the structures are subjected during their service life. Its impact has been investigated accounting for different aspects of problems in various fields of engineering. It may take place as a result of a variety of causes, for example, when civil buildings are exposed to earthquakes or ships are subjected to extreme wave loads (Det Norske Veritas-Germanischer LIyod, DNV-GL [1], Eurocode-3 [2], Eurocode-8 [3], the Federal Emergency Management Agency, FEMA [4]).

When the steel structure is subjected to a cyclic load, their hysteretic behaviour becomes a critical issue to be investigated involving the structural capacity and stiffness reduction. Its impact is magnified with the strain reversals. Ibarra et al., 2005 [5] provided the description, calibration and application of hysteretic models accounting for the strength and stiffness deteriorations for a variety of materials including steel.

Azevedo et al., 1994 [6] provided an overview of experimental methodologies for the cyclic load and analytical methods to simulate the hysteretic behaviour of steel structural components.

Zhou et al., 2015 [7] performed a series of cyclic load tests accounting for several cyclic loading protocols and material properties under considerable inelastic strain exposure and they concluded that the loading history has a considerable influence on the stress-strain response and it is more pronounced at low amplitude loadings. Different cyclic load protocols have also been studied by Shi et al., 2011 [8] and Shi et al., 2012 [9] where the difference between the monotonic and hysteretic curves has been presented.

Krolo et al., 2016 [10] investigated the behaviour of structures, built of mild steel and subjected to the cyclic load accounting for variable strain ranges by applying the displacement controlled load. They compared the hysteretic curves as predicted by the finite element solution and experimental results, showing a good agreement.

J. Mar. Sci. Eng. **2020**, *8*, 65; doi:10.3390/jmse8020065

www.mdpi.com/journal/jmse

Zhao et al., 2019 [11] studied aluminium alloys under low-cycle fatigue loading and they showed that as the number of the cyclic load increases, it gives rise to the load-carrying capacity and stiffness reduction. Wang et al., 2019 [12] analysed steel-reinforced concrete columns subjected to the cyclic load employing a damage assessment approach where the hysteretic and skeleton curves have been developed based on the test results.

Ship hull structures are made up of steel by and largely are exposed to a variety of loads throughout the ship's service life at sea. The imposed loads play a significant role in defining the overall structural capacity of the ship structure. Hence, the structural behaviour and capacity to resist different loads are to be well understood in the first place to enhance the ship and crew safety and also to protect the marine environment in case of structural failure.

The cyclic load is also one of the load types that the ships are subjected to. The degree of the cyclic load exposure may differ depending on the sea state conditions where the ships are operating.

A variety of tools and methods have been developed to estimate the ultimate load-carrying capacity of the ship structure (Smith 1977 [13], Paik et al., 2012 [14], ALPS/HULL [15]).

Smith 1977 [13] proposed a progressive collapse method also widely used by the Common Structural Rules [16], in order to estimate the ultimate ship strength. The ship cross-section is divided into components defined as a unit of plates with associated stiffeners, hard corner or plate elements. Each element is independent and progressively loses its strength and stiffness during the incremental permissible curvature.

Gordo and Guedes Soares 1997 [17] used the progressive collapse method to assess the ultimate load-carrying capacity of the hull girders and verified with the experimental results demonstrating good accuracy. The progressive collapse method was also implemented in Gordo and Guedes Soares 1996 [18] and Paik et al., 2012 [19].

The finite element solution is also being commonly used for the ultimate collapse analysis of marine structures which was initially performed by Chen et al., 1983 [20]. Several authors, Paik et al., 2008 [21], Xu et al., 2013 [22] and Tekgoz et al., 2018 [23], studied the ultimate shipload carrying capacity using the finite element solution which is based on a force-rotation-controlled static load.

In these approaches, the structure is allowed to follow a path under a static pure-bending load with an incrementally increasing curvature.

However, the ship plating is predominantly subjected to the dynamic loads that subsequently leads to the cyclic load attack, which is added to the complexity of the geometrical and material nonlinearity of the structural assessment.

Yao et al., 1990 [24] performed a series of elastic-plastic large deflection analysis on plates under cyclic load. They studied the influence of the cyclic load on the plate in-plane rigidity, re-yielding and ultimate load capacity reduction for a wide plate.

Goto et al., 1995 [25] studied the influence of the localization of the plastic buckling on the steel structures where they concluded that it significantly reduces the loading capacity of the steel structure under the cyclic load.

Komoriyama et al., 2018 [26] studied the influence of the cumulative buckling under the cyclic load on the load capacity of the stiffened panels. They showed that when the cyclic compressive load is around the ultimate capacity of the structure, the cumulative buckling deformation is high. However, its impact on the ultimate load carrying capacity is small.

Yao et al., [27] developed an analytical solution for a plate subjected to a cyclic load in order to simulate the collapse behaviours accounting for the welding induced residual stresses where a simple dynamical model has been introduced presenting a good agreement with the one defined by the finite element solution.

Cui et al., 2018 [28] studied the ultimate load-carrying capacity of the ship hull girder under a cyclic load using the Smith's method and a good agreement has been achieved when compared to the one predicted by the finite element solution.

Li et al., 2019 [29] proposed an analytical solution to predict the buckling and collapse response of both plates and stiffened panels under the cyclic load showing a good agreement with the FEM prediction.

The cyclic response is a complex phenomenon that involves several aspects to be considered both from material non-linearity being a function of the material stress-strain definition and geometrical non-linearity being the buckling, initial imperfections, plastic formation pattern, etc.

Here in this study, the ship is considered to be exposed to an extreme cyclic load. The term extreme cyclic load has been considered in the sense that the ship is already failed and she is in post-collapse stages. In this state, the ship may experience extreme cyclic behaviour, and its post-collapse structural capacity might be lower than what the static approaches predict.

Therefore, the strain ranges considered in this study can be considered within the range of the extreme ones. The plates, as a part of the ship hull structure, have been assumed to be failed in the initial loading and exposed to the multiple cyclic loads in order to see how the structural response of the plate changes.

The objective of this study is to investigate the strength of rectangular plates subjected to cyclic load with varying strain ranges. The finite element solution is implemented to estimate the load-carrying capacity. The influence of the initial imperfections, plate thicknesses and aspect ratio parameters have been accounted for.

2. Finite Element Modelling

2.1. Material and Structural Description

The material properties are assumed as reported in Krolo et al., 2016 [10] for the present study, as can be seen in Table 1.

Table 1. Material property descriptors.

Yield Stress (MPa)	Young's Modulus (MPa)	ν	C_1 (MPa)	γ	C_2 (MPa)	γ	C_3 (MPa)	γ	Q_∞ (MPa)	b
285	207,000	0.3	13,921	765	4240	52	1573	14	25.6	4.4

Due to the complexity of the cyclic behaviour of the structures which may exhibit strain hardening accompanying a structural capacity increase, and material yield stress reduction which is termed as the Bauschinger effect, that leads to the structural capacity reduction, a comprehensive material model that may mimic this complex behaviour under the cyclic load is defined.

Here the Chaboche [30] nonlinear kinematic hardening and the non-linear isotropic hardening rules under the cyclic load has been used, and its descriptors have been shown in Table 1. which have been calibrated based on the experimental cyclic test data as given by Krolo et al., 2016 [10].

The yield surface definition is defined following the Von Mises criterion, ANSYS [31]:

$$F = f(\sigma - \alpha) - \sigma^0 = 0 \tag{1}$$

where σ^0 is the material yield stress and $f(\sigma - \alpha)$ is the equivalent Von Mises stress concerning the back stress α, that equals to:

$$f(\sigma - \alpha) = \sqrt{\frac{3}{2}(S - \alpha^{dev}) : (S - \alpha^{dev})} \tag{2}$$

where σ is the stress tensor, S is the deviatoric stress tensor and α^{dev} is the deviatoric part of the back stress tensor.

The material yield stress definition with the material isotropic hardening rule is defined as, ANSYS [31]:

$$\sigma^0 = \sigma\big|_0 + Q_\infty\left(1 - e^{-b\varepsilon^{pl}}\right) \tag{3}$$

where $\sigma|_o$ is the material initial yield stress at zero plastic strain. Q_∞ and b are the material parameters of the isotropic hardening behaviour of the materials, defined based on the experimental cyclic test data and ε^{pl} is the equivalent plastic strain.

The evolution of each back stress model with the kinematic hardening rule equals to, ANSYS [31]:

$$\Delta\hat{\alpha}_i = \frac{2}{3}C_i\Delta\varepsilon^{pl} - \gamma_i\alpha_i\Delta\varepsilon^{pl} \tag{4}$$

where C_i and γ_i are the material parameters of the kinematic hardening behaviour of materials, defined based on the experimental cyclic test data, and finally, α is the overall back stress defined as, ANSYS [31]:

$$\alpha = \sum_{i=1}^{N}\alpha_i \tag{5}$$

where N is the back stress number which has been set to 3 here.

Three different plates with varying plate aspect ratios and plate thicknesses have been studied here in order to investigate their impact on the structural capacity under the cyclic load, as shown in Table 2.

Table 2. The plate structural definition.

Plates	Length (mm)	Breadth (mm)	Thickness (mm)	Aspect Ratio, Length/Breadth
1	500	500	5,10	1
2 [1]	2610	880	10,15,20	3
3 [1]	4950	830	10,15,20	6

[1] Paik et al., 2012 [19].

2.2. Load, Boundary Condition and Initial Imperfection

The boundary conditions applied to the Finite Element Method, FEM model edges are simply supported conditions. The simply supported boundary conditions have been kept for all FEM studies performed here.

For the plate with an aspect ratio 1, only a quarter part of the plate has been modelled, and the symmetry boundary conditions have been applied to the respective locations, and for the longer plates, the entire plate has been modelled as shown in Figure 1:

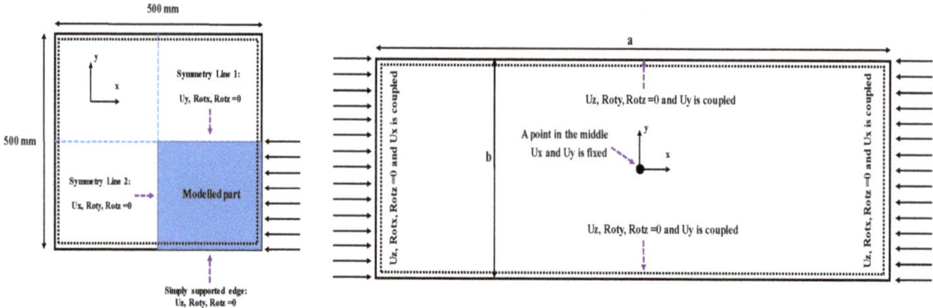

Figure 1. Boundary conditions for plate 1 (**left**) and plate 2 and 3 (**right**).

There are two types of initial imperfections used, namely, the uni-modal and the multi-modal ones (see Figure 2). The uni-modal one, which is termed here as Initial imp_B, takes the general form as defined by Ueda et al., 1985 [32]:

$$W_o = w_{max} \, sin\left(\frac{m\pi x}{a}\right) sin\left(\frac{n\pi y}{b}\right) \tag{6}$$

where a is the length of the plate, b is the breath of the plate, m and n are parameters depending on the number of half-waves considered. For all FE studies here, n is set to 1 and m is calculated as the minimum integer as follows:

$$\frac{a}{b} \leq \sqrt{m(m+1)} \tag{7}$$

As for the mid-plate maximum initial imperfection, W_{max} is given as for the average initial imperfections unless stated otherwise in the respective sections, as defined by Smith et al., 1988 [33]:

$$W_{max} = 0.1\beta_p{}^2 t_p \tag{8}$$

where β_p represents plate slenderness and t_p is the plate thickness.

As for the multi-modal initial imperfection that is labelled as Initial imp_A, it takes the general form of Ueda et al., 1985 [32]:

$$W_o = \sum\left(\sum A_{0mn} \, sin\left(\frac{m\pi x}{a}\right)\right) sin\left(\frac{n\pi y}{b}\right) \tag{9}$$

where A_{0mn} is the maximum magnitude of each component of initial imperfection which is defined concerning each plate aspect ratio and n is set to 1 for each component here.

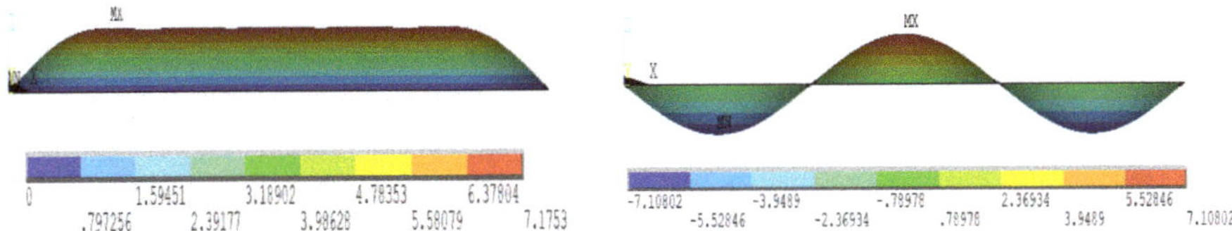

Figure 2. Multi-modal initial imperfection (**left**) and uni-modal initial imperfection (**right**).

As to the cyclic load, a displacement controlled load has been applied in order to avoid sudden load fluctuations that may occur during the plate buckling phenomenon as can be seen in Figure 3.

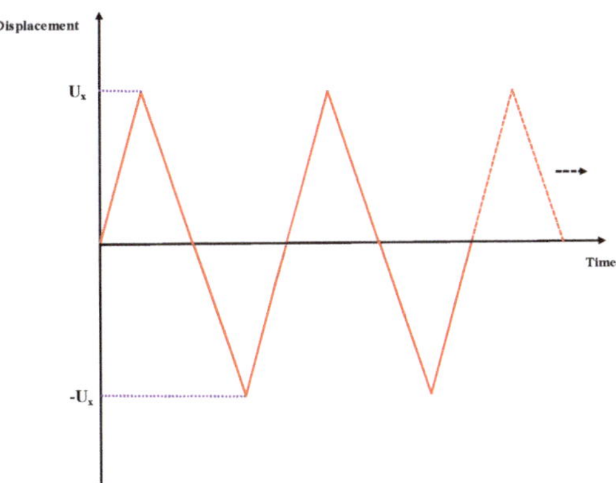

Figure 3. Load application.

The cyclic load response has been estimated by the FEM method using commercial finite element software, ANSYS [31]. The shell element SHELL181 has been used to model the studied plates. The element type has four nodes with six degrees of freedom at each node, including translations and rotations about the x, y, and z-axes.

3. Results

3.1. The Impact of the Cyclic Load on the Load Capacity, Plate 1

Here the impact of the cyclic load on the ultimate strength is analysed. Firstly, a square plate, that is labelled as Plate 1, with a thickness of 5 mm and 10 mm is studied. The amplitude of the initial imperfection, W_{max}, is taken as 10% of the plate thickness. The material property and load definition are provided in Sections 2.1 and 2.2. Under these conditions, the plate may experience the elastic and plastic buckling.

The influence of the varying strain range is studied employing one half-cycle load as can be seen in Figures 4 and 5. Figure 4 shows the normalized strength and strains as a result of the half-cycle load. The load is initially compressive, and the tensile load follows to complete the half cycle. Figure 5 shows the normalized strength and strains under half-cycle load. The load is initially tensile, and the compressive load follows to complete the half cycle.

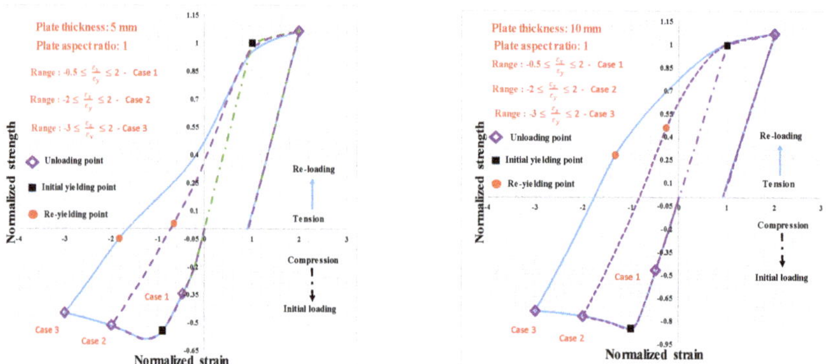

Figure 4. Plate thickness: 5 mm (**left**) and plate thickness: 10 mm (**right**), first compressive followed by a tensile load.

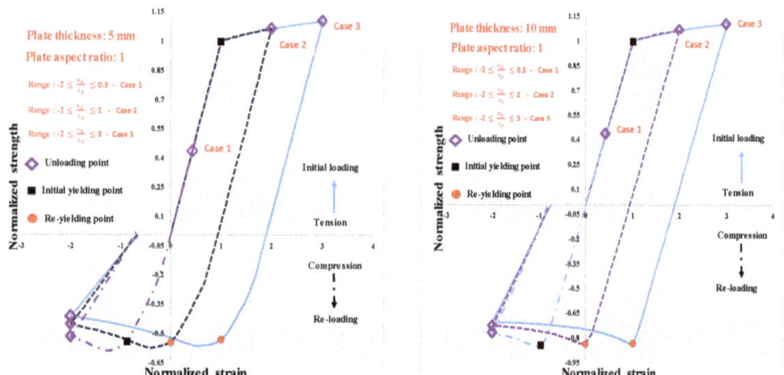

Figure 5. Plate thickness: 5 mm (**left**) and plate thickness: 10 mm (**right**), first tensile followed by compressive load.

The square plate response may differ depending on how the plate is initially loaded. The plate re-yielding points do not differ if it is first loaded under tensile load and followed by a compressive load. However, it significantly changes if it is firstly loaded under compression and followed by a tensile load. This holds true for both plate thickness cases. Similar findings have also been given in Yao et al., 1990 [24] for a wider plate.

This might be explained with the non-uniform residual plastic deformations that occur in previous compressive loading history and due to the buckling phenomenon.

Additionally, the plate with a 5 mm thicknesses creates a local plasticity line forming a partial failure mechanism (see Figure 6). When a failure mechanism occurs in the plate, it governs the plate deformation and causes unloading on the stresses in the other parts of the plate, and this phenomenon, apart from the developed residual plastic strains and Bauschinger effect, may also explain why the re-yielding reduction is more influenced when it is re-loaded in the case of a plate thickness of 5 mm (see Figure 4).

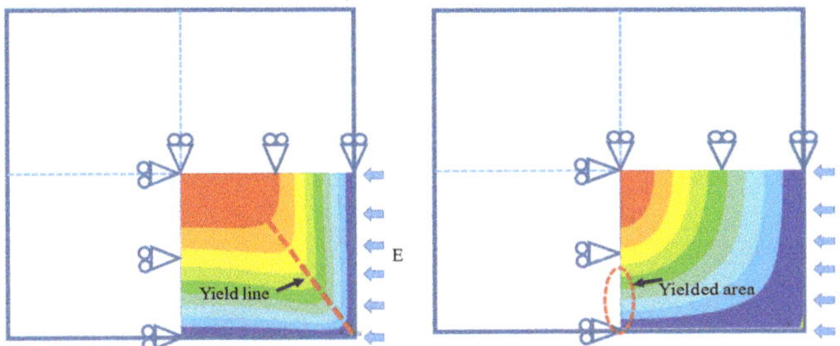

Figure 6. Plate thickness: 5 mm (**left**) and plate thickness: 10 mm (**right**), yielding location.

On the next stage, the square plate is subjected to the multiple cycle load with a variable strain range. Figure 7 shows the normalized strength and strain response of the plate that is subjected to the multiple cyclic loads accounting for the plate thicknesses. For both cases, a plate thickness of 5 mm and 10 mm, the response approaches to converged loop after several cyclic loads.

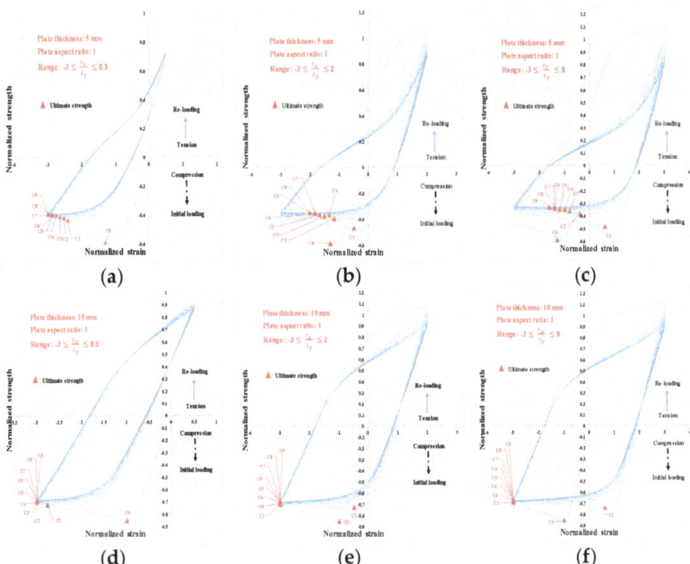

Figure 7. (**a**) $t = 5$ mm, $-3 \leq \frac{\varepsilon_x}{\varepsilon_y} \leq 0.5$; (**b**) $t = 5$ mm, $-3 \leq \frac{\varepsilon_x}{\varepsilon_y} \leq 2$; (**c**) $t = 5$ mm, $-3 \leq \frac{\varepsilon_x}{\varepsilon_y} \leq 3$; (**d**) $t = 10$ mm, $-3 \leq \frac{\varepsilon_x}{\varepsilon_y} \leq 0.5$; (**e**) $t = 10$ mm, $-3 \leq \frac{\varepsilon_x}{\varepsilon_y} \leq 2$; (**f**) $t = 10$ mm, $-3 \leq \frac{\varepsilon_x}{\varepsilon_y} \leq 3$.

As the number of the cyclic load is increasing, this gives a rise of the ultimate strength and initial stiffness reduction which has been observed with the square plate (see Figure 8).

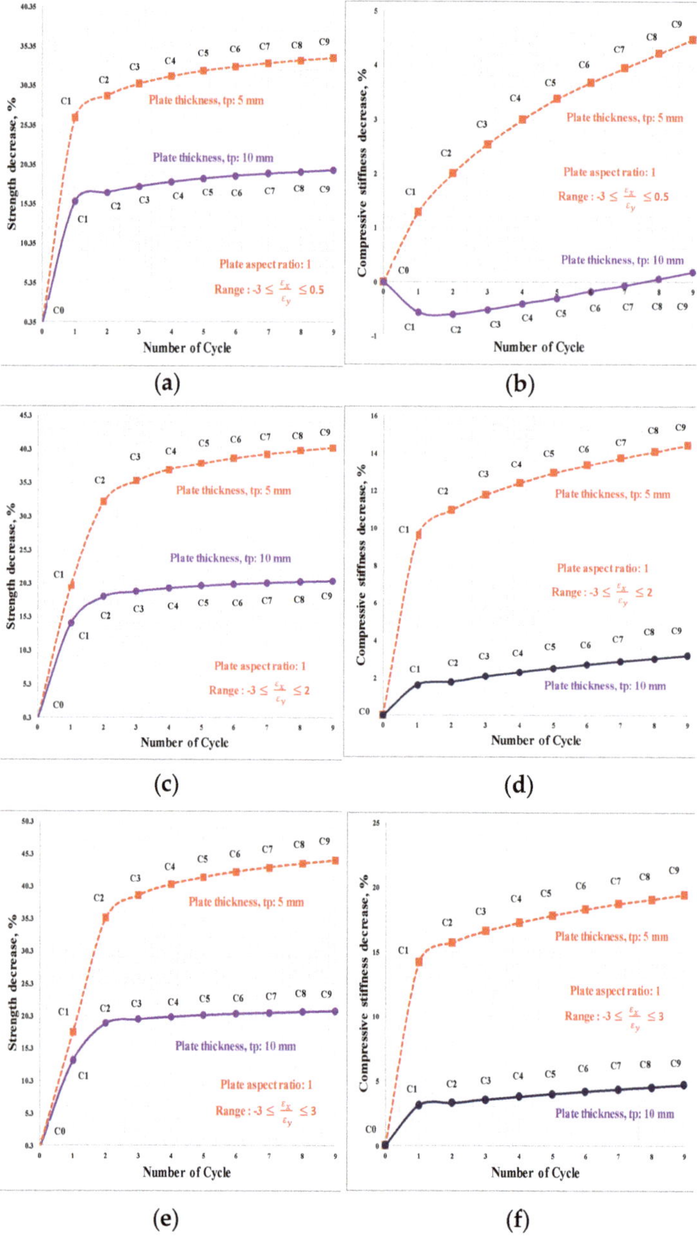

Figure 8. (**a**) Strength decrease, $-3 \leq \frac{\varepsilon_x}{\varepsilon_y} \leq 0.5$; (**b**) Stiffness decrease, $3 \leq \frac{\varepsilon_x}{\varepsilon_y} \leq 0.5$; (**c**) Strength decrease, $-3 \leq \frac{\varepsilon_x}{\varepsilon_y} \leq 2$; (**d**) Stiffness decrease, $-3 \leq \frac{\varepsilon_x}{\varepsilon_y} \leq 2$; (**e**) Strength decrease, $-3 \leq \frac{\varepsilon_x}{\varepsilon_y} \leq 3$; (**f**) Stiffness decrease, $-3 \leq \frac{\varepsilon_x}{\varepsilon_y} \leq 3$.

It has been observed in the case of a square plate that as the plate thicknesses are reduced, the ultimate load-carrying capacity and its stiffness reduction is more pronounced and in addition to this, the reduction is magnified with the increase of the strain range.

3.2. Impact of the Cyclic Load on the Load Capacity, Plate 2 and 3

Firstly, the impact of the plate thickness using the Initial imp_B has been studied under a half load cycle, as shown in Figure 9. This study is performed to see the influence of the order of the loading with long plates by varying the plate thicknesses. A particular case with a plate thickness of 8 mm is also presented to analyse the response under the half-cycle load concerning the order of the cyclic load.

The results show that when the plate thickness is less than 10 mm, it may show a different structural capacity when it is loaded first under tensile load. For the rest of the cases, the order of the loading does not have a significant impact on the structural response.

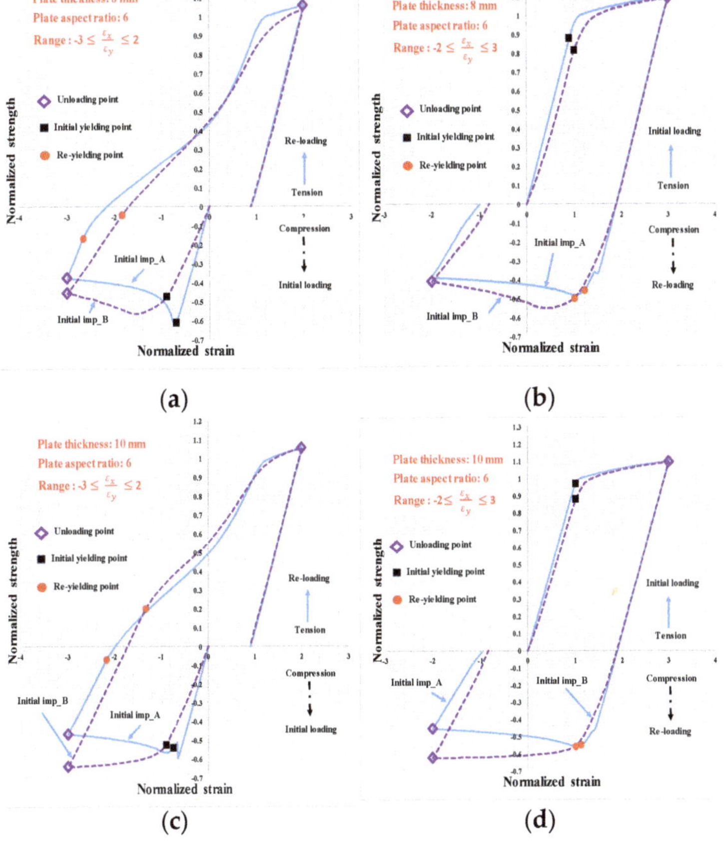

Figure 9. *Cont.*

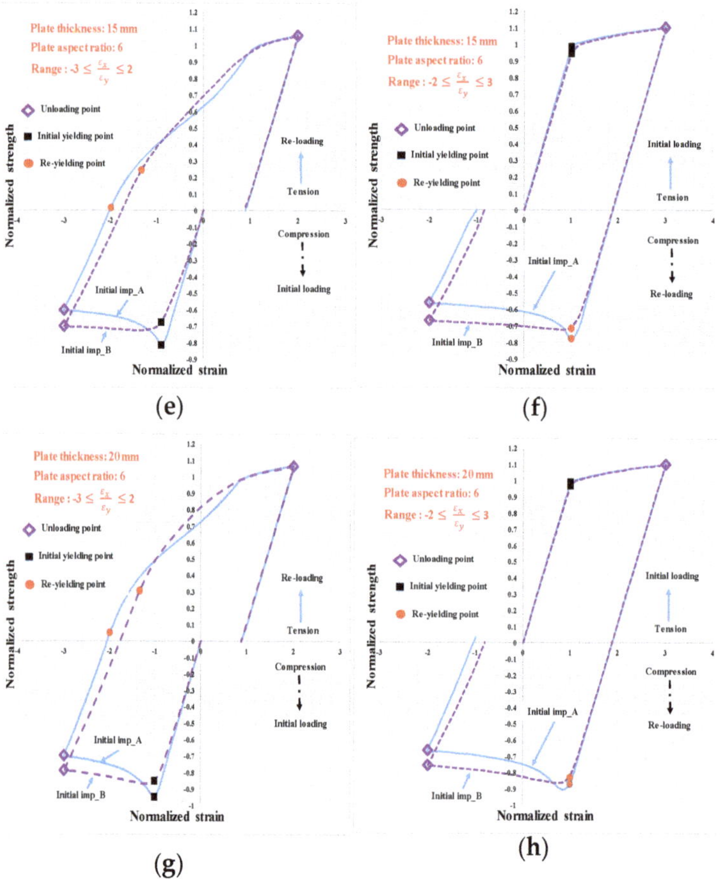

Figure 9. (**a**) $t = 8$ mm, first compressive followed by tensile load; (**b**) $t = 8$ mm, first tensile followed by compressive load; (**c**) $t = 10$ mm, first compressive followed by tensile load; (**d**) $t = 10$ mm, first tensile followed by compressive load; (**e**) $t = 15$ mm, first compressive followed by tensile load; (**f**) $t = 15$ mm, first tensile followed by compressive load; (**g**) $t = 20$ mm, first compressive followed by tensile load; (**h**) $t = 20$ mm, first tensile followed by compressive load.

At the next stage, the impact of the uni-modal and multi-modal initial imperfection is studied under multiple cyclic loads with a constant strain range of $-3 \leq \frac{\varepsilon_x}{\varepsilon_y} \leq 3$ using Plate 2 and 3 that has an aspect ratio of 3 and 6, respectively in order to see the effect of the initial imperfection on the ultimate load-carrying capacity under the multiple cyclic load exposure.

Figure 10 shows the normalized strength and strain prediction of the FEM accounting for the plate initial imperfections under the multiple cyclic loads for the plate with an aspect ratio: 6.

The multi-modal initial imperfection, the initial imp_A, exhibits lower strength performance contrary to the uni-modal initial imperfection, the Initial imp_B, as can be seen in Figure 10.

In the case of the uni-modal case, Initial imp_B, the load transition forms smooth plastic deformation to other parts and follows the pattern during the entire cyclic load exposure. However, in the case of the multi-modal case, Initial imp_A, the load transition is not smooth and very local plastic formation occurs at the early stage of the cyclic load and continues with it throughout the cyclic load exposure. In addition to that, at the final stages of the cyclic load exposure, in both cases, the structure creates a local plastic mechanism by which the structure is governed, and it may finally lead to rupture as can be seen from Figures 11 and 12.

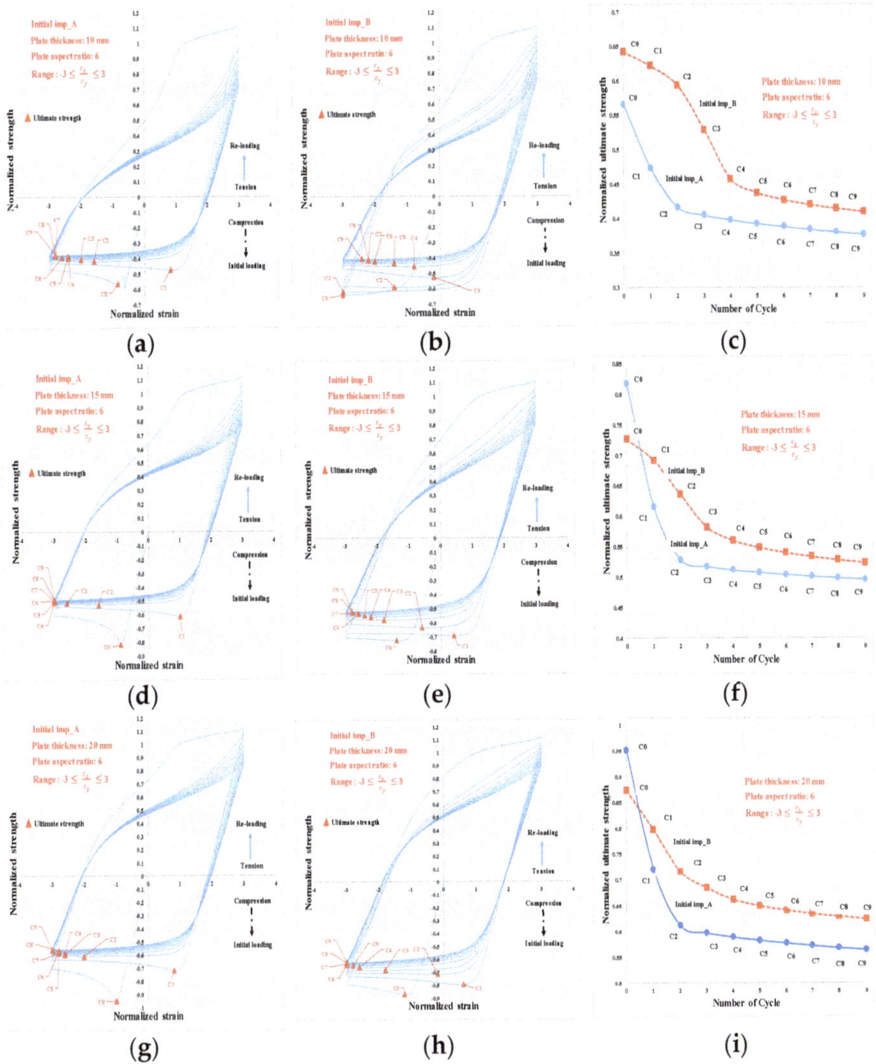

Figure 10. (**a**) t = 10 mm, Initial imp_A $-3 \leq \frac{\varepsilon_x}{\varepsilon_y} \leq 3$; (**b**) t = 10 mm, Initial imp_B, $-3 \leq \frac{\varepsilon_x}{\varepsilon_y} \leq 3$; (**c**) Normalized ultimate strength, $-3 \leq \frac{\varepsilon_x}{\varepsilon_y} \leq 3$; (**d**) t = 15 mm, Initial imp_A, $-3 \leq \frac{\varepsilon_x}{\varepsilon_y} \leq 3$; (**e**) t = 15 mm, Initial imp_B $-3 \leq \frac{\varepsilon_x}{\varepsilon_y} \leq 3$; (**f**) Normalized ultimate strength $-3 \leq \frac{\varepsilon_x}{\varepsilon_y} \leq 3$; (**g**) t = 20 mm, Initial imp_A $-3 \leq \frac{\varepsilon_x}{\varepsilon_y} \leq 3$; (**h**) t = 20 mm, Initial imp_B $-3 \leq \frac{\varepsilon_x}{\varepsilon_y} \leq 3$; (**i**) Normalized ultimate strength $-3 \leq \frac{\varepsilon_x}{\varepsilon_y} \leq 3$.

Figure 11 shows the progress of the equivalent plastic strains at the ultimate compressive capacity under the cyclic load for the plate thicknesses of 10 mm using an aspect ratio of 6 accounting for the different initial imperfection.

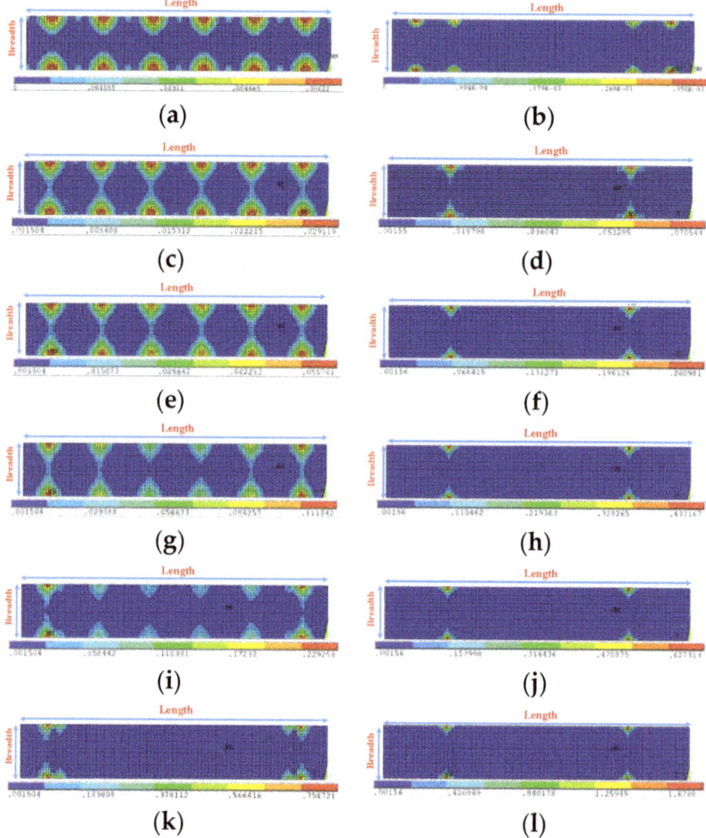

Figure 11. (**a**) Initial imp_B, plastic strains at C0; (**b**) Initial imp_A, plastic strains at C0; (**c**) Initial imp_B, plastic strains at C1; (**d**) Initial imp_A, plastic strains at C1; (**e**) Initial imp_B, plastic strains at C2; (**f**) Initial imp_A, plastic strains at C2; (**g**) Initial imp_B, plastic strains at C3; (**h**) Initial imp_A, plastic strains at C3; (**i**) Initial imp_B, plastic strains at C4; (**j**) Initial imp_A, plastic strains at C4; (**k**) Initial imp_B, plastic strains at C9; (**l**) Initial imp_A, plastic strains at C9.

Figure 12 shows the progress of the equivalent plastic strains at the ultimate compressive capacity for a plate thickness of 20 mm with an aspect ratio of 6.

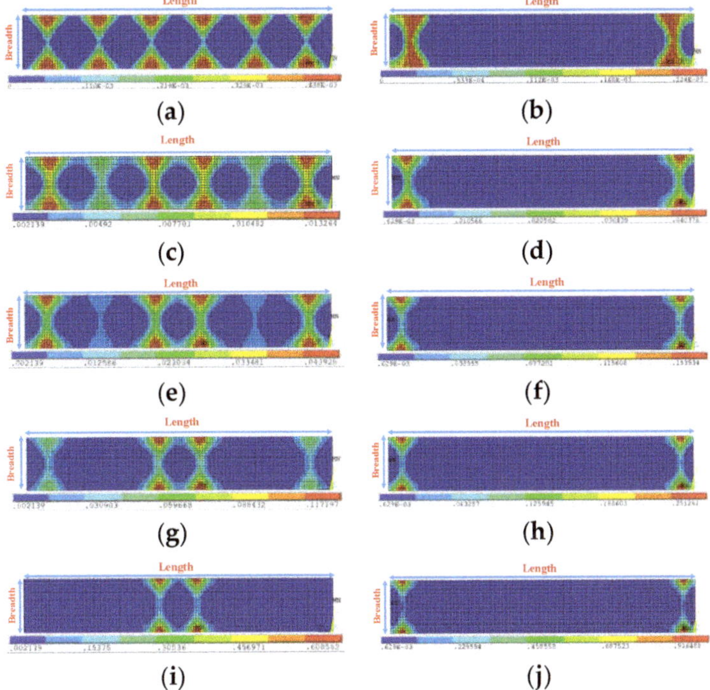

Figure 12. (**a**) Initial imp_B, plastic strains at C0; (**b**) Initial imp_A, plastic strains at C0; (**c**) Initial imp_B, plastic strains at C1; (**d**) Initial imp_A, plastic strains at C1; (**e**) Initial imp_B, plastic strains at C2; (**f**) Initial imp_A, plastic strains at C2; (**g**) Initial imp_B, plastic strains at C3; (**h**) Initial imp_A, plastic strains at C3; (**i**) Initial imp_B, plastic strains at C9; (**j**) Initial imp_A, plastic strains at C9.

It is observed that in the case of a plate thickness of 10 mm, the load transition is faster contrary to the one of 20 mm as shown in Figures 11 and 12 in the case of the uni-modal initial imperfection. This may be one of the reasons why the plate with a 10 mm thickness exhibits more considerable ultimate carrying capacity reduction as the cycle number is increasing.

Figure 13 shows the normalized strength and strain prediction of the FEM accounting for the plate initial imperfections under multiple cyclic loads with an aspect ratio of 3. The multi-modal initial imperfection exhibits a lower strength performance compared to the uni-modal initial imperfection, as can be seen in Figure 13. Similar predictions have also been observed with the plate aspect ratio of 6.

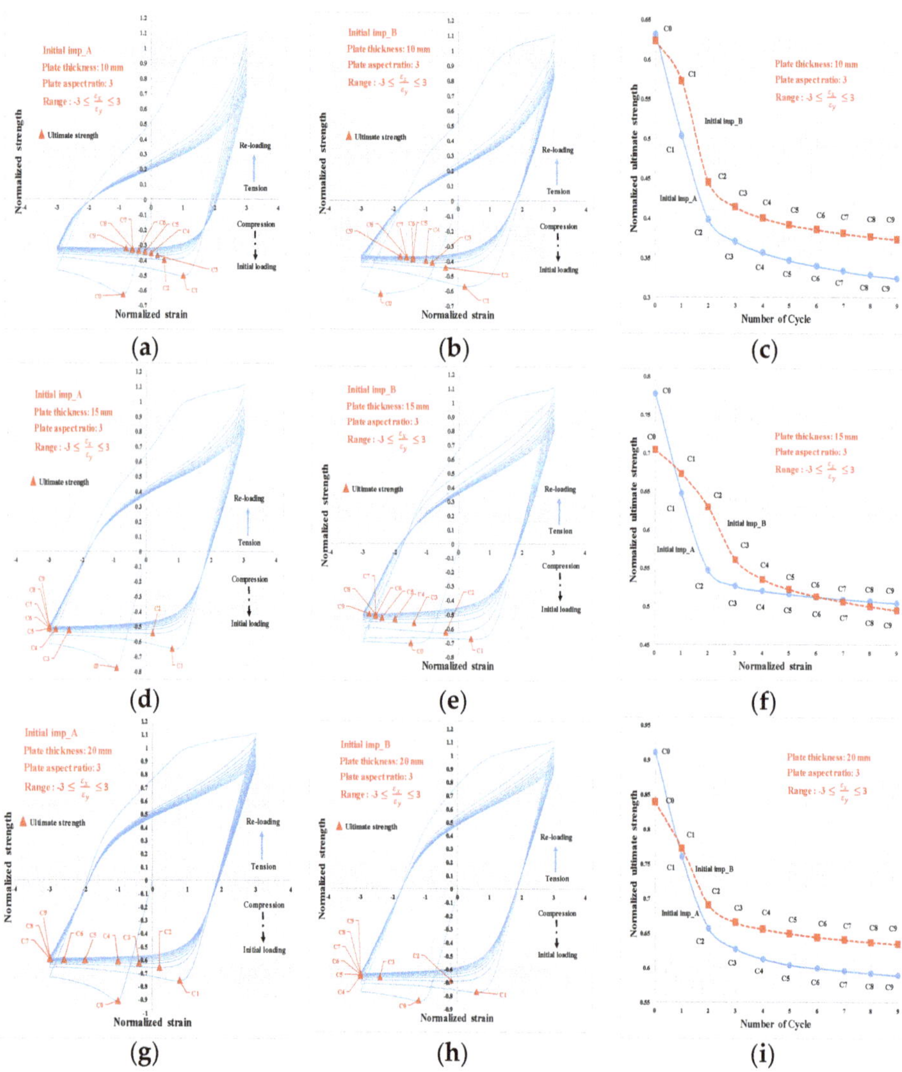

Figure 13. (**a**) $t = 10$ mm, Initial imp_A $-3 \leq \frac{\varepsilon_x}{\varepsilon_y} \leq 3$; (**b**) $t = 10$ mm, Initial imp_B, $-3 \leq \frac{\varepsilon_x}{\varepsilon_y} \leq 3$; (**c**) Normalized ultimate strength, $-3 \leq \frac{\varepsilon_x}{\varepsilon_y} \leq 3$; (**d**) $t = 15$ mm, Initial imp_A, $-3 \leq \frac{\varepsilon_x}{\varepsilon_y} \leq 3$; (**e**) $t = 15$ mm, Initial imp_B $-3 \leq \frac{\varepsilon_x}{\varepsilon_y} \leq 3$; (**f**) Normalized ultimate strength $-3 \leq \frac{\varepsilon_x}{\varepsilon_y} \leq 3$; (**g**) $t = 20$ mm, Initial imp_A $-3 \leq \frac{\varepsilon_x}{\varepsilon_y} \leq 3$; (**h**) 20 mm, Initial imp_B $-3 \leq \frac{\varepsilon_x}{\varepsilon_y} \leq 3$; (**i**) Normalized ultimate strength $-3 \leq \frac{\varepsilon_x}{\varepsilon_y} \leq 3$.

Figure 14 shows the progress of the equivalent plastic strains at the ultimate compressive capacity under the cyclic load for a plate thicknesses of 10 mm with an aspect ratio of 3.

Figure 15 shows the progress of the equivalent plastic strains at the ultimate compressive capacity under the cyclic load for a plate thicknesses of 20 mm with an aspect ratio of 3.

Similar observation as with the plate with an aspect ratio of 6 can be seen in the case of a plate thickness of 20 mm, where the uni-modal initial imperfection has smooth load transition, plastic formation, as can be seen in Figure 15 and in the case of the multi-modal one, the local plastic formation again occurs at early stages during the cyclic load exposure.

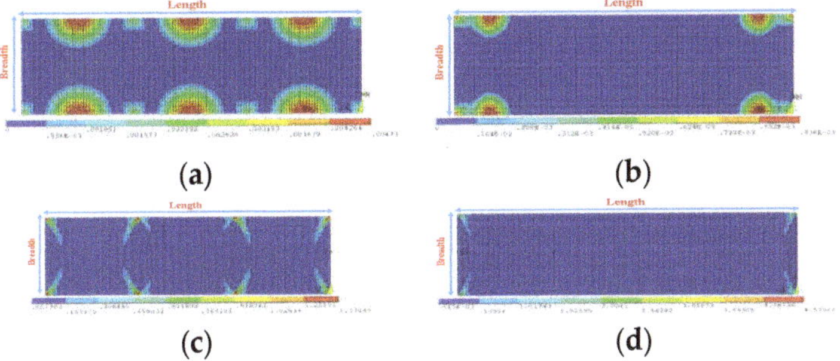

Figure 14. (**a**) Initial imp_B, plastic strains at C0; (**b**) Initial imp_A, plastic strains at C0; (**c**) Initial imp_B, plastic strains at C9; (**d**) Initial imp_A, plastic strains at C9.

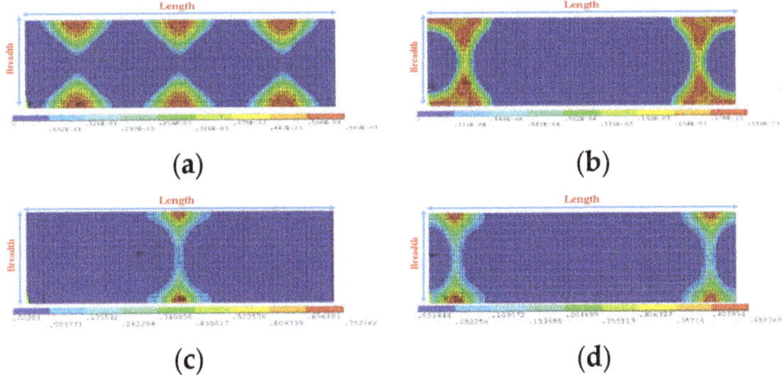

Figure 15. (**a**) Initial imp_B, plastic strains at C0; (**b**) Initial imp_A, plastic strains at C0; (**c**) Initial imp_B, plastic strains at C9; (**d**) Initial imp_A, plastic strains at C9.

The plate with a thickness of 10 mm and aspect ratio:3 shows a different plastic formation which may be one of the reasons why the ultimate load-carrying capacity is more pronounced under the multiple cyclic load exposure. The plate, in this case, might follow the plastic formation pattern, as shown in Figure 16. In this case, the plastic lines govern the plate deformation, which gives rise to more considerable ultimate load capacity reductions during the cyclic load exposure.

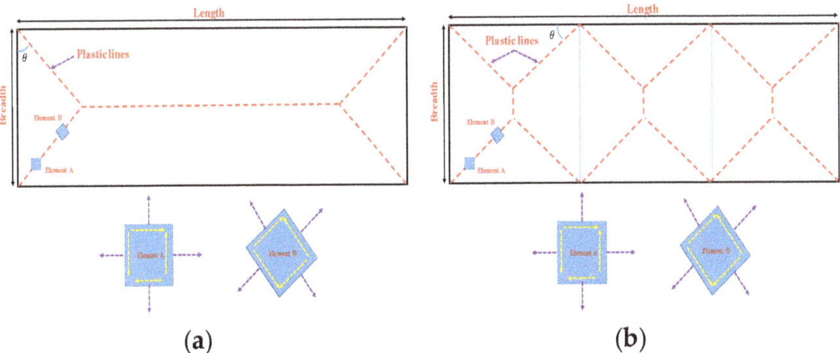

Figure 16. Initial imp_A (**left**) and initial imp_B (**right**), possible plastic formation pattern, *t* = 10 mm, aspect ratlio:3.

In the next study, the longer plates are subjected to a multiple load exposure considering only the uni-modal initial imperfection, the initial imp_B, in order to analyse the ultimate load-carrying capacity reduction accounting for the plate thickness and aspect ratios.

Figures 17 and 18 show the cyclic response of the plate with an aspect ratio of 6 and 3, respectively, by varying the plate thicknesses and strain ranges.

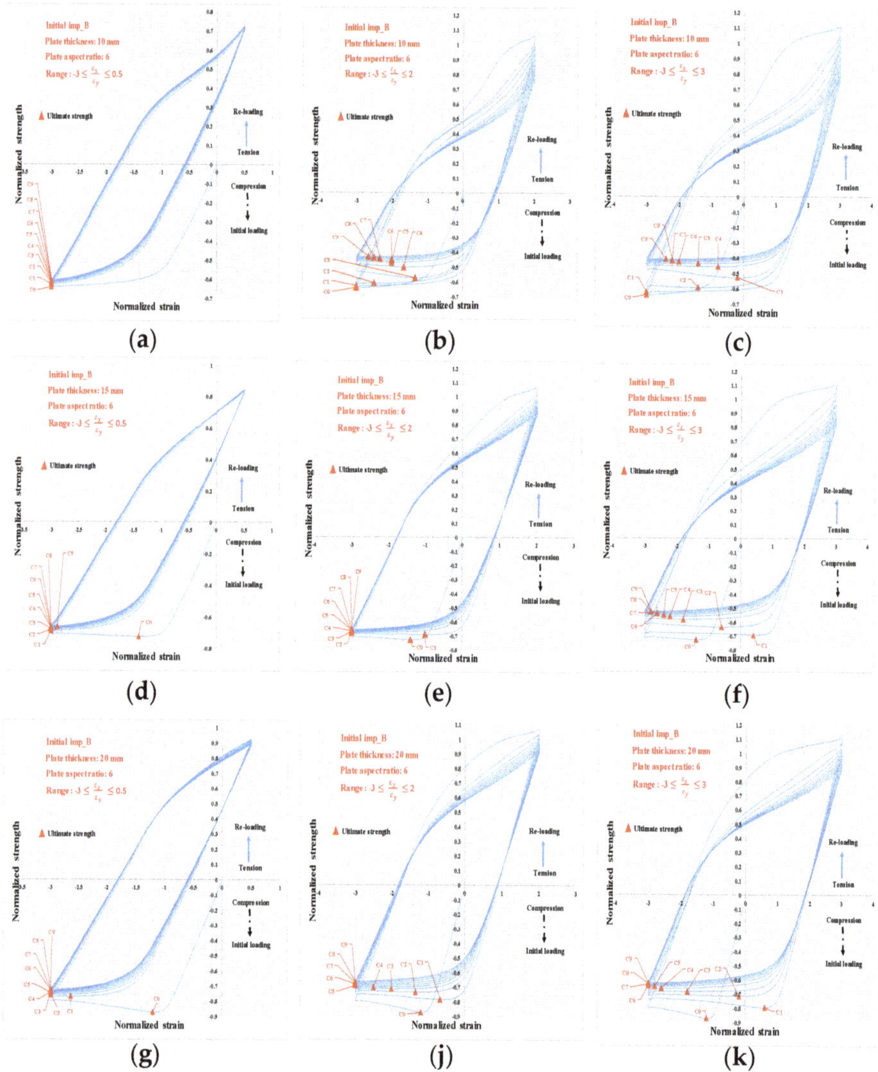

Figure 17. (**a**) $t = 10$ mm, $-3 \leq \frac{\varepsilon_x}{\varepsilon_y} \leq 0.5$; (**b**) $t = 10$ mm, $-3 \leq \frac{\varepsilon_x}{\varepsilon_y} \leq 2$; (**c**) 10 mm, $-3 \leq \frac{\varepsilon_x}{\varepsilon_y} \leq 3$; (**d**) $t = 15$ mm, $-3 \leq \frac{\varepsilon_x}{\varepsilon_y} \leq 0.5$; (**e**) 15mm, $-3 \leq \frac{\varepsilon_x}{\varepsilon_y} \leq 2$; (**f**) 15 mm, $-3 \leq \frac{\varepsilon_x}{\varepsilon_y} \leq 3$; (**g**) 20 mm, $-3 \leq \frac{\varepsilon_x}{\varepsilon_y} \leq 0.5$; (**j**) 20 mm, $-3 \leq \frac{\varepsilon_x}{\varepsilon_y} \leq 2$; (**k**) 20 mm, $-3 \leq \frac{\varepsilon_x}{\varepsilon_y} \leq 3$.

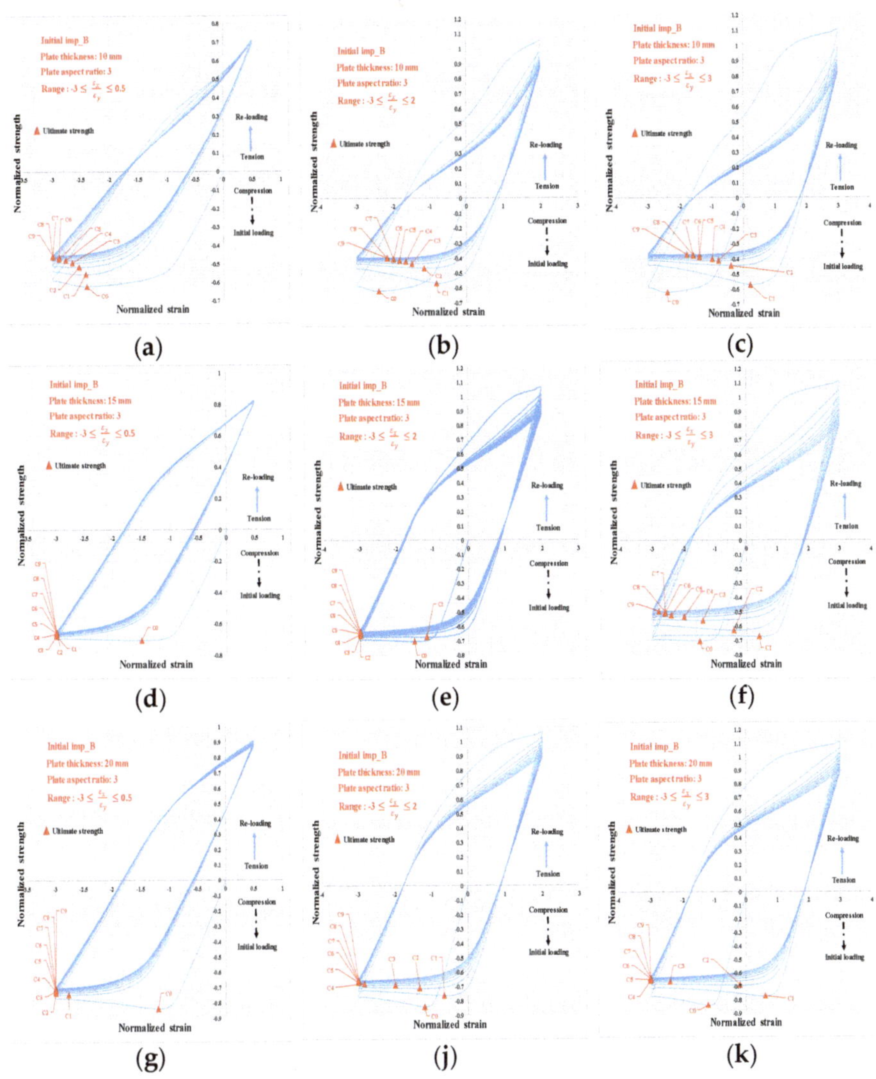

Figure 18. (**a**) $t = 10$ mm, $-3 \leq \frac{\varepsilon_x}{\varepsilon_y} \leq 0.5$; (**b**) $t = 10$ mm, $-3 \leq \frac{\varepsilon_x}{\varepsilon_y} \leq 2$; (**c**) $t = 10$ mm, $-3 \leq \frac{\varepsilon_x}{\varepsilon_y} \leq 3$; (**d**) $t = 15$ mm, $-3 \leq \frac{\varepsilon_x}{\varepsilon_y} \leq 0.5$; (**e**) $t = 15$mm, $-3 \leq \frac{\varepsilon_x}{\varepsilon_y} \leq 2$; (**f**) 15 mm, $-3 \leq \frac{\varepsilon_x}{\varepsilon_y} \leq 3$; (**g**) $t = 20$ mm, $-3 \leq \frac{\varepsilon_x}{\varepsilon_y} \leq 0.5$; (**j**) $t = 20$ mm, $-3 \leq \frac{\varepsilon_x}{\varepsilon_y} \leq 2$; (**k**) $t = 20$ mm, $-3 \leq \frac{\varepsilon_x}{\varepsilon_y} \leq 3$.

Figure 19 shows the ultimate capacity reduction accounting for the plate thickness and aspect ratios subjected to multiple cyclic loads.

It has been observed that in the longer plates, there is a correlation between the plate thickness and ultimate load capacity reduction that as the plate thickness gets larger, it gives rise to more load-carrying capacity reduction during the cyclic load.

However, there are cases where the reduction is more pronounced with the thinner plates. The reason for this is attributed to the pattern of the plastic formation where a partial failure mechanism may form and governs the entire plate collapse. This phenomenon has been observed with a plate thickness of 10 mm, with an aspect ratio of 3, as shown in Figure 14.

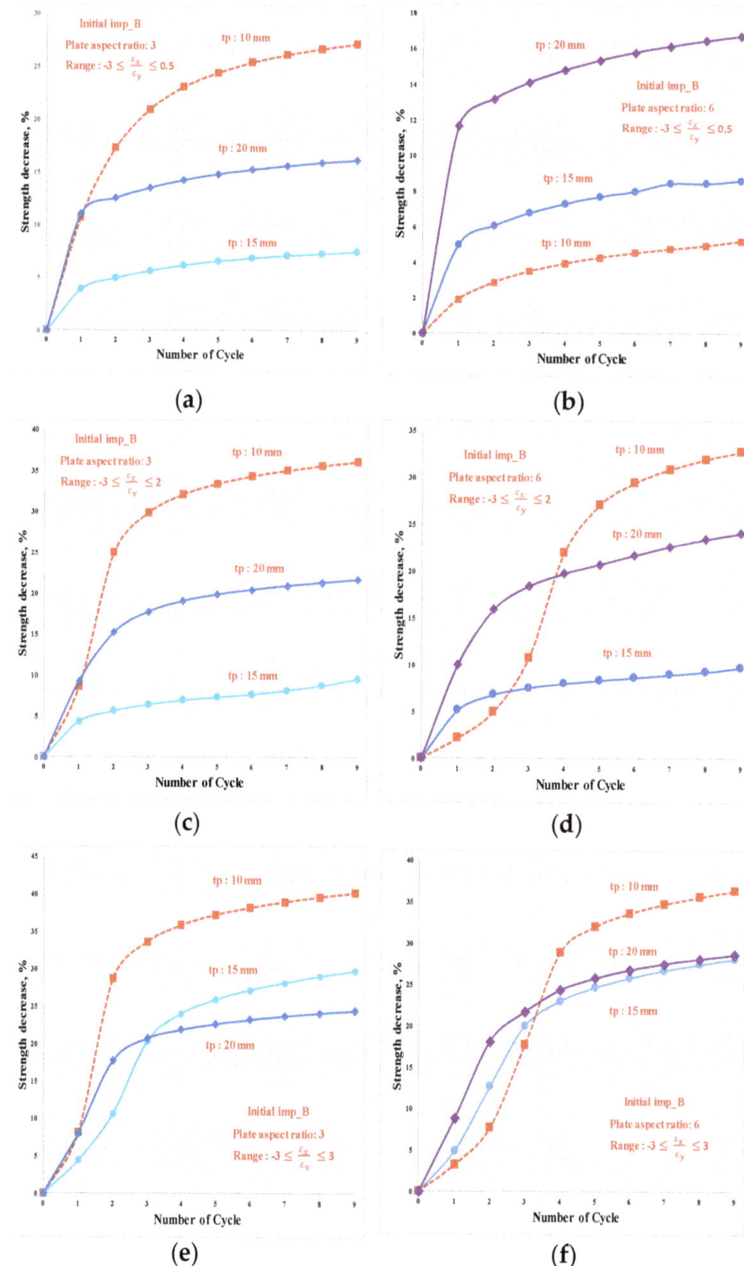

Figure 19. (**a**) Aspect ratio: 3, Initial imp_B, Strength decrease, $-3 \leq \frac{\varepsilon_x}{\varepsilon_y} \leq 0.5$; (**b**) Aspect ratio: 6, Initial imp_B, Strength decrease, $-3 \leq \frac{\varepsilon_x}{\varepsilon_y} \leq 0.5$; (**c**) Aspect ratio: 3, Initial imp_B, Strength decrease, $-3 \leq \frac{\varepsilon_x}{\varepsilon_y} \leq 2$; (**d**) Aspect ratio: 6, Initial imp_B, Strength decrease, $-3 \leq \frac{\varepsilon_x}{\varepsilon_y} \leq 2$; (**e**) Aspect ratio: 3, Initial imp_B, Strength decrease, $-3 \leq \frac{\varepsilon_x}{\varepsilon_y} \leq 3$; (**f**) Aspect ratio: 6, Initial imp_B, Strength decrease, $-3 \leq \frac{\varepsilon_x}{\varepsilon_y} \leq 3$.

4. Conclusions

The strength of rectangular plates subjected to cyclic loads has been analysed by employing the finite element method accounting for different plate aspect ratios and thicknesses. It has been shown that as the plates are subjected to the cyclic load, the ultimate load capacity, along with their stiffness decreases. The reduction is more pronounced as the strain range is increased.

It has been observed that with the square plate, there is a correlation between the ultimate capacity reduction and the plate thickness as the plate thickness gets smaller that gives rise to the capacity reduction under the multiple cyclic loads. The capacity reduction might also be influenced by the type of plastic formation that may lead to local plastic collapse during the cyclic load exposure.

For the plates with higher aspect ratios, as the plate gets thicker, the plate load capacity reduction gets more pronounced under the cyclic load. The response is also highly influenced by the type of plastic formation.

Therefore, the ultimate capacity reduction might be more pronounced as the cyclic load exposure is increased in the case of the thinner plate because of the local plastic formations.

It has been observed that the initial imperfection has a significant influence on the ultimate load reduction during the cyclic load exposure that has been seen with the plate with higher aspect ratios.

The multi-modal initial imperfection is observed to be less stable, as opposed to the uni-modal initial imperfection. The ultimate load capacity reduction is more significant with the multi-modal initial imperfection during the cyclic load exposure because the local plastic formation is earlier and governs the plate deformations.

However, in the case of the uni-modal initial imperfection, during the cyclic load exposure, the load transition and plastic formation are smooth and more balanced, which can be observed with the equivalent plastic strain formation sequence.

Author Contributions: The concept of the problem is developed by M.T. and Y.G. The analysis is performed by M.T. and writing of the original draft manuscript is done by M.T. and Y.G. All authors have read and agreed to the published version of the manuscript.

Funding: This work was performed within the Strategic Research Plan of the Centre for Marine Technology and Ocean Engineering (CENTEC), which is financed by Portuguese Foundation for Science and Technology (Fundação para a Ciência e Tecnologia-FCT) under contract UID/Multi/00134/2013-LISBOA-01-0145-FEDER-007629.

Acknowledgments: In this section you can acknowledge any support given which is not covered by the author contribution or funding sections. This may include administrative and technical support, or donations in kind (e.g., materials used for experiments).

Conflicts of Interest: The authors declare no conflict of interest.

References

1. DNV-GL. *Class Guidelines, Fatigue Assessment of Ship Structures*; DNV-GL: Oslo, Norway, 2015.
2. Eurocode-3. *Design of Steel Structures-Part 1-6, Strength and Stability of Shell Structures*; European Committee for Standardization: Brussels, Belgium, 2007.
3. Eurocode-8. *Design of Structure for Earthquake Resistance, Part 1: General Rules Seismic Actions and Rules for Buildings*; European Committee for Standardization: Brussels, Belgium, 2004.
4. FEMA. *Improvement of Nonlinear Static Seismic Analysis Procedure*; FEMA 440; Applied Technology Council (ATC-55 Project): Washington, DC, USA, 2005.
5. Ibarra, L.F.; Medina, R.A.; Krawinkler, H. Hysteretic models that incorporate strength and stiffness deterioration. *Earthq. Eng. Struct. Dyn.* **2005**, *34*, 1489–1511. [CrossRef]
6. Azevedo, J.; Calado, L. Hysteretic behaviour of steel members: Analytical models and experimental tests. *J. Constr. Steel Res.* **1994**, *29*, 71–94. [CrossRef]
7. Zhou, F.; Chen, Y.; Wu, Q. Dependence of the cyclic response of structural on loading history under large inelastic strains. *J. Constr. Steel Res.* **2015**, *104*, 64–73. [CrossRef]
8. Shi, Y.; Wang, M.; Wang, Y. Experimental and constitutive model study of structural steel under cyclic loading. *J. Constr. Steel Res.* **2011**, *67*, 1185–1197. [CrossRef]

9.   Shi, G.; Wang, M.; Bai, Y.; Wang, F.; Shi, Y.; Wang, Y. Experimental and modelling study of high-strength structural steel under cyclic loading. *Eng. Struct.* **2012**, *37*, 1–13. [CrossRef]

10.   Krolo, P.; Grandi, D.; Smolcic, Z. Experimental and numerical study of mild steel behaviour under cyclic loading with variable strain ranges. *Adv. Mater. Sci. Eng.* **2016**, *2016*, 1–13. [CrossRef]

11.   Zhao, X.; Li, H.; Chen, T.; Cao, B.; Li, X. Mechanical properties of aluminium under low-cycle fatigue loading. *Materials* **2019**, *12*.

12.   Wang, B.; Sun, Y.; Zheng, S. Hysteretic behaviour of steel-reinforced concrete columns based on damage analysis. *Appl. Sci.* **2019**, *9*, 687. [CrossRef]

13.   Smith, C. Influence of local compressive failure on ultimate longitudinal strength of a ship hull. In Proceedings of the International Symposium on Practical Design in Shipbuilding (PRADS), Tokyo, Japan, 18–20 October 1977; pp. 73–79.

14.   Paik, J.K.; Kim, D.K.; Park, D.H.; Kim, H.B.; Mansour, A.E.; Caldwell, J.B. Modified Paik-Mansour formula for ultimate strength calculations of ship hulls. *Ships Offshore Struct.* **2012**, *8*, 245–260. [CrossRef]

15.   ALPS/HULL. *A Computer Program for Progressive Collapse Analysis of Ship Hulls*; Advanced Technology Center, DRS C3 Systems, Inc.: Parsippany, NJ, USA, 2018. Available online: www.proteusengineering.com; www.maestromarine.com; (accessed on 1 November 2019).

16.   CSR. *Common Structural Rules for Bulk Carriers and Oil Tankers*; International Association of Classification Societies, IACS: London, UK, 2017.

17.   Gordo, J.M.; Soares, C.G. Approximate method to evaluate the hull girder collapse strength. *Mar. Struct.* **1997**, *9*, 449–470. [CrossRef]

18.   Gordo, J.M.; Soares, C.G. Approximate assessment of the ultimate longitudinal strength of the hull girder. *J. Ship Res.* **1996**, *40*, 60–69.

19.   Paik, J.K.; Amlashi, H.; Boon, B.; Branner, K.; Caridis, P.; Das, P.; Kim, C.W.; Fujikubo, M.; Huang, C.-H.; Josefson, L.; et al. ISSC committee III.1 ultimate strength. In Proceedings of the 18th International Ship and Offshore Structures Congress, Rostock, Germany, 9–13 September 2012; Fricke, W., Bronsart, R., Eds.; Schiffbautechnische Gesellschaft: Hamburg, Germany, 2012; pp. 285–364.

20.   Chen, K.Y.; Kutt, L.M.; Piaszczyk, C.M.; Bieniek, M.P. Ultimate strength of ship structures. *SNAME Trans.* **1983**, *91*, 149–168.

21.   Paik, J.K.; Kim, B.J.; Seo, J.K. Methods for ultimate limit state assessment of ships and ship-shaped offshore structures. *Ocean Eng.* **2008**, *35*, 281–286. [CrossRef]

22.   Xu, M.; Garbatov, Y.; Soares, C.G. Ultimate strength assessment of a tanker hull based on experimentally developed master curves. *J. Mar. Sci. Appl.* **2013**, *12*, 127–139. [CrossRef]

23.   Tekgoz, M.; Garbatov, Y.; Soares, C.G. Strength assessment of an intact and damaged container ship subjected to asymmetrical bending loadings. *Mar. Struct.* **2018**, *58*, 172–198. [CrossRef]

24.   Yao, T.; Nikolov, P.I. Buckling/Plastic collapse of plates under cyclic loading. *Soc. Nav. Archit. Jpn* **1990**, *168*, 449–462. [CrossRef]

25.   Goto, Y.; Toba, Y.; Matsuoka, H. Localization of plastic buckling patterns under cyclic loading. *J. Eng. Mech.* **1995**, *121*, 493–501. [CrossRef]

26.   Komoriyama, Y.; Tanaka, Y.; Ando, T.; Hashizume, Y.; Tatsumi, A.; Fujikubo, M. Effects of cumulative buckling deformation formed by cyclic loading on ultimate strength of stiffened panel. In Proceedings of the International Conference on Ocean, Offshore and Arctic Engineering, Madrid, Spain, 17–22 June 2018.

27.   Yao, T.; Fujikubo, M.; Nie, C. *Development of a Simple Dynamical Model to Simulate Collapse Behaviour of Plates with Residual Welding Stresses under in-Plane Load*; The Japan Society of Naval Architects and Ocean Engineers: Hiroshima, Japan, 1997; pp. 171–182.

28.   Cui, H.W.; Yang, P. Ultimate strength assessment of hull girder under cyclic bending based on Smith´s method. *J. Ship Res.* **2018**, *62*, 77–88. [CrossRef]

29.   Li, S.; Hu, Z.; Benson, S. An analytical method to predict the buckling and collapse behaviour of plates and stiffened panels under cyclic loading. *Eng. Struct.* **2019**, *199*, 109627. [CrossRef]

30.   Chaboche, J.L. Constitutive equations for cyclic plasticity and cyclic viscoplasticity. *Int. J. Plast.* **1989**, *5*, 247–302. [CrossRef]

31.   ANSYS. *Advanced Analysis Techniques Guide, Southpointe, 2600 ANSYS Drive*; Ansys, Inc.: Canonsburg, PA, USA, 2019.

32. Ueda, Y.; Yao, T. The influence of complex initial deflection modes on the behaviour and ultimate strength of rectangular plates in compression. *J. Constr. Steel Res.* **1985**, *5*, 265–302. [CrossRef]
33. Smith, C.S.; Davidson, P.C.; Chapman, J.C. Strength and stiffness of ship´s plating under in-plane compression and tension. *R. Inst. Nav. Arch. Trans.* **1988**, *130*.

Journal of
*Marine Science
and Engineering*

Article

Effect of Residual Stresses on the Elastoplastic Behavior of Welded Steel Plates

José Manuel Gordo

CENTEC, Instituto Superior Técnico, University of Lisbon, 1649-004 Lisboa, Portugal; jose.gordo@centec.tecnico.ulisboa.pt

Received: 15 August 2020; Accepted: 8 September 2020; Published: 10 September 2020

Abstract: A robust methodology to simulate virtually the residual stresses pattern in welded steel plates is presented. The methodology is applied to the structural analysis of typical welded plates belonging to ship structures, and the effect of residual stresses on the elastoplastic behavior of plates loaded axially is analyzed in comparison to the residual stress free case, both for tension and compression and including initial imperfections. Residual stresses affect in different manner plates with different geometries; thus a parametric study is performed covering the usual range of variation of the most important plate parameters that control the strength of the plates, more precisely the slenderness and the aspect ratio. The results from finite elements analysis are compared with codes and most established formulations and recommendations of applicability in the prediction of load-shortening curves for hull's bending strength evaluation, ultimate strength and ultimate strain of plate elements are made.

Keywords: ultimate strength of plates; residual stresses; initial imperfections

---

1. Introduction

The study of unstiffened plates under different loading conditions has been object of very deep and extensive analysis along time in relation to the effect of the most important parameters that influence the ultimate strength of such plates.

Plate elements are a fundamental component of stiffened plates that are the basis of structural strength of thin-walled structures evaluation. The hull girder of a ship is subjected to different loadings, like bending moment, shear force and lateral pressure. Plate elements are very important in all cases contributing greatly to resist those loadings. In the particular case of the hull girder strength assessment, the structure may be considered to be formed by a set of stiffened plates each of them contributing to resist the applied bending moment and other loads [1–3]. The response of each stiffened element depends greatly from the contributions of the associated plate attached to the stiffeners and frames [4].

The evaluation of the residual stresses distribution along the structure has been concentrated in three main fields: development of non-destructive and destructive techniques [5], use of finite element methods and analytical studies associated with the mechanical characteristics of the different material and welding techniques. Leggatt [6] explains the residual stresses in welded structures, i.e., how the different factors affect the magnitude, direction and distribution of residual stresses in welded joints and structures. There exist today several commercial finite elements (FE) codes that may be employed for detailed nonlinear simulations of the development of the temperature and stress fields present during welding.

Lindgren [7], Runesson et al. [8] and Dong [9] discussed methods for numerical simulation of the welding process taking into consideration actual thermal, mechanical and microstructure developments. The focus is on material modelling, coupling effects between thermal, mechanical fields and microstructure, numerical techniques and modelling aspects.

The evaluation of the residual stresses level in 3-D structures may be done indirectly. It requires the imposition of some hypotheses to establish the residual stress pattern and may be obtained by two methods: the structural tangent modulus method and the total energy dissipation method [10]. The first one considers the variation of the tangent modulus under alternate loading at a point of a cycle corresponding to the maximum loading point in the previous cycle. The variation of the tangent modulus is the result of the variation on the effective inertia of the section due to the development of local plasticity at points where residual stresses are still high. The second method considers the dissipation of energy in a structure with residual stresses when an external load is applied [11].

Tekgoz et al. [12] analyzed the effect of residual stress on the ultimate strength assessment of a stiffened panel by a nonlinear finite element method. They develop modified stress strain curves in order to include a model of residual stresses on material properties. An improved approach was proposed by Launert et al. [13] by prescribing initial deformations and simplified residual stress pattern manually in the numerical model applied to welded plate girders. The present methodology allows to implement residual stresses in direct way by using the concept of heat affected zone (HAZ) and mechanical and thermal properties of the material.

2. Methodology

This work studies the effect of residual stresses on the behavior of unstiffened rectangular plates subjected to axial compression. In order to achieve this objective, a methodology to implement residual stresses in structural plate elements due to welding process is presented. The internal state of stresses obtained in this first stage is used as initial condition for the evaluation of the unstiffened plate structural performance by finite element analysis.

2.1. Implementation of Residual Stresses due to Welding

The plate elements are divided into different regions according to the thermal process that they are subjected during the welding process. A typical plate element belonging to a ship's panel has a length a and a width b and is welded on its borders to stiffeners and frames, thus there are strips of the plate in the borders that suffer the effect of heating and cooling during the welding process, developing permanent plastic strains that affect the initial internal state of stresses in the whole plate in the end of the welding process.

2.1.1. Theoretical Model of Residual Stresses for Perfect Plates

It is commonly accepted that these residual stresses patterns may be simply represented by a strip of plate near each border with a width ηt at yield stress (σ_0) in tension and a central region under compression at a stress σ_r with a width of b-$2\eta t$ that equilibrates the plate. η is the factor that relates the width of the heat affected zone with the thickness of the plate t.

The equilibrium equation of loads for the plate quantifies the residual stress as:

$$\sigma_r = -\sigma_0 \frac{2\eta t}{b - 2\eta t} \tag{1}$$

In the transverse direction the transversal residual stress in the middle of the plate is:

$$\sigma_{rt} = -\sigma_0 \frac{2\eta t}{a - 2\eta t} \tag{2}$$

Figure 1 presents the distribution of residual stresses in the flat plate, after welding, both in longitudinal and transversal directions, for the simplified approach (left) [11] and the one obtained by thermal analysis (right) [14]. A similar approach was present by Yi et al. [15] to represent the biaxial state of residual stresses in welded panels.

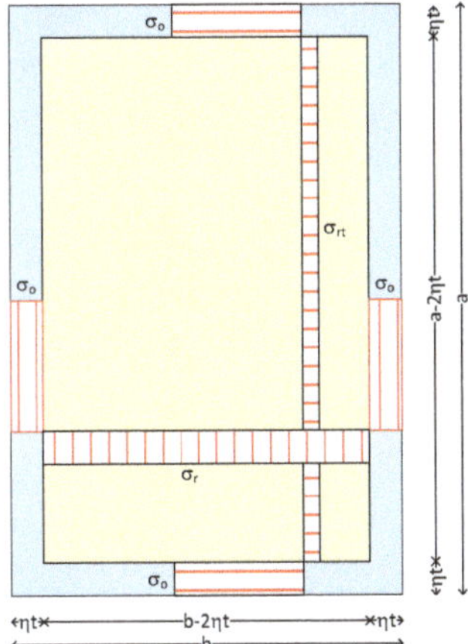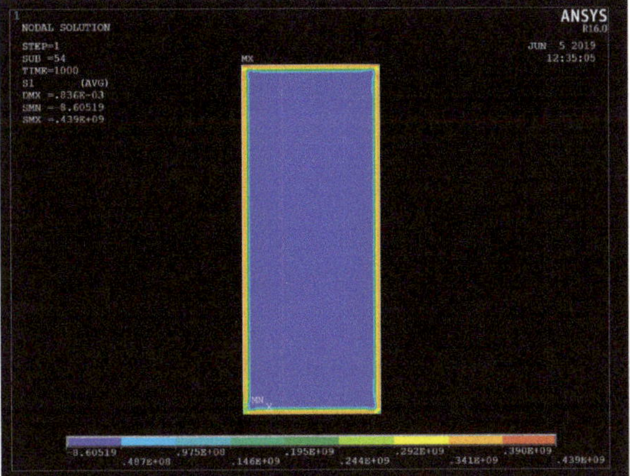

Figure 1. Idealized residual stress model for an unstiffened perfect plate welded in the edges (**left**) and results by FE thermal analysis (**right**).

2.1.2. Implementation of Residual Stresses by FE Modelling

The study of plates with initial residual stresses requires, as a first step, the introduction of the residual stresses pattern as described in Figure 1. There are several ways to achieve such objective [16–18], some of them very complex and computationally costly.

The present method allows to perform the thermal analysis to introduce residual stresses, followed by structural analysis without intermediate steps and been computationally very efficient.

The plate model in finite elements (FE) is divided by parts as described in Figure 1 where the lateral strips representing the heat affected zone (HAZ) have a width ηt. The characterization of the material for such analysis requires in general the information about the mechanical properties for different temperatures and the respective coefficient of thermal expansion.

The present method only uses the modulus of elasticity, the yield stress at room temperature and the respective coefficient of thermal expansion, considering an elastic-perfectly plastic deformation without hardening. The thermal loading is inputted by rising the temperature into the HAZ strips to a reference temperature, in this study 500 °C, followed by a decrease to room temperature, while the temperature in the rest of the plate is kept at room temperature during the whole process [14].

In the first step, the temperature rises to 500 °C in HAZ and it generates plasticity in compression in the lateral strips and some tensile stresses in the central part of the plate. During the second step, cooling of the HAZ strips, the state of internal stresses reverses and, in the end, one has tensile yield stress in the lateral strips and residual compressive stresses in the central part.

This simplified procedure allows to obtain the required residual stress pattern and proceed directly to a structural analysis in a sequential FE analysis load step. It is recommended to use a mesh size less than half of the HAZ width, preferably $\eta t/3$. The peak temperature needs to be bigger than a reference value given by:

$$\Delta T = \frac{2\sigma_0}{\alpha_T E} \tag{3}$$

which is the minimum difference of temperature required to pass from yield stress in compression to yield stress in tension on HAZ strips. For mild steel, reference values in equation are σ_0 of 240 MPa, E is 210 GPa and the coefficient of thermal expansion α_T is 1.1×10^{-5} K^{-1} resulting in a minimum difference of temperatures ΔT of 208 °C.

2.1.3. Implementation of Procedure in FE Modelling

The complete procedure is composed by four stages: 1. Heating of HAZ strips up to 500 °C; 2. Cooling of HAZ strips down to room temperature (T = 20 °C); 3. Initial shortening; 4. Compressive displacement in top of plate up to collapse and beyond. The full procedure is presented in Figure 2.

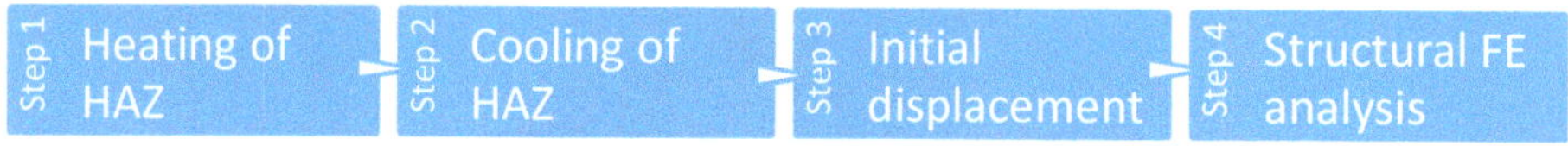

Figure 2. FE loading procedure and steps.

The step 3 is not mandatory but it is very important and it guarantees that part of the residual stresses are not dissipated in the beginning of step 4. Its value must be bigger than the residual longitudinal displacement at the top in the end of thermal analysis, step 2, and the structural analysis on this step 3 should be performed without sub-steps. In more detail and taking as example a plate with residual stresses σ_r after the thermal analysis, one has a residual longitudinal shortening in the end of step 2 given by $\sigma_r / E \cdot a$. If the initial increment of displacement in the structural analysis is smaller than the residual shortening, the plate suffers residual stresses relaxation by plasticity in tensile strips with a decrease in the level of residual stresses. So one should guarantees that the initial increment of load avoids this in the structural analysis. In this work, it decided to do it independently (step 3) and restart the analysis from that point with better control of the increment of load (step 4).

Figure 3 shows the distribution of stresses in the end of steps 1, 2 and at collapse in step 4 for the top layer. As it can be seen, in the end of 1st step the HAZ strips are in compression at yield stress while they are in tension after plasticity in the end of the 2nd step. In the whole process the residual stress pattern varies from point to point due to the presence of initial imperfections and deformations developed during the loading.

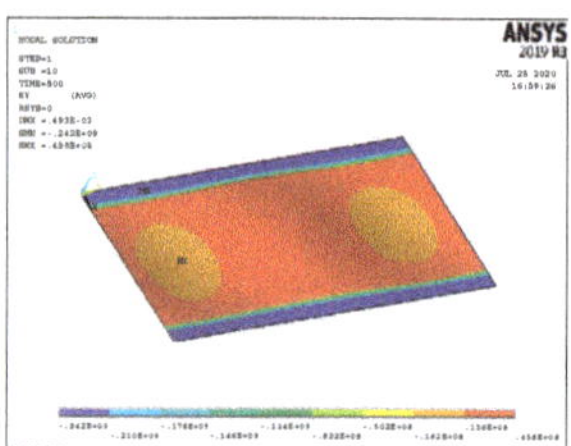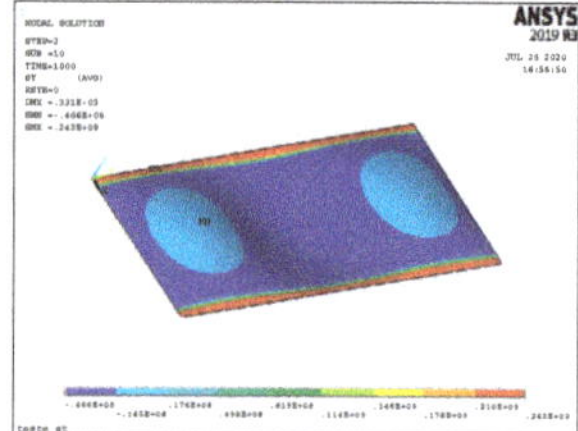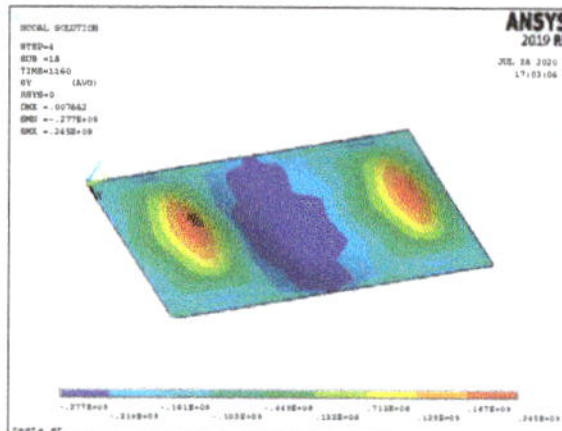

Figure 3. Example of state of stresses of a plate after HAZ heating (**left**), HAZ cooling (**middle**) and at collapse (**right**).

The commercially available finite element software ANSYS [19] has been used for this study. Eight-node shell elements SHELL281 are used which are well-suited for large strain nonlinear applications. Full Newton-Raphson solution procedure was implemented to perform the nonlinear finite element analysis. The mesh of each plate was associated with the tensile strip width as mentioned previously, mesh size less than half of the HAZ width, preferably $\eta t/3$ in order to obtain a reliable representation of residual stresses pattern.

2.2. Parameters Affecting the Strength of Unstiffened Plates

The parameters that most affect the structural behavior in compression, represented by the load-shortening curves (LSC), and the ultimate strength, are the plate slenderness, initial geometric imperfections, residual stresses, boundary conditions and complexity of loading.

2.2.1. Plate Slenderness

The definition of the plate slenderness β results directly from the resolution of rectangular plate's buckling formulation where the elastic critical Euler's stress of a plate under axial compression is:

$$\frac{\sigma_{cr}}{\sigma_0} = k_{cr}\frac{\pi^2}{12(1-\nu^2)\beta^2}$$

(4)

β considers both the geometry of the plate and its material properties and is given by Equation (5) while k_{cr} accounts for the mode of buckling having as minimum 4 when $m = a/b$.

$$\beta = \frac{b}{t}\sqrt{\frac{\sigma_0}{E}}$$

(5)

$$k_{cr} = \left(\frac{mb}{a} + \frac{a}{mb}\right)^2$$

(6)

When the critical stress approaches or overlaps the yield stress, some plasticity is developed in some points of the plate and the ultimate stress of the plate becomes less than the critical one. Several proposals have been presented in the past to estimate the ultimate strength, some of them having the plate slenderness as unique parameter. Two important examples are Faulkner's (Equation (7)) and Frankland's (Equation (8)) equations that are used as well to introduce the concept of effective width. The former is representative of the strength of simply supported welded plates [20] and the latter is used in International Association of Classification Societies (IACS) formulation in the definition of the effective width of the attached plate in a stiffened structural element [3,21].

$$\phi_u = \frac{\sigma_u}{\sigma_0} = \frac{2}{\beta} - \frac{1}{\beta^2}(for \beta > 1)$$

(7)

$$\phi_u = \frac{\sigma_u}{\sigma_0} = \frac{2.25}{\beta} - \frac{1.25}{\beta^2}(for \beta > 1.25)$$

(8)

Figure 4 compares these formulas showing that IACS equation predicts higher strength under in-plane compression than Faulkner's approach, which is eventually a consequence of the effect of residual stresses on the strength of the plates used in the database. It also presents in the same figure a qualitative classification of unstiffened plates according to relative importance of plasticity and elastic instability on the behavior and strength of the plates, as follows: stocky plates, dominated by plasticity and yield stress; intermediate plates, where elastoplastic effects are very important with a critical stress above yield stress; slender plates, having high degradation of the ultimate strength with increase of slenderness and a critical stress above 0.5 yield stress: and very slender plates with very low ultimate strength due to very low critical stress. The range of slenderness covered in this classification is the most common in ship's structures, which varies from 1 to 2.5 according Zhan [22].

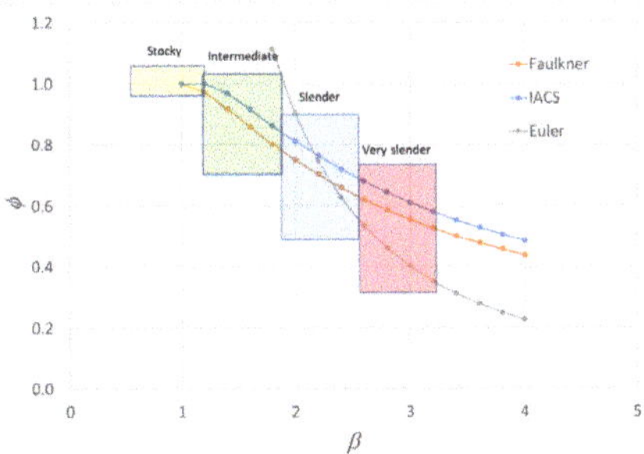

Figure 4. Strength of rectangular plates according to Faulkner, IACS and Euler equations.

2.2.2. Geometric Imperfections

Several studies [23–29] proposed different values and formulas to estimate the amplitude of initial imperfections, most of them pointing out to the maximum value without consideration to the mode of imperfections.

In this work a reference value given by Equation (9) is considered and a wide range of amplitude around that value is used to estimate the relevancy of the parameter.

$$\frac{w_i}{t} = 0.1\beta^2 \tag{9}$$

Different amplitudes and modes of imperfections are considered, corresponding to the expansion of the shape of imperfections into Fourier series that can be represented by:

$$w(x,y) = \sum\sum w_i.sin\frac{m\pi x}{a}.sin\frac{\pi y}{b} \tag{10}$$

The importance of the mode of imperfections is analyzed by considering the fundamental mode ($m = 1$) which normally leads to stronger plates, and the critical one corresponding to the weaker plate. Since a/b is 2.5 in the models of this study, the critical mode m is 3.

2.2.3. Boundary Conditions

The adoption of appropriate boundary conditions (BC) on each type of structure is of vital importance because the differences on structural behavior and strength can be very large between them. A plate element works normally associated with stiffeners and frames and its contribution to the strength of the structure depends on the restraining action of these elements, both in terms of displacements and rotations.

One may consider 4 main classes of boundary conditions for plate elements belonging to panels:

- Unrestrained; all lateral edges and tops, respectively long and short edges in a rectangular plate, are supported perpendicularly to the plate's plane, but in-plane movements along the lateral edges are allowed and rotations are free.
- Restrained; all lateral edges and tops are supported perpendicularly to the plate's plane, but in-plane movements along the edges are not allowed, called as fixed condition, and rotations are free.
- Constrained; all lateral edges and tops are supported perpendicularly to the plate plane, in-plane movements of the lateral edges are allowed but kept straight and rotations are free.
- Clamped; displacements and rotations are null in all edges.

Each of the above classes leads to differences on the load-elongation curves, ultimate strength and ultimate strain, modes of collapse both in compression and tension. In terms of ultimate strength, the difference may be bigger than 20% in compression, leading to completely different modes of collapse due to the development of transversal stresses induced by Poisson effect [30,31].

Unrestrained and constrained conditions guarantee that the loading in uniaxial compression do not generate globally transversal loading in the lateral edges. Restrained conditions generate a biaxial state of stresses leading normally to stronger response of the plate, with higher ultimate strength and difference collapse mode in certain cases [30]. Clamped conditions can be representative for very rigid framing system in panels, which is not usual in ship's structures, at least on the hull girder.

Due to the nature of stiffening on ship's panels, it is considered that the constraint condition is the most appropriate to represent the behavior of plate elements belonging to ship's structures. Unrestrained are rather conservative and restrained conditions are too optimistic. Figure 5 shows the boundary conditions on the FE model used in this study.

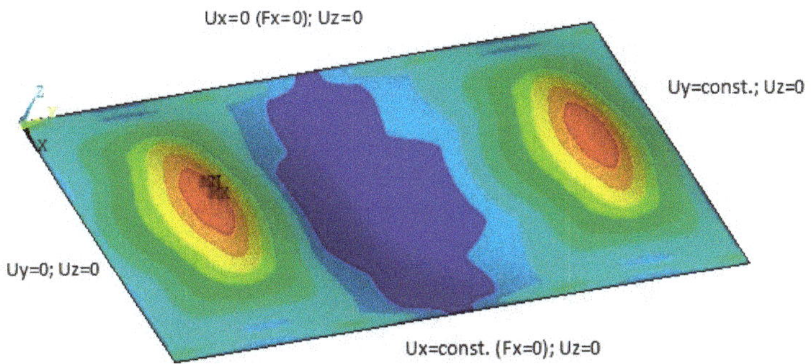

Figure 5. Boundary conditions on a rectangular plate loaded longitudinally (constrained conditions).

The rectangular plate is loaded longitudinally in y direction applying a displacement (Uy) on the right top edge, that remains straight; the left top is fixed (Uy = 0). The upper lateral edge is fixed in the direction perpendicular to the loading (Ux = 0) and the lower lateral edge remains straight but it is free to move in-plane, Ux = constant for each load step, satisfying the condition of a null total net load in the lateral edges during the load path (Fx = 0). These BC's guarantee a pure uni-axial loading of the plate. Similar BC for the plate element were used by Li et al. [32] in the study of ultimate compressive strength of welded stiffened plates.

3. Results

The analysis of the influence of residual stresses on plate's behavior is performed by FE analysis considering a rectangular plate with length a of 1.5 m and a width b of 0.6 m resulting in an aspect ratio α of 2.5 as the basic model.

Four different thicknesses of the plate are considered: 20, 15, 10 and 8 mm. These thicknesses are representative of different classes of plates in terms of elastoplastic behavior and buckling, respectively stocky, intermediate, slender and very slender plates with increase in the elastic instability in compression. The width to thickness ratio, b/t, is respectively 30, 40, 60 and 75.

At least 2 levels of initial geometric imperfections are considered for each group in order to evaluate its effect on the degradation of the ultimate strength in compression. Since the plate's structural behavior is very sensitive to the mode of initial imperfections, 2 modes are considered and compared, the fundamental ($m = 1$) and the critical one ($m = 3$).

The material is considered to have an elastic, perfectly plastic behavior with a Young modulus E of 210 GPa and yield stress, σ_0, of 240 MPa.

The boundary conditions in this study are simply supported in all edges. The edges remain straight during the loading path and the net load in the lateral edges is null, which means that the loading is globally uniaxial, and a global transversal movement of lateral edges is allowed (constrained condition).

3.1. Stocky Plates

In this group the plates have a slenderness β of 1.014 ($b/t = 30$). The reference value for the initial imperfections of these plates, w_i, is 2 mm according to Equation (9). Three different levels of imperfections are considered: 2, 5 and 10 mm, which corresponds to w_i/t ratio of 0.1, 0.25 and 0.5, respectively. Two modes of imperfections are analyzed: the fundamental, $m = 1$, and the critical one, $m = 3$. The effect of residual stresses for each plate is evaluated for three levels of width of the tensile strip at the edges: $\eta = 0$, 2 and 3 to which corresponds a normalized residual stress level of 0.0, 0.15 and 0.25.

The load shortening curves are presented in Figure 6, where the strength is the average compressive stress normalized by the yield stress. The results of the ultimate strength and the corresponding strain normalized by yield strain are presented in Table 1 for the different cases.

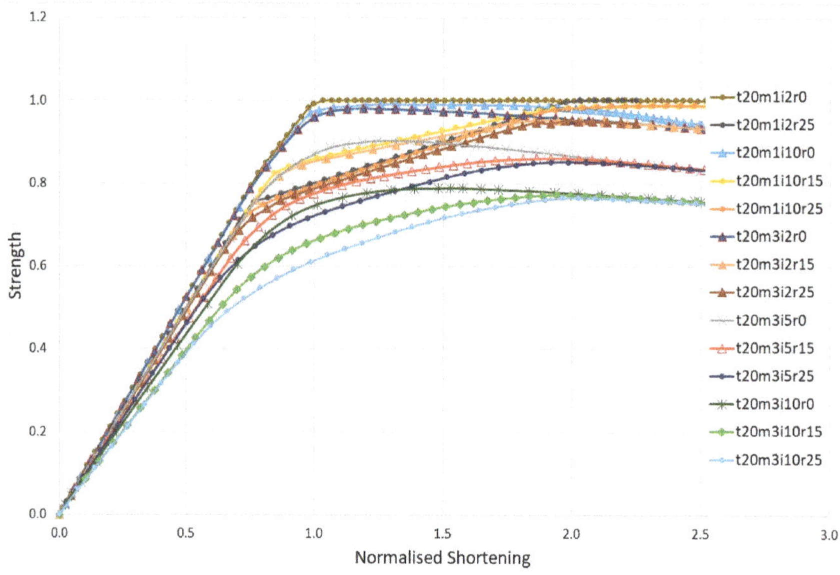

Figure 6. Load shortening curves for stocky plates with different initial imperfections in amplitude (i = 2, 5 and 10 mm) and mode (m = 1, 3), and several levels of residual stresses (r = 0, 0.15 and 0.25%).

Table 1. Summary of data and main results for plates with a = 1.5 m, b = 0.6 m and t = 20 mm.

Identification	Mode i	w_i (mm)	w_i/t	η	Ultimate Stress	Ultimate Strain	Residual Stress %
t20m1i2r0	1	2	0.10	0	1.001	1.00	0.00
t20m1i2r25	1	2	0.10	3	1.002	2.00	0.25
t20m1i10r0	1	10	0.50	0	0.991	1.46	0.00
t20m1i10r15	1	10	0.50	2	0.989	2.36	0.15
t20m1i10r25	1	10	0.50	3	0.990	2.37	0.25
t20m3i2r0	3	2	0.10	0	0.982	0.98	0.00
t20m3i2r15	3	2	0.10	2	0.954	1.94	0.15
t20m3i2r25	3	2	0.10	3	0.952	1.97	0.25
t20m3i5r0	3	5	0.25	0	0.904	1.31	0.00
t20m3i5r15	3	5	0.25	2	0.861	1.92	0.15
t20m3i5r25	3	5	0.25	3	0.851	1.99	0.25
t20m3i10r0	3	10	0.50	0	0.789	1.53	0.00
t20m3i10r15	3	10	0.50	2	0.772	1.94	0.15
t20m3i10r15	3	10	0.50	3	0.765	2.03	0.25

The analysis of the curves and their maximum values allows to withdraw several conclusions.

Plates with fundamental mode (m = 1) have a behavior very similar to material behavior and do not suffer almost any degradation of strength with the increase of amplitude of the imperfections. The ultimate strength maintains constant for large shortening. Residual stresses change the shape of residual stress free LSC in the range of normalized shortening from $1 - \sigma_r/\sigma_0$ to 2 by an almost straight cut. The ultimate stress passes to occur at normalized ultimate shortening of 2.

One the other hand, plates with critical mode ($m = 3$) suffer a large degradation of strength with the increase of the amplitude of initial imperfections. Nevertheless, the less imperfect one ($w_i = 2$ mm) has an ultimate strength close to the values found in plates with fundamental mode of imperfections.

The degradation of the ultimate strength due to increase on amplitude of imperfections is very important for this group of plates ($m = 3$) and the ultimate strength can be quantified as:

$$\phi_u(\beta = 1.01; rs = 0\%) = 1.027 - 0.479\frac{w_i}{t} \tag{11}$$

The ultimate strain of the residual stresses free plate also suffers a shift with the increase of the amplitude of initial imperfections towards a normalized strain of 1.5, as shown in Figure 7.

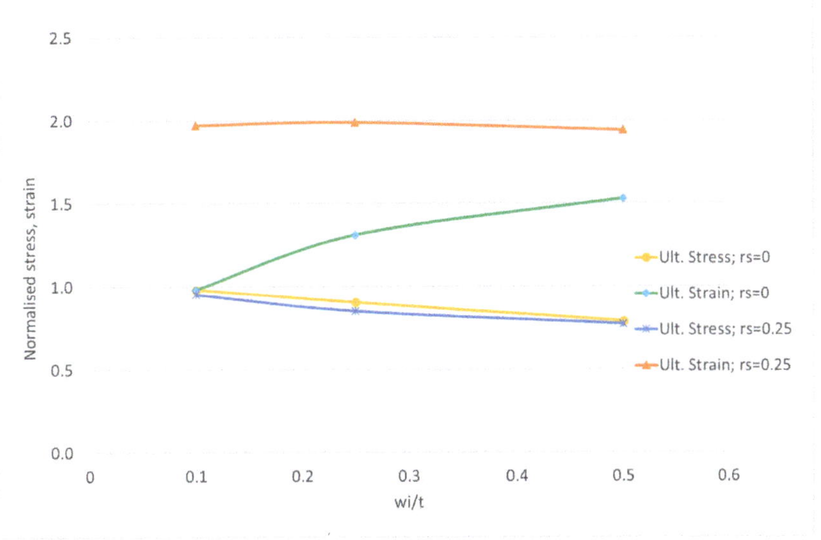

Figure 7. Variation of ultimate strength and corresponding normalized strain for plates with $t = 20$ mm, $m = 3$ and no residual stresses.

Residual stresses have a similar effect to the one descripted previously for plate with $m = 1$ by cutting the LSC to lower values in the same range of shortening. Finally, initial imperfections and residual stress seems to reduce the structural modulus of the plate in elastic range given by the slope of LSC in Figure 6. The ultimate normalized strain in plates with residual stresses is close to 2, which is the limit of the effect of the tensile strips due to welding.

The load-shedding pattern, after the ultimate load on plates with residual stresses, follows the path of the similar residual stress free plate.

The ultimate strength of plates with residual stresses is given by:

$$\phi_u(\beta = 1.01; rs = 25\%) = 0.983 - 0.438\frac{w_i}{t} \tag{12}$$

The degradation of ultimate strength due to the level of imperfections is lower by 8.6% in the plates with residual stresses (-0.438) than in plates free of residual stresses (-0.479).

3.2. Intermediate Plates

The plates in this group have a slenderness, β, of 1.352 ($b/t = 40$). The reference value for the initial imperfections of these plates, w_i, is 2.74 mm according to Equation (9). Three different levels of imperfections are considered: 2, 3 and 5 mm, which correspond to w_i/t ratios of 0.13, 0.2 and 0.33, respectively. The structural analysis concentrates on the critical mode, $m = 3$. The effect of residual stresses for each plate is evaluated for three levels of the width of the tensile strip at the edges: $\eta = 0, 2$ and 3 to which corresponds a normalized residual stress level of 0, 0.11 and 0.18.

The load shortening curves are presented in Figure 8. The results of the ultimate stress normalized by the yield stress and the corresponding strain normalized by yield strain are presented in Table 2 for the different cases. The results for a plate with imperfections in the fundamental mode for comparison are also included.

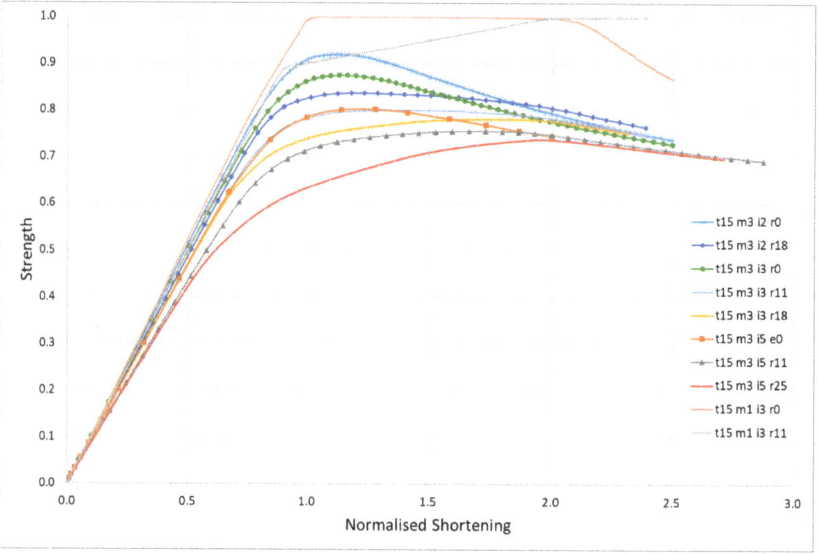

Figure 8. Load shortening curves for intermediate plates ($\beta = 1.352$) with different initial imperfections in fundamental and critical mode and residual stresses.

Table 2. Summary of data and main results for plates with $a = 1.5$ m, $b = 0.6$ m and $t = 15$ mm.

Identification	Mode i	w_i (mm)	w_i/t	η	Ultimate Stress	Ultimate Strain	Residual Stress %
t15m1i3r0	1	3	0.20	0	1.000	1.402	0.00
t15m1i3r11	1	3	0.20	2	0.999	2.114	0.11
t15m3i2r0	3	2	0.13	0	0.919	1.113	0.00
t15m3i2r18	3	2	0.13	3	0.837	1.231	0.18
t15m3i3r0	3	3	0.20	0	0.874	1.129	0.00
t15m3i3r11	3	3	0.20	2	0.801	1.449	0.11
t15m3i3r18	3	3	0.20	3	0.782	1.742	0.18
t15m3i5r0	3	5	0.33	0	0.802	1.276	0.00
t15m3i5r11	3	5	0.33	2	0.755	1.660	0.11
t15m3i5r25	3	5	0.33	4	0.737	1.969	0.25

Qualitatively, the conclusions are very much in agreement with the ones described for stocky plates in relation to the increase in imperfections and residual stresses: degradation of ultimate strength, reduction of initial structural modulus and increase in ultimate strain, as it may be seen in Figure 9.

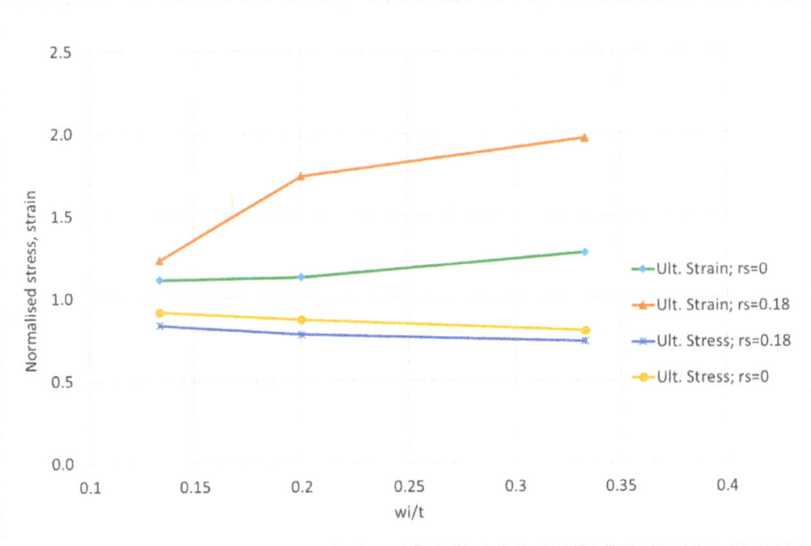

Figure 9. Variation of ultimate strength and corresponding normalized strain for plates with *t* = 15 mm, *m* = 3 and no residual stresses.

The degradation of the ultimate strength due to increase on amplitude of imperfections is even bigger than for the stocky ones for this group of plates (*m* = 3) and the ultimate strength of the residual stresses free plate can be quantified as:

$$\phi_u(\beta = 1.35; rs = 0\%) = 0.994 - 0.580\frac{w_i}{t} \tag{13}$$

The equation for the ultimate strength for plates with residual stress of 18% of the yield stress is given by:

$$\phi_u(\beta = 1.35; rs = 18\%) = 0.891 - 0.474\frac{w_i}{t} \tag{14}$$

One should note that the level of residual stresses does not affect much the ultimate strength of the plate that achieved at a normalized strain close to 2 and after that the LSC follows the residual stresses free plate LSC.

3.3. Slender Plates

The plates in this group have a slenderness β of 2.03 (*b/t* = 60). The reference value for the initial imperfections of these plates, w_i, is 4.11 mm according to Equation (9) corresponding to a w_i/t ratio of 0.411. Three levels of imperfections, w_i = 2, 5 and 6 mm and residual stresses of 0, 0.07 and 0.15 σ_y are considered. The LSC's of this group of plates are shown in Figure 10. Table 3 presents the main results of the ultimate strength and the corresponding normalized shortening for the different cases.

The effects of residual stresses are the reduction the initial structural modulus of the plate, the delay of collapse to normalized shortening in the range of 1.7 to 1.9, as presented in Figure 11. It is also marked the existence of the almost straight line due to the action of HAZ strips in the range 0.8–1.8 of normalized shortening.

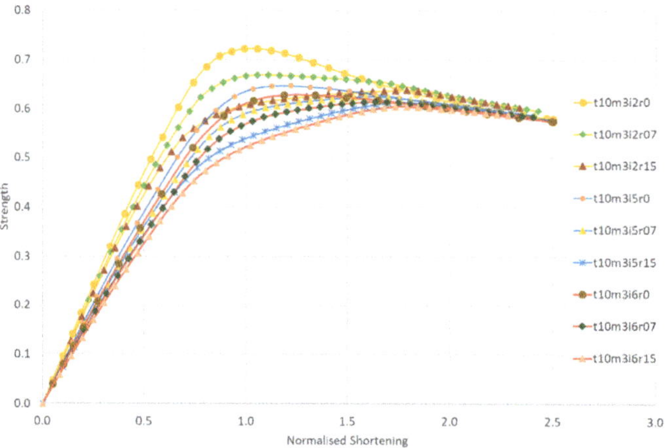

Figure 10. Load shortening curves for intermediate plates (β = 2.03) with different initial imperfections in critical mode and residual stresses.

Table 3. Summary of data and main results for plates with a = 1.5 m, b = 0.6 m and t = 10 mm.

Identification	Mode i	w_i (mm)	w_i/t	η	Ultimate Stress	Ultimate Strain	Residual Stress %
t10m3i2r0	3	2	0.2	0	0.724	1.052	0.00
t10m3i2r07	3	2	0.2	2	0.670	1.105	0.07
t10m3i2r15	3	2	0.2	4	0.639	1.787	0.15
t10m3i5r0	3	5	0.5	0	0.648	1.215	0.00
t10m3i5r07	3	5	0.5	2	0.626	1.641	0.07
t10m3i5r15	3	5	0.5	4	0.615	1.748	0.15
t10m3i6r0	3	6	0.6	0	0.629	1.183	0.00
t10m3i6r07	3	6	0.6	2	0.615	1.635	0.07
t10m3i6r15	3	6	0.6	4	0.606	1.733	0.15
t10m3i8r0	3	8	0.8	0	0.598	1.432	0.00
t10m3i8r15	3	8	0.8	4	0.588	1.756	0.15

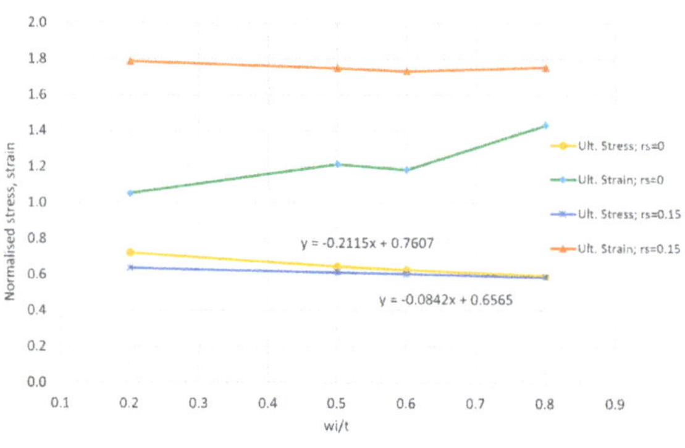

Figure 11. Variation of ultimate strength and corresponding normalized strain for plates with t = 10 mm, m = 3 and no residual stresses.

The degradation of the ultimate strength due to increase on amplitude of imperfections is much smaller than in the previous groups, stocky and intermediate ones and the ultimate strength of residual stress-free plate can be quantified as:

$$\phi_u(\beta = 2.03; rs = 0\%) = 0.761 - 0.212\frac{w_i}{t} \tag{15}$$

The ultimate strength for plates with residual stress of 15% of the yield stress is given by:

$$\phi_u(\beta = 2.03; rs = 15\%) = 0.657 - 0.084\frac{w_i}{t} \tag{16}$$

Imperfections degrade the ultimate strength by a factor of 0.212 in the case of residual stress-free plate, while this factor is 0.58 in the same case for intermediate plates, which is 2.7 times more.

The ultimate strength of the plates with residual stresses is almost constant with the variation of residual stresses and initial imperfections in critical mode.

3.4. Very Slender Plates

The plates in this group have a slenderness, β, of 2.54 (b/t = 75). The reference value for the initial imperfections of these plates, w_i, is 5.14 mm according to Equation (9) and a w_i/t ratio of 0.643. Three levels of imperfections, w_i = 2, 5 and 6 mm and residual stresses of 0, 0.06 and 0.12 σ_y are considered. The LSC's of this group of plates are shown in Figure 12.

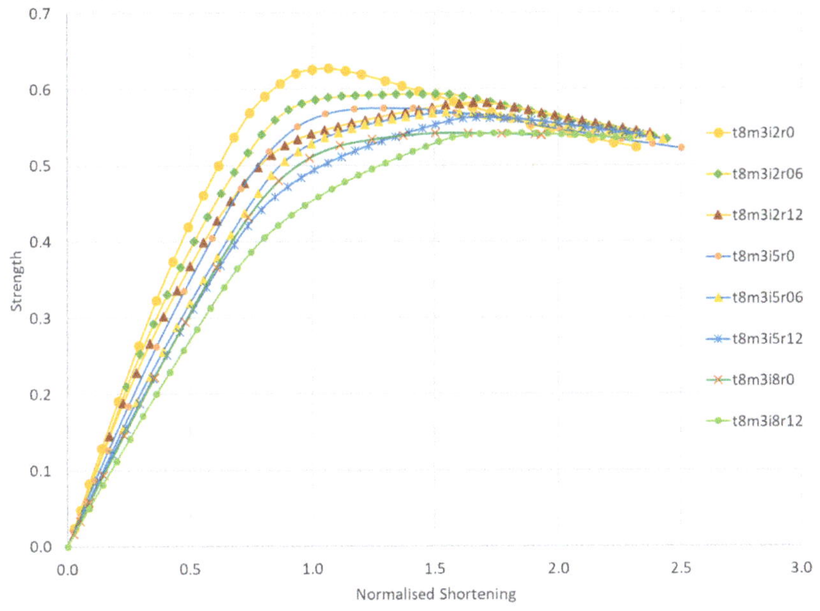

Figure 12. Load shortening curves for very slender plates with different amplitude of initial imperfections in critical mode and residual stresses.

Table 4 presents the main results of the ultimate strength and the corresponding normalized shortening for the different cases.

Table 4. Summary of data and main results for plates with a = 1.5 m, b = 0.6 m and t = 8 mm.

Identification	Mode i	w_i (mm)	w_i/t	η	Ultimate Stress	Ultimate Strain	Residual Stress %
t8m3i2r0	3	2	0.25	0	0.627	1.063	0.00
t8m3i2r07	3	2	0.25	2	0.593	1.505	0.06
t8m3i2r15	3	2	0.25	4	0.582	1.654	0.12
t8m3i5r0	3	5	0.63	0	0.575	1.289	0.00
t8m3i5r07	3	5	0.63	2	0.569	1.544	0.06
t8m3i5r15	3	5	0.63	4	0.563	1.668	0.12
t8m3i6r0	3	8	1.00	0	0.542	1.633	0.00
t8m3i6r07	3	8	1.00	4	0.542	1.792	0.12

Since the slenderness of this group is very high the ultimate strength of the plates is small but does not degrade much with increase in imperfections and residual stresses, as presented in Figure 13. In fact, the ultimate strength of plates with residual stresses is almost insensitive to imperfections as confirmed by the very low coefficient 0.044 in Equation (18), but the pre-collapse behavior is rather dissimilar for each of them in terms of structural modulus and softening until collapse, Figure 12.

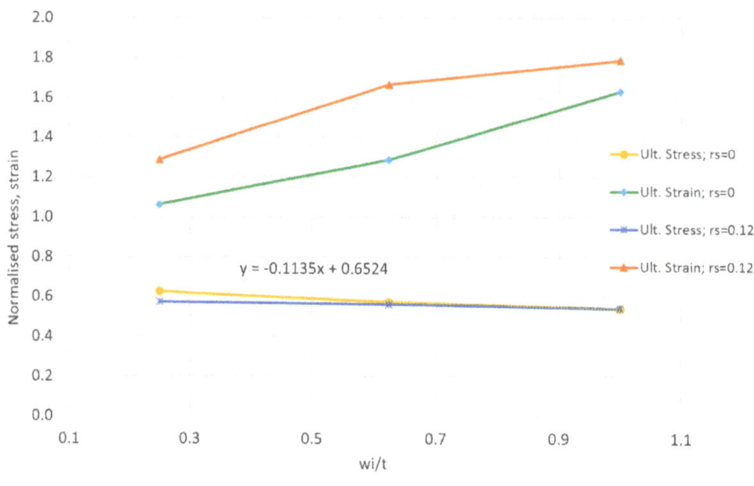

Figure 13. Variation of ultimate strength and corresponding normalized strain for plates with t = 8 mm, m = 3 and residual stresses with η = 0, 4.

The ultimate normalized shortening of residual stress-free plates tends to be greater than in previous groups and the one for plates with residual stresses is much smaller than the values between 1.8 and 2 found previously. This is the result of the very low level of loading carrying capacity of these very slender plates.

The degradation of the ultimate strength due to increase on amplitude of imperfections is the smallest of all groups of plates with critical mode considered in this study and the ultimate strength can be quantified as:

$$\phi_u(\beta = 2.54; rs = 0\%) = 0.652 - 0.114\frac{w_i}{t} \tag{17}$$

The equation for the ultimate strength for plates with residual stress of 12% of the yield stress is given by:

$$\phi_u(\beta = 2.54; rs = 12\%) = 0.587 - 0.044\frac{w_i}{t} \tag{18}$$

These results confirm the conclusion of Ueda et al. [33] that welding residual stresses reduce the buckling strength remarkably but have a little effect on the ultimate strength when the plate is thin.

4. Discussion

The present methodology proved to be very efficient on the structural analysis of plate elements with residual stresses. The method is only based on mechanical and thermal properties of the material and allows to obtain the analysis in short time of computation with high accuracy.

The results obtained show how the main parameters already indicated in Section 3 affect the load-shortening curves (LSC), the ultimate strength and the corresponding ultimate shortening. The results were partially analyzed for each group of plates and formulas were presented for the ultimate strength of imperfect plates with and without residual stresses. Three main aspects need deeper discussion:

- Effect of slenderness on ultimate strength and correlation with imperfections and residual stresses;
- Dependency of structural tangent modulus from initial conditions of plate, i.e., geometry, imperfections and residual stresses;
- Effect of residual stresses in the LSC's.

The analysis performed for each group of plates indicates that the main parameters may affect the ultimate strength largely, in some cases by more than 20%. Two modes of imperfections ($m = 1, 3$) were considered in some groups of plates to demonstrate this variation. For the stocky plates ($t = 20$ mm) without residual stresses, one may find a difference of 2% or 20% depending on the amplitude of imperfections, 2 mm and 10 mm respectively.

4.1. Effect of Slenderness on Ultimate Strength and Correlation with Imperfections and Residual Stresses

The dependency of ultimate strength on slenderness may be obtained by treating the strength equations for each group in integrated manner. Each equation has a constant term that represents the strength of virtual 'perfect' plate in terms of imperfections and a term related to the degradation of strength due to amplitude of imperfections. Collecting such information, one may present the relationship between strength and slenderness, as plotted on Figure 14, where some usual strength formulations are also plotted for comparison.

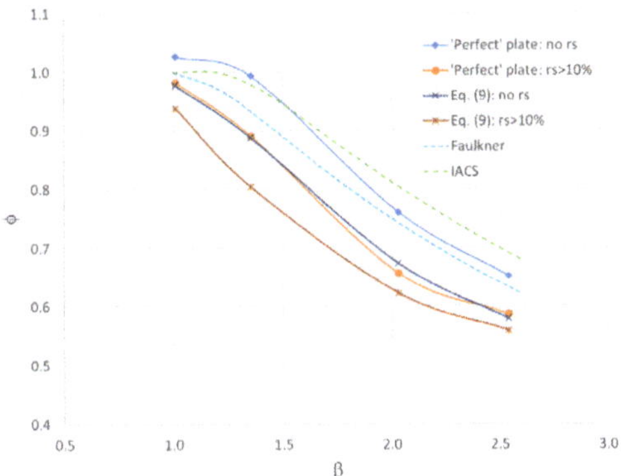

Figure 14. Comparison of results for constraint plates considering residual stress and initial imperfections and standard formulations.

The curves for the influence of imperfections (Equation (9): no *rs*) and for effect of residual stresses alone ('Perfect plate: *rs* > 10%') are very close, which means that the effect of imperfections or residual stresses individually are of the same order in comparison to virtual 'perfect' plate ('Perfect' plate: no *rs*). Nevertheless, the corresponding ultimate shortening and the LSC for each case are completely different.

The comparison of the plates with average imperfections but no residual stresses (Equation (9): no *rs*) or with residual stresses (Equation (9): *rs* > 10%) allows to conclude that the degradation of strength due to residual stresses is more marked in intermediate plates in the range of slenderness from 1.2 to 1.7. Thin plates are poorly affected by residual stresses as mentioned previously.

Curve 'Equation (9): *rs* > 10%' is representative of real welded plate and may be given by:

$$\phi_u\left(\beta; \frac{w_i}{t} = 0.1\beta^2; rs > 10\%\right) = 0.309 + \frac{0.645}{\beta} \tag{19}$$

One aspect of concern in relation to structural codes is that all curves with exception for the "Perfect' plate: no *rs*' curve, are well below the Faulkner and IACS formulation. This may be result of boundary conditions of the tests that are the database for such formulations, inducing fixed conditions or some degree of rotational restraining.

4.2. Dependency of Structural Tangent Modulus from Initial Conditions

The comparison of initial stage of LSC for different initial imperfections with the same slenderness leads to the conclusion that plates present different structural tangent modulus in elastic range, where structural tangent modulus is the slope of the LSC, $\delta\sigma/\delta\varepsilon$. This means that the plate suffers a loose of effectiveness in early stages of loading due to out-of-plane of initial imperfections. With the increase in the compressive loading, the out-of-plane geometry of plate increases, generating additional loss in effectiveness. The effectiveness depends not only on the level of imperfections but also on the slenderness of the plate. Figure 15 plots the normalized tangent modulus of the plate for different thicknesses, t = 15 mm at left and t = 8 mm at right, for different amplitudes of initial imperfections and residual stresses.

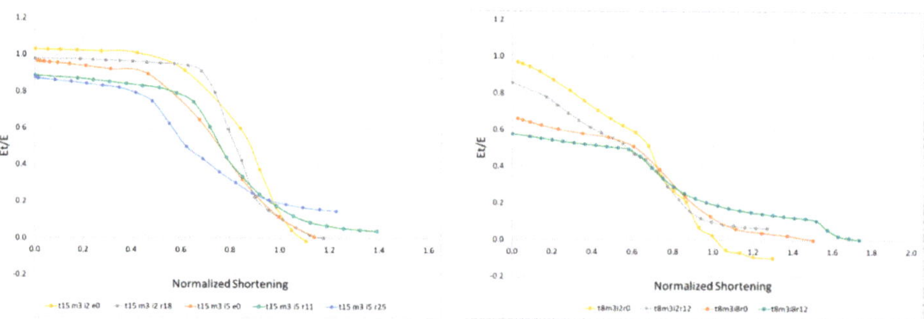

Figure 15. Variation of tangent modulus for intermediate (**left**) and very slender plates (**right**).

The reduction in effectiveness at early stage of load is larger in slenderer plates and increases with w_i in both cases, as seen by comparison of yellow and red curves in both graphics. The presence of residual stresses represents a further reduction on initial effectiveness measured as E_t/E. The total reduction can be very high, representing a great softening of the structural element in elastic range.

This aspect requires further investigation in future research because it may have consequences on the response of 3-D structures, like the hull girder, under longitudinal bending.

4.3. Effect of Residual Stresses in the LSC's

The results obtained for imperfect plates with residual stresses confirms the hypotheses adopted by Gordo and Guedes Soares [4] to generate predictive LSC's for unstiffened and stiffened plates, $\phi_{pl}(\varepsilon)$. In such work, LSC are obtained by the product of the effective width of the plate without residual stresses at each level of loading, $\phi_{ef}(\varepsilon)$, and a modified material behavior that includes the effect of the tensile strips in HAZ, $\phi_{mat+rs}(\varepsilon)$, reading as:

$$\phi_{pl}(\varepsilon) = \phi_{ef}(\varepsilon) \cdot \phi_{mat+rs}(\varepsilon) \tag{20}$$

Figure 16 shows $\phi_{mat+rs}(\varepsilon)$ for compression and tensile loading. This response is very well reproduced by the stocky plates presented in this study, Figure 6, where the effectiveness $\phi_{ef}(\varepsilon)$ is close to 1. In all other groups of plates, the behavior reproduced by the red dot line is present, but curves in the range of strain affected by residual stresses are not totally straight due to the reduction on $\phi_{ef}(\varepsilon)$.

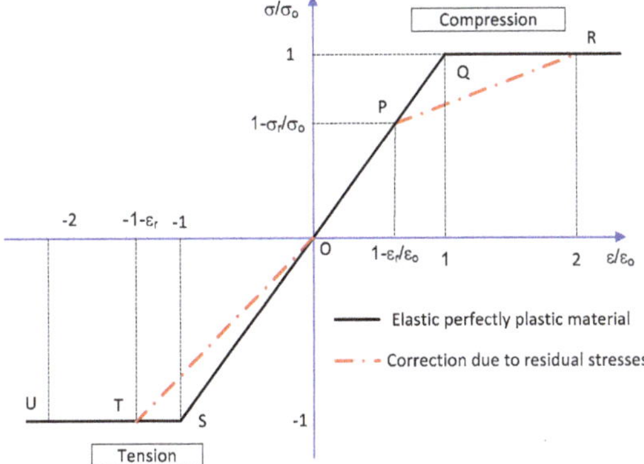

Figure 16. Average stress-strain curve for an unstiffened plate with and without residual stresses.

Furthermore, $\phi_{ef}(\varepsilon)$ proposed in [4] was based on Faulkner's formulation. For the prediction of LSC of unstiffened plates with average imperfections in critical mode and constrained edges, curves from Figure 14 should be used, in particular the curve 'Equation (9): no *rs*' representative of plates with average imperfections and no residual stresses.

5. Conclusions

A methodology to evaluate the performance of welded structural element was presented and shows to be efficient and reliable in the study of unstiffened plate elements under compression.

The parameters used in this study, slenderness, residual stresses, mode and amplitude initial imperfections, prove to be of high importance in the prediction of the ultimate strength but also in the response prior to collapse. The variation in ultimate strength of plates due to the change of magnitude of one parameter alone may reach high values.

Residual stresses affect the pre-collapse behavior of the plates very much, but the ultimate strength tends to be the same, keeping other parameters constant. The corresponding ultimate strain is much higher than in the case of residual stresses free plate, which means that one has a softer response with increase in residual stresses level.

Both residual stresses and initial imperfections affect the initial slope of LSC, structural tangent modulus, and this may affect the performance of 3-D structures, so it should be considered in the future.

Future application of this methodology involves the study of the structural behavior of welded stiffened plates and 3-D structures, considering different loading conditions.

J. Mar. Sci. Eng. **2020**, *8*, 702

Funding: This research received no external funding.

Conflicts of Interest: The author declares no conflict of interest.

References

1. Gordo, J.; Guedes Soares, C.; Faulkner, D. Approximate assessment of the ultimate longitudinal strength of the hull girder. *J. Ship Res.* **1996**, *40*, 60–69.
2. Yao, T.; Nikolov, P. Progressive Collapse Analysis of a Ship's Hull under Longitudinal Bending. *J. Soc. Nav. Archit. Jpn.* **1991**, 449–461. [CrossRef]
3. IACS UR S-Strength of Ships. Available online: http://www.iacs.org.uk/publications/unified-requirements/ur-s/ (accessed on 1 May 2020).
4. Gordo, J.; Soares, G. Approximate load shortening curves for stiffened plates under uniaxial compression. *Integr. Offshore Struct. 5* **1993**, 189–211.
5. Okada, T.; Caprace, J.D.; Estefen, S.F.; Han, Y.; Josefson, L.; Kvasnytskyy, V.F.; Papazoglou, V.; Race, J.; Roland, F.; Schipperen, I.; et al. Committee V.3-Materials and Fabrication Technology. In Proceedings of the 17th International Ship and Offshore Structures Congress, Seoul, Korea, 16–21 August 2009; Volume 2, pp. 137–200.
6. Leggatt, R.H. Residual stress in welded structures. *Int. J. Press. Vessel. Pip.* **2008**, *85*, 144–151. [CrossRef]
7. Lindgren, L.-E. Finite element simulation of welding Part 3: Efficiency and integration. *J. Therm. Stress.* **2001**, *24*, 305–334. [CrossRef]
8. Runesson, K.; Skyttebol, A.; Lindgren, L.-E. Nonlinear finite element analysis and applications to welded structures. In *Comprehensive Structural Integrity*; Elsevier: Oxford, UK, 2003; Volume 3, pp. 255–320.
9. Dong, P. Residual stresses and distortions in welded structures: A perspective for engineering applications. *Sci. Technol. Weld. Join.* **2005**, *10*, 389–398. [CrossRef]
10. Gordo, J.M.; Guedes Soares, C. Experimental evaluation of the ultimate bending moment of a box girder. *Mar. Syst. Ocean Technol.* **2004**, *1*, 33–46. [CrossRef]
11. Gordo, J. Residual stresses relaxation of welded structures under alternate loading. In *Developments in Maritime Transportation and Exploitation of Sea Resources*; Taylor & Francis: London, UK, 2013; pp. 321–328.
12. Tekgoz, M.; Garbatov, Y.; Soares, C. Finite element modelling of the ultimate strength of stiffened plates with residual stresses. In *Analysis and Design of Marine Structures*; Romanoff, J., Ed.; CRC Press: Leiden, The Netherlands, 2013; pp. 309–317. ISBN 978-1-138-00045-2.
13. Launert, B.; Li, Z.; Pasternak, H. Development of a new method for the direct numerical consideration of welding effects in the component design of welded plate girders. In Proceedings of the 7th International Conference on Structural Engineering, Mechanics and Computation (SEMC 2019), Cape Town, South Africa, 2–4 September 2019.
14. Gordo, J.M.; Teixeira, G. A simplified method to simulate residual stresses in plates. In Proceedings of the Martech2020; Lisbon, Portugal, 16–19 November 2020.
15. Yi, M.S.; Hyun, C.M.; Paik, J.K. An empirical formulation for predicting welding-induced biaxial compressive residual stresses on steel stiffened plate structures and its application to thermal plate buckling prevention. *Ships Offshore Struct.* **2019**, *14*, 18–33. [CrossRef]
16. Chen, B.-Q.; Hashemzadeh, M. Numerical analysis of the effects of weld parameters on distortions and residual stresses in butt welded steel plates. In *Developments in Maritime Transportation and Exploitation of Sea Resources*; Soares, C., Peña, F., Eds.; CRC Press: Leiden, The Netherlands, 2013; pp. 309–320. ISBN 978-1-138-00124-4.
17. Pasternak, H.; Launert, B.; Kannengießer, T.; Rhode, M. Advanced Residual Stress Assessment of Plate Girders Through Welding Simulation. *Procedia Eng.* **2017**, *172*, 23–30. [CrossRef]
18. Hashemzadeh, M.; Garbatov, Y.; Guedes Soares, C. Numerical stress-strain analysis of butt-welded plates during the welding process. In *Developments in the Collision and Grounding of Ships and Offshore Structures*; Guedes Soares, C., Ed.; CRC Press: Leiden, The Netherlands, 2019; pp. 157–162. ISBN 978-1-00-300242-0.
19. *ANSYS 2019R3*; Ansys, Inc.: Canonsburg, PA, USA, 2019.
20. Faulkner, D. A Review of Effective Plating for use in the Analysis of Stiffened Plating in Bending and Compression. *J. Ship Res.* **1975**, *19*, 1–17.
21. Frankland, J.M. *The Strength of Ship Plating under Edge Compression*; US EMM Report: Washington, DC, USA, 1940.

22. Zhang, S. A review and study on ultimate strength of steel plates and stiffened panels in axial compression. *Ships Offshore Struct.* **2015**, *11*, 1–11. [CrossRef]

23. Guedes Soares, C. Design equation for the compressive strength of unstiffened plate elements with initial imperfections. *J. Constr. Steel Res.* **1988**, *9*, 287–310. [CrossRef]

24. Bonello, M.A.; Chryssanthopoulos, M.K.; Dowling, P.J. Probabilistic strength modelling of unstiffened plates under axial compression. In Proceedings of the 10th International Conference on Offshore Mechanics and Arctic Engineering (OMAE), ASME, Stavanger, Norway, 23–28 June 1991; Volume 2, pp. 255–264.

25. Carlsen, C.A.; Czujko, J. The specification of tolerances for post welding distortion of stiffened plates in compression. *Struct. Eng.* **1978**, *56*, 133–141.

26. Antoniou, A.C. On the maximum deflection of plating in newly built ships. *J. Ship Res.* **1980**, *24*, 31–39.

27. Antoniou, A.C.; Lavidas, M.; Karvounis, G. On the shape of post-welding deformations of plate panels in newly built ships. *J. Ship Res.* **1984**, *28*, 1–10.

28. Kmiecik, M.; Jastrzębski, T.; Kuźniar, J. Statistics of ship plating distortions. *Mar. Struct.* **1995**, *8*, 119–132. [CrossRef]

29. Yi, M.S.; Lee, D.H.; Lee, H.H.; Paik, J.K. Direct measurements and numerical predictions of welding-induced initial deformations in a full-scale steel stiffened plate structure. *Thin-Walled Struct.* **2020**, *153*, 106786. [CrossRef]

30. Gordo, J.M. Effect of initial imperfections on the strength of restrained plates. *J. Offshore Mech. Arct. Eng.* **2015**, *137*. [CrossRef]

31. Gordo, J.M.; Soares, C.G. Degradation of long plate's ultimate strength due to variation on the shape of initial imperfections. In *Towards Green Marine Technology and Transport*; CRC Press: Leiden, The Netherlands, 2015; pp. 365–374.

32. Li, C.; Dong, S.; Wang, T.; Xu, W.; Zhou, X. Numerical Investigation on Ultimate Compressive Strength of Welded Stiffened Plates Built by Steel Grades of S235–S390. *Appl. Sci.* **2019**, *9*, 2088. [CrossRef]

33. Ueda, Y.; Yasukawa, W.; Yao, T.; Ikegami, H.; Ohminami, R. Effect of Welding Residual Stresses and Initial Deflection on Rigidity and Strength of Square Plates Subjected to Compression (Report II). *Trans. JWRI* **1997**, *6*, 33–38.

Journal of
*Marine Science
and Engineering*

Article

Multi-Pass Welding Distortion Analysis Using Layered Shell Elements Based on Inherent Strain

Jaemin Lee [1], Diego Perrera [2] and Hyun Chung [3,*]

[1] Department of Naval Architecture and Ocean Engineering, Chonnam National University, Yeosu 59626, Korea; jae27v@jnu.ac.kr
[2] Department of Mechanical Engineering, Korea Advanced Institute of Science and Technology, 291 Daehak-ro, Yuseong-gu, Daejeon 34141, Korea; die.perrera@gmail.com
[3] Department of Naval Architecture & Ocean Engineering, Chungnam National University, Daejeon 34134, Korea
* Correspondence: hchung@cnu.ac.kr

Abstract: In this article, a layered shell element-based, elastic finite element method for predicting welding distortion in multi-pass welding is developed. The welding distortion generated in each pass can be predicted by employing layer-by-layer equivalent plastic strains as thermal expansion coefficients and using the heat-affected zone (HAZ) width as the mesh size. The final distortion can be expressed as the sum of the distortions for each pass. This study focuses on extraction of the equivalent plastic strain and HAZ width through 3D thermal elastic plastic analysis (TEPA) for each pass. The input variables extracted from each pass can be converted and added to simulate the final distortion of the multi-pass welding. A 10 mm thick, multi-pass butt-welded joint, subjected to three passes, is simulated via the proposed method. The predicted welding distortion is compared with the 3D TEPA results and the measured experimental data. The outcome indicates that good agreement can be obtained.

Keywords: welding distortion; inherent strain; multi-pass welding deformation prediction

Citation: Lee, J.; Perrera, D.; Chung, H. Multi-Pass Welding Distortion Analysis Using Layered Shell Elements Based on Inherent Strain. *J. Mar. Sci. Eng.* **2021**, *9*, 632. https://doi.org/10.3390/jmse9060632

Academic Editors: Joško Parunov and Yordan Garbatov

Received: 19 May 2021
Accepted: 4 June 2021
Published: 6 June 2021

Publisher's Note: MDPI stays neutral with regard to jurisdictional claims in published maps and institutional affiliations.

1. Introduction

Welding distortion inevitably occurs in welded structures. Uneven temperature distribution occurs due to welding, resulting in residual stress and permanent deformation [1,2]. This causes problems such as a decrease in dimensional accuracy and a decrease in productivity. If welding deformation can be predicted through computer simulation, production plans, such as structural changes to reduce welding distortion, can be established [3]. The biggest problem in implementing welding distortion analysis by computer simulation is the fact that welding is a very complex multi-physics phenomenon [4]. Actual welding is a multi-physical phenomenon in which thermal, mechanical, and metallurgical effects occur, and so, the more accurately reflected these are, the longer the calculation time takes exponentially [5–9]. Since the analysis time takes several hours for unit specimens with a welding length of several hundred millimeters, it is practically impossible to apply 3D TEPA to a welded structure with a length of several tens of meters [1].

Therefore, since the 1980s, to overcome this problem of high calculation cost, simplified methods involving reduced computation time, based on inherent strain theory, have been developed [10,11]. Inherent strain refers to the permanent deformation generated in the heat-affected zone (HAZ) [12]. The inherent strain value can be determined by adding the inelastic strain values in the HAZ by using numerical analysis, such as 3D TEPA, and experimental validation [13]. The estimated inherent strain value can be assigned as an elastic load, such as an equivalent nodal load or equivalent thermal strain load to a finite element (FE) model [14]. With regards to the practical application of welding analysis to large welded structures, welding distortion analysis methods based on equivalent thermal strain and shell elements have been proposed [15,16]. In particular, the method developed by Ha [16], which is known as the "strain as direct boundary (SDB) method," uses the virtual thermal expansion coefficient and virtual temperature distribution in an elastic FEM-based shell model to simulate the plastic deformation that occurs in the HAZ. Unlike conventional shell element-based welding distortion analysis techniques, such as the equivalent load method [17], the SDB method uses scalar input variables for reduced modeling time. Chung et al. [18] also developed a layered shell-based welding distortion analysis method, which can estimate the welding distortion of both the plate and stiffener in fillet welds; this cannot be predicted using the conventional SDB method. Welding distortion analysis methods, based on thermal strain and shell elements, can be effectively used for welding distortion analysis of large welding structures because they use both scalar input variables, to reduce the modeling time, and shell model-based FEM, which can also reduce the computation time effectively. However, most previous research has focused on single-pass welding only. As multi-pass welding is widely used to join thick plates, welding distortion analysis methods based on thermal strain and shell elements should be extended to include multi-pass welding.

In this paper, by exploiting the advantage of the layered shell element-based method, which can provide different thermal strain values per layer, we propose a layered shell element-based elastic FEM for predicting welding distortion in multi-pass welding. We focus on extracting the equivalent strain and HAZ width through 3D TEPA, along with its application to layered, shell element-based elastic FEM. A 10 mm thick, multi-pass butt-welded joint, subjected to three passes, is simulated using the proposed method and the predicted welding distortion is then compared with that obtained from a conventional method and experimentally measured data.

2. SDB Method

The core principle of the SDB method is that the inherent strain can be used as the equivalent thermal strain. In commercial FEM codes, the thermal expansion coefficient can be used as a tool to simulate thermal strain when the temperature variances at specific nodes are given. In the welding distortion analysis method developed by Ha [16], artificial top and bottom temperatures at the nodes, the mesh size at the welding region, and the thermal expansion coefficient are taken as input parameters. The shrinkage can be estimated from the average value obtained from the artificial top and bottom temperatures, whereas the angular deformation can be estimated from the mean difference between the artificial top and bottom temperatures. The model mesh size is equivalent to the maximum width of the inherent strain region [18]. This methodology is based on an experimental case study in which the inherent strain, which was used as the thermal coefficient value, was measured. The artificial temperature values were estimated by analyzing the HAZ shape by conducting 3D TEPA and experimental validation. For more detailed explanations, please refer to Ha [16] and Chung et al. [18].

For multi-pass welding analysis, Ha and Yang [19] extended the conventional SDB method. In the case of multi-pass welding, when the welding of a specific pass is performed, the welding deformation is determined by the thickness up to the stage accumulated in the weld. However, in shell element-based FEM, the bending stiffness (which affects the degree of deformation) is determined by the thickness of the adjacent plate, which is similar to the final thickness following welding completion. Thus, shell element-based welding distortion analysis is performed for a state with considerably higher stiffness than the actual condition encountered in multi-pass welding. Considering these issues, Ha and Yang [19] idealized the problem by employing the following major assumptions:

A.   Each pass has the same cross-sectional area;
B.   All passes are stacked in the layer direction;
C.   The area of the HAZ generated by each pass is ignored, but the bead reinforcement is considered;

D.   The deformation due to the internal residual stress caused by the temperature differences between passes is neglected.

By applying this method, it is possible to implement shell element-based welding distortion analysis in the case of multi-pass welds of the same joint shape, in which the deformation increases with the number of passes. According to a case study by Ha and Yang [19], the proposed method can qualitatively predict that the total amount of angular distortion increases as the number of passes increases, but actual experimental values show an error of 40–50%. This suggests that a limit exists when the idealized assumptions are actually applied, with the main problem being the difficulty in specifying the inherent strain value, which is equal to the thermal expansion coefficient in the SDB method. As the welding passes accumulate, the residual stress and inherent strain values are also influenced by each pass. Ultimately, the inherent strain value generated in each pass is different. Furthermore, it cannot be confidently assumed that the obtained value represents the entire inherent strain region. For a more detailed explanation of this problem, please refer to Ha and Yang [19].

3. Proposed Method

3.1. Layered Shell Element-Based Welding Distortion Analysis Method

Chung et al. [18] introduced a novel approach using layered shell element-based FEM. The main contribution of this method is that composite shell elements can be used for the different thermal expansion coefficients along the joint thickness in fillet welding; hence, it is possible to represent the deformations of both members simultaneously, which cannot be achieved using the conventional SDB method. In the shell element model, the intersecting region at which the base plate and fillet member are attached share the same node; thus, the temperature degree of freedom is shared at the intersection nodes. However, in the composite shell element-based method, different thermal coefficients are employed for each layer representing the inherent region; thus, the HAZ area can be separately modeled for both members, and it is possible to represent the distortions of both members simultaneously.

Reviewing Section 2, it is apparent that the conventional SDB method for multi-pass welding involves major assumptions that may not apply to actual welding conditions, possibly generating prediction errors. Note that assumptions (C) and (D) are critical and may decrease prediction accuracy. First, in the case of V-groove multi-pass butt welding, for which the number of passes is relatively small, it is common to fill in the lateral, rather than layer, direction. This corresponds to adherence to the conventional third assumption, which generates greater angular distortion than the actual state. Second, based on a literature survey [20,21], the temperature difference between passes has a significant effect on the final weld deformation. Thus, following the conventional multi-pass SDB method, it is possible to predict the qualitative results regarding the increase in deformation with an increased number of passes; however, it is difficult to accurately predict the angular distortion. Moreover, in the case of multi-pass welding, it is difficult to specify the inherent strain value (representing the final deformation) based on an experiment. As a substitute method, multi-pass welding analysis can be performed through numerical analysis. Then, the inherent strain value for each layer can be calculated in each pass and directly applied to layered shell element-based welding distortion analysis. As noted above, a previous study on fillet welding [18] focused on modeling the HAZ shapes of both members to simulate their distortions simultaneously. However, in this study, we focus on extraction of the layer-by-layer inherent strain values according to the 3D TEPA results for each pass of the multi-pass welding simulations, followed by accumulation of these distortions to predict the final distortion.

3.2. Proposed Analysis Procedure Based on 3D TEPA Results

As discussed in the previous sections, in order to perform the layered shell element-based welding distortion analysis, definition of the input variables, i.e., the thermal coefficients and mesh size, is required. The thermal coefficients for each layer correspond to the equivalent plastic strain, and the mesh size in the welding region is identical to the equivalent HAZ width. The welding deformation is determined by the thickness up to the stage accumulated in the weld; thus, the element birth and death technique [22] is used for both heat transfer analysis and elasto-plastic mechanical analysis. In the pre-processing stage for the 3D TEPA, the weldment for each pass is modeled following the bead shape and excluding the reinforcement. For the heat transfer analysis, a double ellipsoidal heat source model [23] is used and the heat source parameters are calibrated through comparison with the experimentally measured temperatures. The elasto-plastic analysis is performed next. The heat transfer analysis results, i.e., the time-temperature distribution data, are used for the elasto-plastic analysis. Following each welding pass, the equivalent plastic strains at the HAZ area are extracted. In conventional composite shell element-based welding analysis, the HAZ shape is depicted in order to extract the inherent strain value. In this study, however, we use the element plastic strain value obtained from the elasto-plastic analysis directly when obtaining the equivalent strain value in the HAZ area.

The assumptions employed in the proposed method are as follows:

A. The deformation occurring in each pass is caused by the inherent strain region occurring below the minimum equivalent thickness accumulated in each pass;
B. Only the plastic strain in the inherent strain region in the transverse direction is considered;
C. Reinforcement is neglected;
D. In the 3D TEPA analysis, the strain of each pass is calculated after cooling for each pass is complete.

The shrinkage generated by the inherent strain distributed in the elements can be replaced by the inherent deformation introduced as the discontinuity of the nodal displacements [24]. As the angular distortion is generated by the transverse shrinkage difference in the thickness direction, we consider an inherent strain region containing nine elements with three layers (Figure 1). Furthermore, we assume that each element has a unique transverse-direction plastic strain following the 3D TEPA. The final deformation of each layer in the inherent strain zone can be presented as the sum of the product of the element size (transverse direction) and the plastic strain, such that:

$$\delta_{x_1} = a_1\varepsilon^*_{x_{1,1}} + a_2\varepsilon^*_{x_{2,1}} + a_3\varepsilon^*_{x_{3,1}},$$
$$\delta_{x_2} = a_1\varepsilon^*_{x_{1,2}} + a_2\varepsilon^*_{x_{2,2}} + a_3\varepsilon^*_{x_{3,2}}, \qquad (1)$$
$$\delta_{x_3} = a_1\varepsilon^*_{x_{1,3}} + a_2\varepsilon^*_{x_{2,3}} + a_3\varepsilon^*_{x_{3,3}},$$

where

$\delta_{xi} = displacement\ of\ i^{th}\ layer\ in\ inherent\ strain\ region,$

$\varepsilon^*_{xi,j} = plastic\ strain\ (x\ direction)\ of\ i^{th}\ element\ of\ j^{th}\ layer\ in\ inherent\ strain\ region,$

$a_i = equivalent\ size\ (transverse\ direction)\ of\ i^{th}\ element\ in\ each\ layer.$

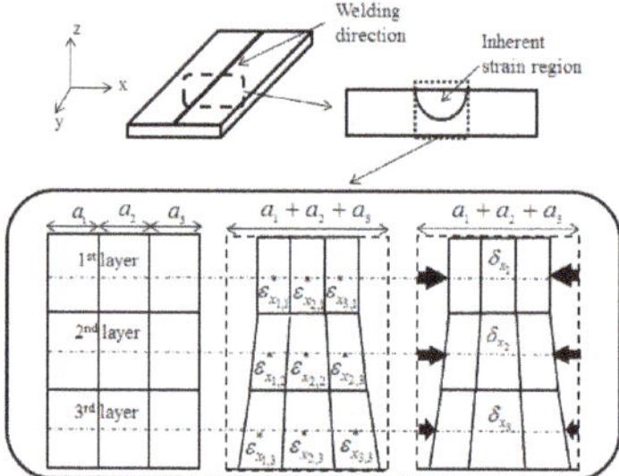

Figure 1. Inherent strain concept and inherent deformation due to element plastic strain.

The 3D elasto-plastic result can be used to determine the transverse-direction displacement according to the thickness in the inherent strain region. It is possible to use this result directly in the composite shell analysis. The thermal coefficients for each layer, which are used as input variables in the composite shell method, are obtained as follows [18]:

$$
\alpha_{i,1} = \frac{\sum_{k=1}^{n} \varepsilon^*_{k,i,1,1} \cdot b_{k,i,1}}{B_1},
$$

$$
\alpha_{i,j}(j \geq 2) = \frac{\sum_{k=1}^{n} (\varepsilon^*_{k,i,j,j} - \varepsilon^*_{k,i,j,j-1}) \cdot b_{k,i,1}}{B_j},
$$

$$(2)$$

where

$\alpha_{i,j}$ = *thermal expansion of ith layer of jth pass*,
$\varepsilon^*_{k,i,j,l}$ = *platic strain (transverse direction) of kth element of ith layer within j^{th} pass's HAZ width after l^{th} pass welding*,
$b_{k,i,j}$ = *equivalent length (transverse direction) of kth element of ith layer within j^{th} pass's HAZ width*,
B_j = *equivalent HAZ width of jth pass*.

One additional step added in the approach proposed in this paper is the subtraction of the amount of plastic strain generated in the previous pass for the elements of the overlapping inherent strain region; this additional step is performed for each pass. In this manner, the input variables for the welding deformation induced in each pass can be obtained separately, and the effect of the residual stress generated in each pass can be reflected. The overall procedure for extracting the equivalent plastic strain and equivalent HAZ width for each pass is summarized in Figure 2.

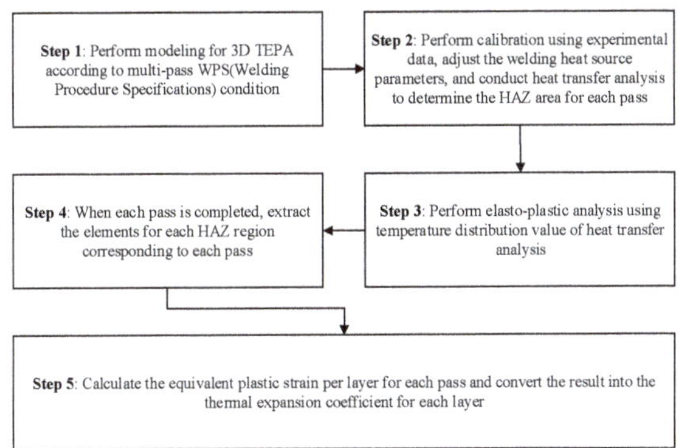

Figure 2. General procedure for extracting input variables in proposed method.

The next step is converting input variables generated by each pass. The process for obtaining the required thermal expansion coefficient values and the HAZ widths through 3D TEPA is described above. These values are used as factors of the equivalent load to simulate the welding deformation. The final deformation is determined from the sum of the equivalent loads, which cause deformation in each pass. According to the FEM theory for layered shell elements, the stress resultants can be obtained by integrating the stress components per layer in the thickness direction, such that:

$$
N_x = \int_{-t/2}^{t/2} \sigma_x dz = \frac{t}{2} \sum_{i=1}^{n} \sigma_x^i \Delta \zeta^i,
$$

$$
M_x = \int_{-t/2}^{t/2} \sigma_x z dz = \frac{t^2}{4} \sum_{i=1}^{n} \sigma_x^i \zeta^i \Delta \zeta^i,
$$

$$(3)$$

where

$$N_x = normal\ force\ (x\ direction)\ resultants,$$
$$z = coordinates\ in\ the\ thickness\ direction\ from\ the\ neutral\ axis,$$
$$\sigma_x = normal\ stress\ (x\ direction)\ resultants,$$
$$\zeta^i = location\ of\ the\ i^{th}\ layer\ in\ the\ natural\ coordinate\ system,$$
$$\Delta\zeta^i = increment\ of\ the\ i^{th}\ layer\ in\ the\ natural\ coordinate\ system.$$

Figure 3 shows the layered shell model and stress distribution diagram conforming to Equation (3). Temperature 1 is assigned to the node corresponding to the weld zone and the stress resultant is obtained using the different thermal expansion coefficient of each layer. The equivalent load, which is calculated from the HAZ size obtained from each pass and the thermal expansion coefficient of each layer, can be calculated from:

$$N_{x,j} = \frac{B_j E t_j}{2} \sum_{i=1}^{n} \alpha^i \Delta\zeta^i,$$
$$M_{x,j} = \frac{B_j E t_j^2}{4} \sum_{i=1}^{n} \alpha^i \zeta^i \Delta\zeta^i, \tag{4}$$

where

$$N_{x,j} = normal\ force\ (x\ direction)\ resultants\ of\ j^{th}\ pass,$$
$$M_{x,j} = bending\ moment\ (x\ direction)\ resultants\ of\ j^{th}\ pass,$$
$$\alpha^i = thermal\ expansion\ coefficient\ in\ the\ i^{th}\ layer,$$
$$B_j = equivalent\ HAZ\ width\ of\ j^{th}\ pass,$$
$$t_j = equivalent\ thickness\ of\ j^{th}\ pass.$$

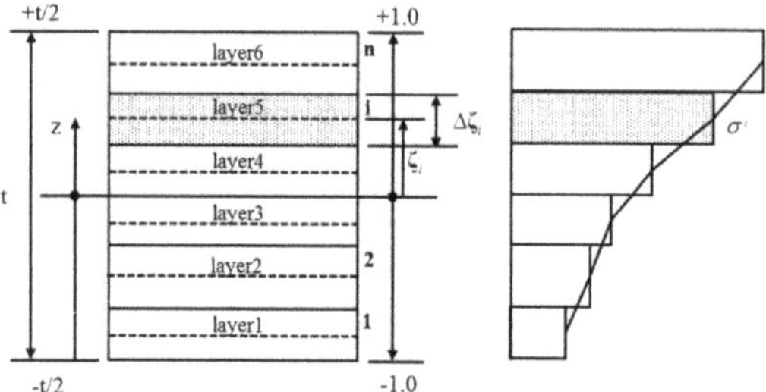

Figure 3. Layered shell model and stress distribution diagram.

The final deformation is obtained from the sum of the equivalent loads of each pass. To simulate the final deformation using composite shell-based welding analysis, it must be possible to simulate the sum of these equivalent loads by adjusting the thermal expansion coefficients of each layer. This can be achieved by introducing a virtual thermal expansion coefficient value and a virtual shrinkage force. The first step is to convert the equivalent load calculated, based on the equivalent thickness of each pass, to the final thickness standard. As the bending stiffness of a plate is proportional to the third power of the thickness, the calculated equivalent bending moment should be proportional to the third power of the value obtained by dividing the final thickness by the equivalent thickness of each pass. Further, the shrinkage force should be proportional to the value obtained by dividing the final thickness by the equivalent thickness of each pass. The final equivalent load can be expressed as:

$$M_{x,final} = \sum_{j=1}^{n} \left(\frac{t}{t_j}\right)^3 M_{x,j},$$
$$N_{x,final} = \sum_{j=1}^{n} \left(\frac{t}{t_j}\right) N_{x,j}, \tag{5}$$

where

$$N_{x,final} = final\ normal\ force\ (x\ direction)\ resultants,$$
$$M_{x,final} = final\ bending\ moment\ (x\ direction)\ resultants,$$
$$t = final\ thickness.$$

The next step is to adjust the thermal expansion coefficient to produce the equivalent bending moment and equivalent shrinkage force. The entire process is outlined in Figure 4.

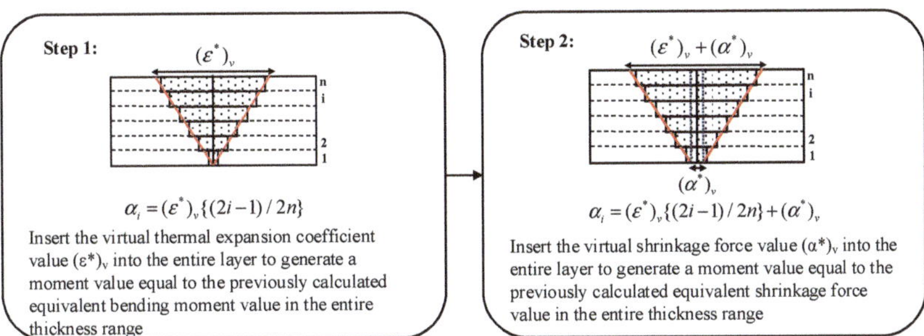

Figure 4. Procedure for adjusting input variables by introducing virtual thermal expansion coefficient value $((\varepsilon^*)v)$ and virtual shrinkage force value $((\alpha^*)v)$.

First, the equivalent bending moment value is adjusted by calibrating the virtual thermal expansion coefficient value $((\varepsilon^*)v)$. Next, by adding the same virtual shrinkage force value $((\alpha^*)v)$ to the entire layer, it is possible to adjust the equivalent shrinkage force. The equivalent bending moment value is first adjusted, considering the fact that adding the same coefficient of thermal expansion to the entire layer does not affect the calculation of the previously calculated moment value. Using the input variables, composite shell element-based welding distortion analysis can be performed.

4. Verification Using Experimental Models

4.1. Experimental Procedure

For validation purposes, multi-pass welding of a butt-joint was conducted. The objective of this experiment was to measure the distortion of the welding specimen and compare the results with the predictions given by the proposed method. For the detailed experiment methods please refer to Perrera [25]. A plate comprised of structural steel SS400 10 mm thick was used for the experiment. The width and length of the plate were 300 and 500 mm, respectively. Figure 5 shows the plate dimensions for the butt-joint.

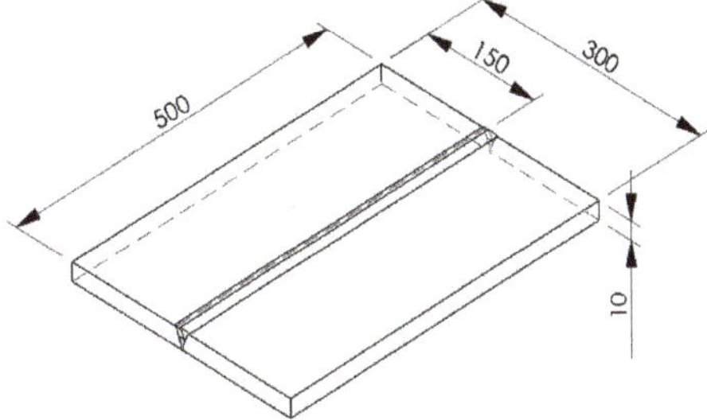

Figure 5. Plate dimensions for butt-joint (mm unit).

Three passes were performed. The welding sequence and bead dimensions for the butt-joint are shown in Figure 6. The welding was performed with a programmable machine and at a constant welding speed.

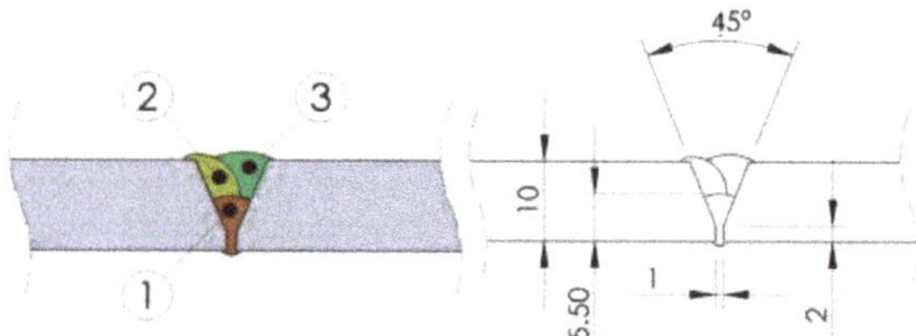

Figure 6. Welding sequence and bead dimensions for butt-joint (mm unit).

Table 1 lists the welding conditions. A pulsed-gas metal arc welding (P-GMAW) machine was used for welding. A DW-300 OTC Daihan digital inverter (OTC Daihen Inc., Tipp City, OH, USA) [26] was used as the power source. Filler wire was 1.2 mm diameter ER70S-6 mild steel and the wire feeding speed was 6 mm/min. The torch was held perpendicular to the workpiece and the distance from the contact tip to the workpiece was kept at 25 mm with an average extension of 18 mm. The gas flow rate was set to 20 L/min using a composition of Ar and 20% CO_2. Figure 7 shows the specimen and equipment for the butt-joint welding.

Table 1. Welding conditions for multi-pass butt-joint.

Pass Number	Current (A)	Voltage (V)	Traveling Speed (mm/s)	Interpass Temperature (°C)
1	220	25.2	8.3	300–350
2	240	27.2	9.0	300–350
3	240	27.2	9.0	300–350

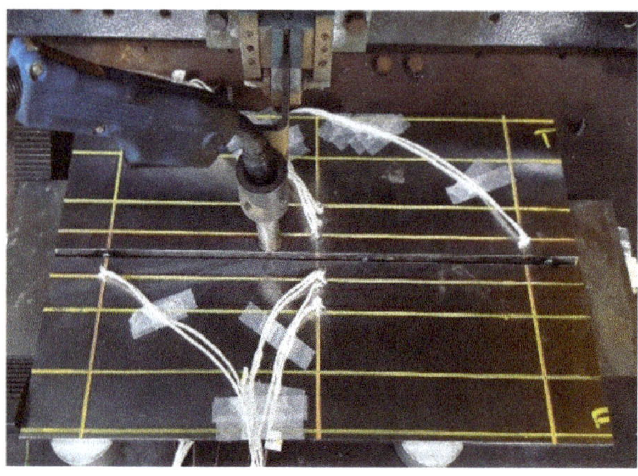

Figure 7. Specimen and equipment for butt-joint welding.

4.2. Numerical Analysis: 3D TEPA

Sequentially coupled 3D TEPA was implemented using Abaqus 6.12 in order to analyze the thermal history of the specimen as well as its welding distortion. The element birth and death technique [19] was used for both heat transfer analysis and elasto-plastic mechanical analysis. First, heat transfer analysis was implemented to obtain the temperature history using a 3D eight-node solid element (DC3D8). The welding heat source was modeled as a double ellipsoidal heat source [20].

The heat intensity distribution of the front and rear half ellipsoids can be expressed as:

$$q_1(x,y,z) = \frac{6\sqrt{3}(f_1 Q)}{abc_1\pi\sqrt{\pi}} \exp\left(-\frac{3x^2}{a^2} - \frac{3y^2}{b^2} - \frac{3z^2}{c_1^2}\right)[\text{W}/\text{m}^3],$$

$$q_2(x,y,z) = \frac{6\sqrt{3}(f_2 Q)}{abc_2\pi\sqrt{\pi}} \exp\left(-\frac{3x^2}{a^2} - \frac{3y^2}{b^2} - \frac{3z^2}{c_2^2}\right)[\text{W}/\text{m}^3],$$

(6)

where a, b, c_1, and c_2 are the heat flow distribution parameters, f_1 and f_2 represent the heat input fractions for the front and rear ellipsoids, respectively, and Q is the effective heat input. Here, $Q = \eta VI$, where η, V, and I represent the efficiency, voltage, and current, respectively. Note that η is usually determined empirically. In this study, the heat source parameters were calibrated using experimental results (Table 2). Considering the characteristics of the multi-pass welding procedure, the center of the moving heat source was modified for each pass (Figure 8). The mechanical analysis was performed by importing the transient temperature as a thermal load, with the C3D8R element being applied. The solid mesh system and boundary conditions for the butt-joint are presented in Figure 9.

Table 2. Heat source parameters.

	1st Pass	2nd Pass	3rd Pass
a (mm)	2.2	3.5	3.5
b (mm)	6.0	5.5	2.5
c_1 (mm)	5	5	5
c_2 (mm)	10	10	10
f_1	0.2	0.2	0.2
f_2	1.8	1.8	1.8

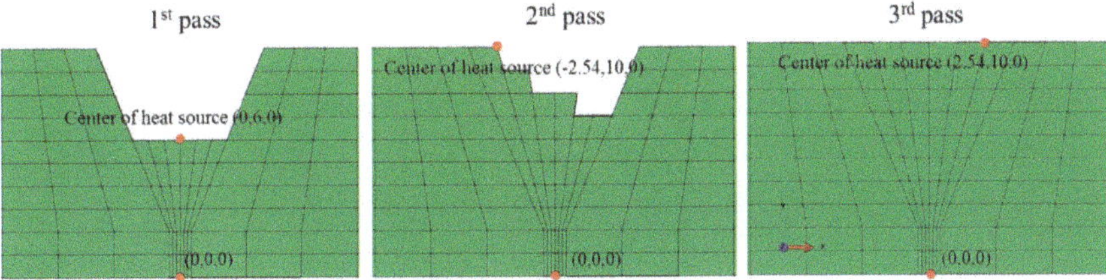

Figure 8. Heat source center modification for each pass (mm unit).

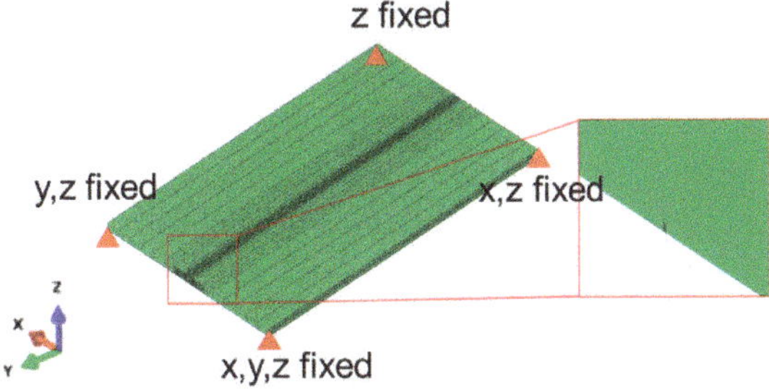

Figure 9. Solid mesh system and boundary conditions for butt-joint model mechanical analysis.

The arc efficiency was assumed to be 0.85. The heat losses due to radiation and convection were considered together using the constant film coefficient. The latent heat was assumed to be 273,790 W, with the solidus and liquidus temperatures being taken as 1427 °C and 1482 °C, respectively. The emissivity coefficient was taken to be 0.32. The other thermal and mechanical properties [18] used in this analysis are presented in Figure 10.

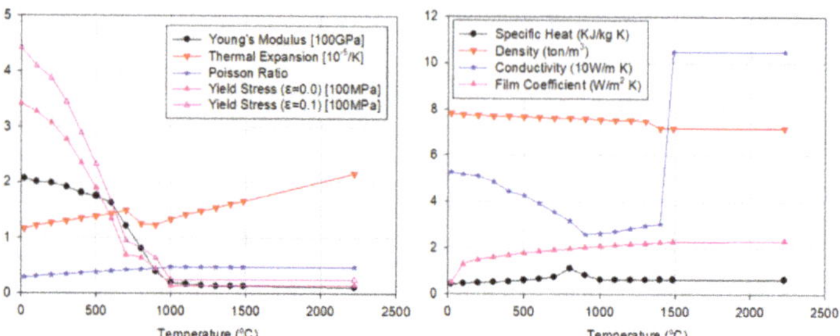

Figure 10. Temperature-dependent material properties.

5. Results and Discussion

5.1. Heat Transfer Analysis Results

The thermal analysis was validated by comparing the simulated cross-sectional profile of each weld pass with the experimental results. In addition, the numerically predicted HAZ was compared with the experimental data. The HAZ is an area of the base metal that has not been melted by the high-temperature heating, but for which the chemical properties are altered. The high temperature from the welding process and the subsequent re-cooling causes these changes from the weld interface to the end of the sensitizing temperature in the metal. To evaluate the width of the HAZ, the phase transformation temperature Ac1 of SS400 was considered as the reference temperature. The isothermal contour of Ac1 temperature is 725 °C [12,15]. It was found that the simulated macro-section HAZ for each weld pass agrees reasonably well with the experimentally obtained HAZ. Further, the modeled sectional profile of the bead is in good agreement with the experimental profile. Figure 11 shows a comparison between the simulated and experimental macro-sections for butt-joint welding. These results establish confidence in the obtained thermal solution, which is used as an input for the mechanical analysis.

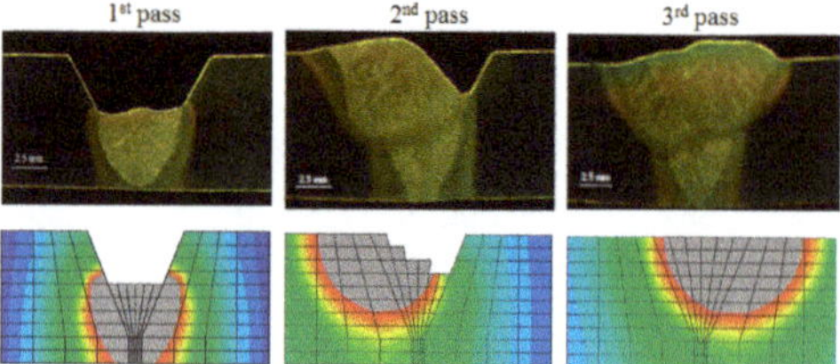

Figure 11. Comparison of simulated weld profile HAZ with macro-section analysis of butt-joint.

5.2. 3D Elasto-Plastic Analysis and Thermal Expansion Values Extraction

By performing elasto-plastic analysis using the temperature distribution of the heat transfer analysis, it is possible to extract the thermal expansion coefficient value for each pass according to the layer in the HAZ region (Figure 12). Through application of Equation (3), the extracted layer-by-layer thermal expansion coefficients are listed in Table 3. Through application of Equation (4), the equivalent loads generated by the inherent strain in each pass are listed in Table 4. Through application of Equation (5), the target moment and target force that generate the final distortion are obtained, as listed in Table 5. The minus direction of the target shrinkage forces means the direction in which the butt-joint shrinks is in the in-plane direction. The minus direction of the target moments means the direction in which the angular distortion occurs is in the upward direction. Next, by introducing $(\varepsilon^*)v$ and $(\alpha^*)v$, by following the procedure described in the Figure 4, it is possible to extract the input variables that are used for the final distortion prediction (Table 6).

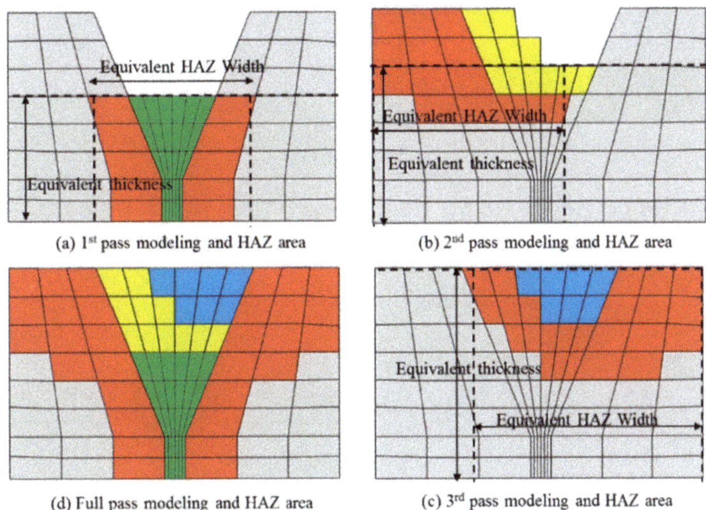

(a) 1st pass modeling and HAZ area

(b) 2nd pass modeling and HAZ area

(d) Full pass modeling and HAZ area

(c) 3rd pass modeling and HAZ area

Figure 12. Equivalent HAZ width and equivalent thickness for each pass.

Table 3. Parameters of each pass for layered shell-element welding analysis method.

	1st Pass	**2nd Pass**	**3rd Pass**
Equivalent thickness (mm)	6	7	10
Equivalent HAZ width (mm)	7.94	10.37	13.04
Thermal expansion in plate ($^\circ\text{C}^{-1}$)	−0.0196 −0.0182 −0.0171 −0.0159 −0.0146 −0.0139	−0.0181 −0.0180 −0.0152 −0.0121 −0.0085 −0.0039 0.0010	−0.0258 −0.0239 −0.0208 −0.0187 −0.0154 −0.0127 −0.0098 −0.0065 −0.0003 −0.0007

Table 4. Equivalent load by pass.

	1st Pass	**2nd Pass**	**3rd Pass**
Thickness (mm)	6	7	10
Moment (N·mm)	−40,067	−191,525	−634,002
Force (N)	−353,794	−154,962	−353,903

Table 5. Target equivalent loads for final distortion prediction.

Target Moment (N·mm)	**Target Force (N)**
−1,377,878	−1,164,933

Table 6. Converted input variables for final distortion prediction.

ε^*	-0.05218
α^*	0.06018
Thickness (mm)	10
Equivalent HAZ width (mm)	16
Thermal expansion in plate ($^\circ$C^{-1})	-0.0129 -0.0181 -0.0233 -0.0286 -0.0338 -0.0390 -0.0442 -0.0494 -0.0546 -0.0599

5.3. Comparison of Various Methods

Here, the results from the various numerical and experimental analyses are presented and compared. Figure 13 shows the deformation results from the experiment.

Figure 13. Experiment result of multi-pass butt-joint.

Figure 14 shows the deformation results from the proposed method. For comparison, we also present the results of the multi-pass SDB method (Figure 15). Artificial temperature values were calculated following the procedure of Ha and Yang [16]. The inherent strain value used here was calculated from the equivalent plastic strain values of the elements in the final HAZ region of the 3D TEPA result (Table 7).

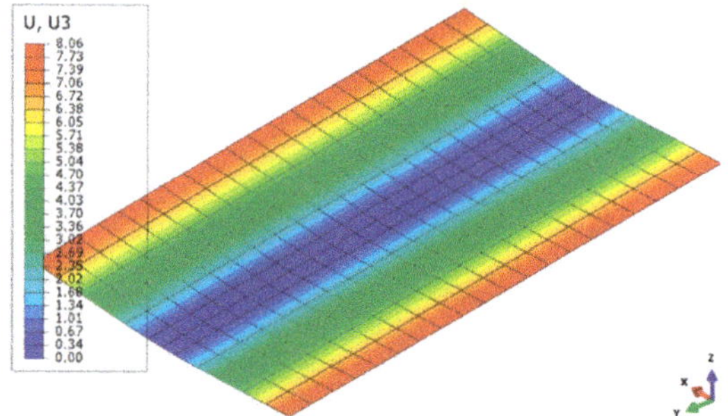

Figure 14. Angular distortion: proposed method (mm unit).

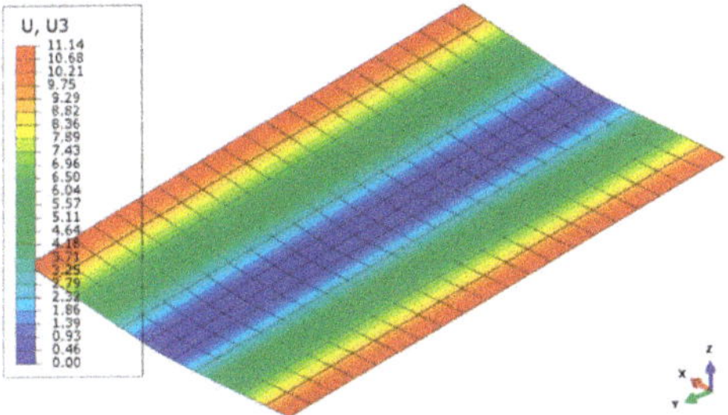

Figure 15. Angular distortion: multi-pass SDB method (mm unit).

Table 7. Input variables for multi-pass SDB method.

Equivalent HAZ Width	Temperature Distribution	Inherent Strain Value
16 mm	Ttop = 1.51 Tbottom = −1.51	−0.0251

Finally, Figure 16 presents a comparison between the existing and proposed methods. It is apparent that the proposed method for the modeling of butt-joint welding can describe the angular deformation precisely. The proposed method is more effective than the existing methods, as the result is closest to the experimental data.

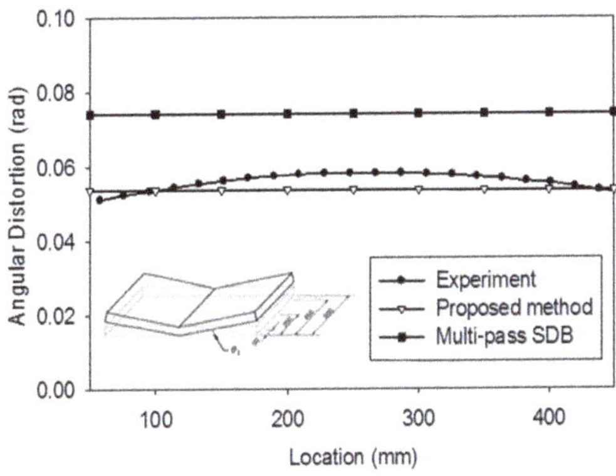

Figure 16. Comparison of angular deformations yielded by each method with the experimental result.

5.4. Discussion

For the existing multi-pass SDB method, although the inherent strain value was extracted from the same 3D TEPA and applied to the shell element-based method, the multi-pass SDB method showed a relatively high difference when compared with the experimental result. It is thought that the accuracy of the result could be improved by supplementing the assumptions of the multi-pass SDB method (i.e., neglect of the effect of the residual stress due to the interpass temperature difference and neglect of the HAZ area for each pass). Nevertheless, the proposed method yielded a difference of approximately 10% from the experiment result. However, this is a limitation of the simplified method. The experiment result showed the difference in each deformation degree depending on the longitudinal bending at each position. However, in the simplified method proposed herein, constant angular distortion could be simulated because of the employed assumptions. At present, only the plastic strain in the transverse direction was considered; thus, the proposed method can be improved by considering a more accurate inherent strain extraction method.

6. Conclusions

In this paper, a layered shell element-based elastic FEM for predicting welding distortions in multi-pass welding was developed. The existing layered shell element-based method depicts the HAZ region and extracts the layer-by-layer thermal strain values. However, in the method proposed in this study, we extracted the layer-by-layer thermal strain values using the element plastic strain values in the HAZ region. Through application of this method, it was possible to consider the influence of the overlapping inherent strain region between each pass. In addition, the influence of the bead shape could also be considered sufficiently. For validation, a 10 mm thick, multi-pass butt-welded joint, subjected to three passes, was simulated using the proposed method; the resultant welding distortion predictions were compared with those given by 3D TEPA and the measured experimental data. The results showed that good agreement can be achieved. Further, compared with the existing method, considerable improvement in accuracy was noted.

Author Contributions: Conceptualization, J.L. and H.C.; methodology, J.L; investigation, D.P.; resources, H.C.; data curation, J.L.; writing—original draft preparation, J.L.; writing—review and editing, J.L. and H.C.; visualization, J.L.; supervision, H.C. All authors have read and agreed to the published version of the manuscript.

Funding: This research was financially supported by the research fund of Chungnam National University.

Institutional Review Board Statement: Not applicable.

Informed Consent Statement: Not applicable.

Data Availability Statement: Not applicable.

Acknowledgments: This article is an addition based on part of J.L.'s doctoral dissertation.

Conflicts of Interest: The authors declare no conflict of interest.

References

1. Lee, J.M.; Seo, H.D.; Chung, H. Efficient welding distortion analysis method for large welded structures. *J. Mater. Process. Technol.* **2018**, *256*, 36–50. [CrossRef]
2. Xin, H.; Correia, J.A.F.O.; Veljkovic, M.; Berto, F.; Manuel, L. Residual stress effects on fatigue life prediction using hardness measurements for butt-welded joints made of high strength steels. *Int. J. Fatigue* **2021**, *147*, 106175. [CrossRef]
3. Lee, J.; Chung, H. Modified Equivalent Load Method for Welding Distortion Analysis. *J. Mar. Sci. Eng.* **2020**, *8*, 794. [CrossRef]
4. Lindgren, L.E. Numerical modelling of welding. *Comput. Methods Appl. Mech. Eng.* **2006**, *195*, 6710–6736. [CrossRef]
5. Lee, J. Development of Efficient Welding Distortion Analysis Method for Large Welded Structure. Ph.D. Dissertation, Korea Advanced Institute of Science and Technology, Daejeon, Korea, 2019.
6. Ding, J. Thermo-Mechanical Analysis of Wire and Arc Additive Manufacturing Process. Ph.D. Dissertation, Cranfield University, Cranfield, UK, 2012.
7. Chen, B.-Q.; Guedes Soares, C. Experimental and numerical investigation on welding simulation of long stiffened steel plate specimen. *Mar. Struct.* **2021**, *75*, 102824. [CrossRef]
8. Hammad, A.; Churiaque, C.; Sánchez-Amaya, J.M.; Abdel-Nasser, Y. Experimental and numerical investigation of hybrid laser arc welding process and the influence of welding sequence on the manufacture of stiffened flat panels. *J. Manuf. Process.* **2021**, *61*, 527–538. [CrossRef]
9. Perić, M.; Garašić, I.; Nižetić, S.; Dedić-Jandrek, H. Numerical Analysis of Longitudinal Residual Stresses and Deflections in a T-joint Welded Structure Using a Local Preheating Technique. *Energies* **2018**, *11*, 3487. [CrossRef]
10. Ueda, Y.; Kim, Y.C.; Yuan, M.G. A predicting method of residual stress using source of residual stress (report I): Characteristics of inherent strain (source of residual stress). *Trans. Jpn. Weld. Res. Inst.* **1989**, *18*, 135–141.
11. Ueda, Y.; Yuan, M.G. A predicting method of welding residual stress using source of residual stress (Report II): Determination of standard inherent strain. *Trans. Jpn. Weld. Res. Inst.* **1989**, *18*, 143–150.
12. Ueda, Y.; Yuan, M.G. A predicting method of welding residual stress using source of residual stress (Report III): Prediction of residual stresses in T- and I-joints using inherent strains. *Trans. Jpn. Weld. Res. Inst.* **1993**, *22*, 157–168.
13. Deng, D.; Murakawa, H.; Ma, N. Predicting welding deformation in thin plate panel structure by means of inherent strain and interface element. *Sci. Technol. Weld. Join.* **2012**, *17*, 13–21. [CrossRef]
14. Park, J.; An, G. Prediction of the welding distortion of large steel structure with mechanical restraint using equivalent load methods. *Int. J. Naval Archit. Ocean Eng.* **2017**, *9*, 315–325. [CrossRef]
15. Jung, H.; Tsai, C.L. Plasticity-based distortion analysis for fillet welded thin-plate T-joints. *Weld. J.* **2004**, *83*, 177–187.

16. Ha, Y.S. Development of thermal distortion analysis method on large shell structure using inherent strain as boundary condition. *J. Soc. Nav. Archit. Korea* **2008**, *45*, 93–100. [CrossRef]
17. Park, J.U.; Lee, H.W.; Bang, H.S. Effects of mechanical constraints on angular distortion of welding joints. *Sci. Technol. Weld. Join.* **2002**, *7*, 232–239. [CrossRef]
18. Kim, K.; Kang, M.; Chung, H. Simplified welding distortion analysis for fillet welding using composite shell elements. *Int. J. Archit. Ocean Eng.* **2015**, *7*, 452–465. [CrossRef]
19. Ha, Y.S.; Yang, J.H. Development of distortion analysis method for multi-pass butt-welding based on shell element. *J. Weld. Join.* **2010**, *28*, 54–59. [CrossRef]
20. Ramjaun, T.; Stone, H.J.; Karlsson, L.; Kelleher, J.; Moat, R.J.; Kornmeier, J.R.; Dalaei, K.; Bhadeshia, H.K.D.H. Effect of interpass temperature on residual stresses in multipass welds produced using low transformation temperature filler alloy. *Sci. Technol. Weld. Join.* **2014**, *19*, 44–51. [CrossRef]
21. Fu, G.; Lourenço, M.I.; Duan, M.; Estefen, S.F. Effects of Preheat and Interpass Temperature on the Residual Stress and Distortion on the T-Joint Weld. In Proceedings of the ASME 2014 33rd International Conference on Ocean, Offshore and Arctic Engineering, San Francisco, CA, USA, 8–13 June 2014.
22. Fanous, I.F.Z.; Younan, M.Y.A.; Wifi, A.S. 3-D Finite Element Modeling of the Welding Process Using Element Birth and Element Movement Techniques. *J. Press. Vessel. Technol.* **2003**, *125*, 144–150. [CrossRef]
23. Goldak, J.; Chakravarti, A.; Bibby, M. A new finite element model for welding heat sources. *Metall. Mater. Trans. B* **1984**, *2*, 299–305. [CrossRef]
24. Murakawa, H.; Ma, N.; Ohsuga, Y. Concept of inherent strain, inherent deformation and inherent force for prediction of welding distortion. *Trans. Jpn. Weld. Res. Inst.* **2011**, *39*, 103–105.
25. Perrera, D. Simplified Distortion Analysis for Multi-Pass Welding Using Layered Shell Elements Based on Inherent Strain. Masters's Dissertation, Korea Advanced Institute of Science and Technology, Daejeon, Korea, 2016.
26. OTC Daihen Inc. Available online: http://www.daihen-usa.com (accessed on 10 May 2021).

Journal of
*Marine Science
and Engineering*

Article

Fatigue Characteristic of Designed T-Type Specimen under Two-Step Repeating Variable Amplitude Load with Low-Amplitude Load below the Fatigue Limit

Jin Gan [1], Di Sun [2], Hui Deng [1,3], Zhou Wang [4], Xiaoli Wang [4,*], Li Yao [4] and Weiguo Wu [3]

[1] Department of Naval Architecture, Ocean and Structural Engineering, School of Transportation, Wuhan University of Technology, Wuhan 430063, China; ganjinwut@163.com (J.G.); denghui@whut.edu.cn (H.D.)
[2] China Ship Development and Design Center, Wuhan 430064, China; sd1069587728@163.com
[3] Green & Smart River-Sea-Going Ship, Cruise and Yacht Research Center, Wuhan University of Technology, Wuhan 430063, China; mailjt@163.com
[4] Department of Automotive Engineering, School of Automotive Engineering, Wuhan University of Technology, Wuhan 430070, China; wangzhou@whut.edu.cn (Z.W.); yaoli1206@163.com (L.Y.)
* Correspondence: xwa@whut.edu.cn

Abstract: In order to investigate the non-linear fatigue cumulative damage of joints in ocean structural parts, one type of low carbon steel Q345D was employed to prepare designed T-type specimens, and a series of fatigue experiments were carried out on the specimens under two-step repeating variable amplitude loading condition. The chosen high cyclic loads were larger than the constant amplitude fatigue limit (CAFL) and the chosen low cyclic loads were below the CAFL. Firstly, the S-N curve of designed T-type specimen was obtained via different constant amplitude fatigue tests. Then, a series of two-step repeating variable load were carried out on designed T-type specimens with the aim of calculating the cumulative damage of specimen under the variable fatigue load. The discussions about non-linear fatigue cumulative damage of designed T-type specimens and the interaction effect between the high and low amplitude loadings on the fatigue life were carried out, and some meaningful conclusions were obtained according to the series of fatigue tests. The results show that fatigue cumulative damage of designed T-type specimens calculated based on Miner's rule ranges from 0.513 to 1.756. Under the same cycle ratio, the cumulative damage increases with the increase of high cyclic stress, and at the same stress ratio, the cumulative damage increases linearly with the increase of cycle ratio. Based on the non-linear damage evaluation method, it is found that the load interaction effect between high and low stress loads exhibits different damage or strengthening effects with the change of stress ratio and cycle ratio.

Keywords: variable amplitude loading; cyclic load below fatigue limit; low carbon steel; load interaction; strengthening effect

Citation: Gan, J.; Sun, D.; Deng, H.; Wang, Z.; Wang, X.; Yao, L.; Wu, W. Fatigue Characteristic of Designed T-Type Specimen under Two-Step Repeating Variable Amplitude Load with Low-Amplitude Load below the Fatigue Limit. *J. Mar. Sci. Eng.* **2021**, *9*, 107. https://doi.org/10.3390/jmse9020107

Academic Editor: Joško Parunov
Received: 4 December 2020
Accepted: 12 January 2021
Published: 20 January 2021

1. Introduction

Fatigue failure is a common phenomenon in steel structures of the ships and ocean assemblies, which are always subjected to cyclic loading during service [1,2]. Fatigue failure always appears at the local region of mechanical parts or components. In most cases, the cyclic loads applied on the structure of ship and ocean engineering are not constant but variable. Prediction of the fatigue life of structures under variable cyclic loads accurately is crucial for the engineers in the field of ocean and structural engineering, and that requires comprehensive understanding the damage mechanism of structures under the different types of variable fatigue load blocks.

Fatigue damage in metals is a process of irreversible accumulation, which is manifested by progressive internal deterioration of material due to nucleation and growth of micro-cracks, debonding, voids, and so on [3,4]. In general, fatigue failure contains three stages: crack initiation, crack propagation and final catastrophic failure. The difference between high cycle fatigue and low cycle fatigue lies in the percentage of the total life spent in each of three stages. As to the high cycle fatigue, much of the fatigue life is spent in the crack initiation stage [4]. Crack initiation is a highly random event which is strongly affected by the initial state of material properties and microstructural defects [5,6].

In order to predict the fatigue life of metallic components, it is necessary to find a reasonable method to calculate the fatigue damage accumulation in variable amplitude cyclic loads. For a long time, most of design engineers in industry always use Palmgren–Miner law to estimate the fatigue lives of metallic components due to relatively accurate and easy operation of this method. The Palmgren–Miner law is a linear damage rule (LDR). The value of cumulative damage *(D)* can be calculated by the Equation (1):

$$D = \sum_{i=1}^{n} \frac{n_i}{N_i} = \sum r_i \tag{1}$$

where n_i is the number of cycles at a given stress amplitude load, N_i is the cycles to failure at the same stress amplitude load. Miner's rule suggests that any structure with $D < 1.0$ during a fatigue test is safe.

The assumption in Miner's rule is that the fatigue damage due to each of a particular cyclic load in a fatigue test with variable amplitude (VA) loads is exactly the same as that in the fatigue test with a constant amplitude (CA) load [7]. However, there are a large number of previous works [8–14] indicating that VA load fatigue tests induce larger fatigue damage than the calculated results by the Miner's law, which leads to non-conservation of Miner's rule in actual VA load fatigue life predictions (i.e., fatigue failure when $D < 1.0$). The main deficiencies of LDR are no consideration of the effect of load sequence, the effect of the damage of low amplitude load below fatigue limit and the interaction effect between the high and low amplitude loadings on the fatigue life [15]. All these factors that are not considered may lead to the deviation of fatigue life prediction. Therefore, the fatigue life predictions are unsafe in some cases [16]. Many researchers tried to modify the form of Miner's rule with the aim of predicting fatigue life precisely. Marco and Starkey [17] proposed a non-linear load-dependent damage rule firstly as follows:

$$D = \sum r_i^{x_i} \tag{2}$$

where x_i is a coefficient depending on the *i*th load. The main novelty of this rule is that the effects of different loading sequences are taken into account in calculation. However, the actual predicted result is usually unsatisfactory, and the experimental results showed that the good agreement only occurred in some cases and some materials [18].

Another approach is making use of the concept of fatigue limit reduction to measure the damage accumulation, when a multi-level load is applied and many rules based on the S-N curve modification have been developed [19]. For all these theories, the original curve is replaced by modified curves. Although these approaches are non-linear damage rules, few of these approaches take into account the load interaction effect.

There are some different methods of treating variable cyclic loading spectra containing the constant amplitude fatigue limit (CAFL) and calculating by the Miner's law [20,21]. The most widely used assumption is that the S-N curve extrapolated beyond the load cycles at 10^7 with a relatively flat slope [22–24], and the damage accumulation rule is still Miner' rule. However, a large number of experiments [9,25–31] show that in fact the actual fatigue damage is larger than in the assumption method. For example, it was reported [12] that the use of a 2-slope S-N curve with the slope change at a stress range above 10.1 N/mm^2 (corresponding to about 5.5×10^8 cycles) was potentially unsafe, particularly for loading spectra containing a large number of small cyclic stresses. Moreover, this was especially true for fatigue life of components under the tensile load.

In ship and ocean engineering, there are many joints in structural parts, such as the intersection between longitudinal and transverse bulkhead [32–34] which is one of the typical ship structures and is prone to fatigue failure during ship voyages. There are many techniques to perform structural health monitoring to guarantee the structural parts' safety during service. In the process of structural health monitoring, it is important to accurately capture or predict the deformation of structures. The stresses and displacements of the structural components in the ships can be calculated by different analytical theories and finite element methods. The Carrera unified formulation (CUF)-based finite element analysis (FEA) model has been validated to accurately simulate the deformation of complex structures by laser-based experiments [35]. Nondestructive testing (NDT) are usually employed to assess structural health and secure structure integrity. With the measurements of laser tracker and terrestrial laser scanning, an optimized surface model can be established as a high-accuracy NDT metrology tool [36]. To obtain accurate deformed state of a given composite thin-walled structures, the geometric nonlinear effects of the composite structures can be analyzed by the CUF-based FEA model [37]. Meanwhile, with the consideration of geometrical nonlinear relations in beam and shell-like structures, the deformed state of a given structure can be correctly captured by refined theories based on the CUF [38]. Although the developed theoretical methods promote the structural health monitoring of the structural parts, the fatigue experiments are still necessary to improve the service life of structural parts.

There is a special type of ship called river-sea going ship, which repeats voyaging between the inland river and coast sea along a special voyage line, and the loading environment from which the ship suffers, is much like the two-step repeating load block [39]. When the ship voyages on the sea, the loading on the ship by the sea wave is different from the river wave in the inland river. The loading on the ships can be variable whether the ships voyage on the sea or the inland river. Usually, the high loading on the ship by the sea wave is larger than that by the inland river. Here in the paper, the variable loading on the river-sea going ship is approached by the model of two-step repeating variable amplitude load with low-amplitude load below the fatigue limit. Therefore, a study on fatigue characteristic of designed T-type specimen under the two-step repeating amplitude loading block was carried out in this work. The loading block contained high cyclic load and low cyclic load levels. The high cyclic load was above the CAFL and the fatigue life produced by the high cyclic load lied in the high cycle fatigue range, and the low cyclic load was below the CAFL. Firstly, tensile tests were conducted on the designed T-type specimens to get the material mechanical properties. Then a series of fatigue experiments with both the CA loading and the two-step repeating amplitude load block were conducted on designed T-type specimens. Furthermore, as a comparison, Miner's rule was used to calculate the fatigue damage accumulation of designed T-type specimen, under every two-step repeating cyclic load condition. Finally, the nonlinear damage parameter was introduced to evaluate the load interaction effect and some conclusions were drawn.

2. Experimental Setup

All the fatigue experiments were carried out via an MTS 322 250 kN Dynamic Fatigue Testing System at the Ship Structure Laboratory of Wuhan University of Technology (Wuhan, China). Figure 1a shows the overall layout of the experiment. The MTS equipment consists of a servo-hydraulic actuator (1) and controller (2). One end of the fatigue specimen was fixed in the platform (3) and the other end was actuated via the servo-hydraulic actuator (1). Figure 1b shows the clamping state of the specimen in fatigue tests. The fatigue test specimen was installed between the grippers (4). All fatigue tests were conducted at room temperature (20 °C).

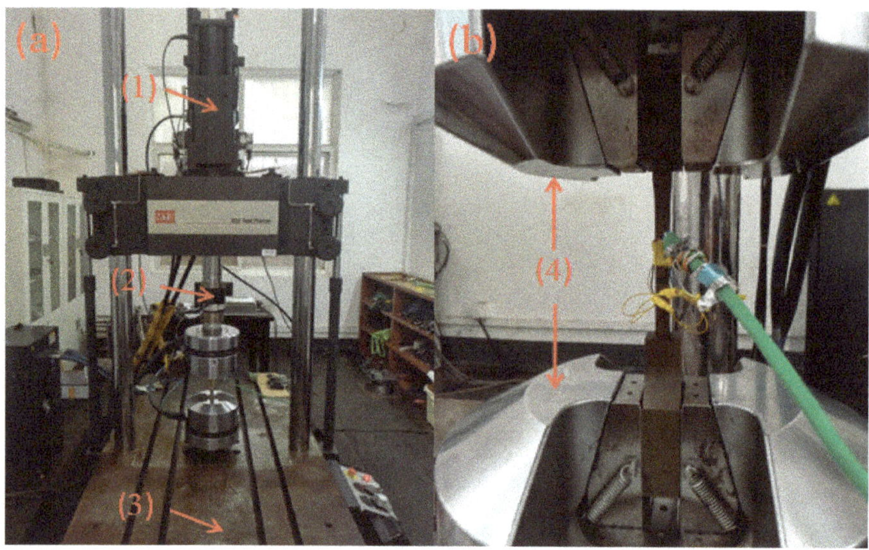

Figure 1. Load frame of the test (**a**), the installation of designed T-type specimen (**b**). A servo-hydraulic actuator (1), a controller (2), the platform (3), grippers (4).

2.1. Material

The material used for the fatigue test was one type of low carbon steel, Q345D steel. Its chemical compositions (in Wt%) were listed in Table 1. In order to obtain the mechanical properties of Q345D steel, tensile tests were conducted on the standard specimen. Figure 2 shows the standard specimen of the tensile test. The tensile tests were conducted in accordance with the published standard [40]. The test results of mechanical properties are shown in Table 2.

Table 1. Chemical compositions of Q345D steel.

Q345D	Element	C	Mn	Si	S	P	Ni	Cr	Mo	V	Cu	Fe
	Wt.-%	0.15	1.39	0.28	0.003	0.015	0.01	0.05	0.007	0.004	0.04	Balance

Figure 2. The fractured specimen after the tensile test.

Table 2. Mechanical properties of Q345D steel.

Ultimate tensile strength, σ_{UTS} [MPa]	539
Monotonic yield strength, σ_{YS} [MPa]	384
Young's modulus, E [GPa]	206
Poisson's coefficient, v	0.26
elongation, δ [%]	30.5

2.2. Designed T-Type Specimen

Figure 3 shows the specimen and its dimensions employed in a fatigue test. The thickness in the middle of uniform cross section was 8 mm. The specimen was prepared with the method of wire-electrode cutting to guarantee its size precision. The specimen had a bilateral symmetrical T-type step (labelled as "A" in Figure 3a) which caused highly localized stress concentration. When the test specimen was subjected to cyclic loading, failure occurred at the T-type step ("A" in Figure 3a).

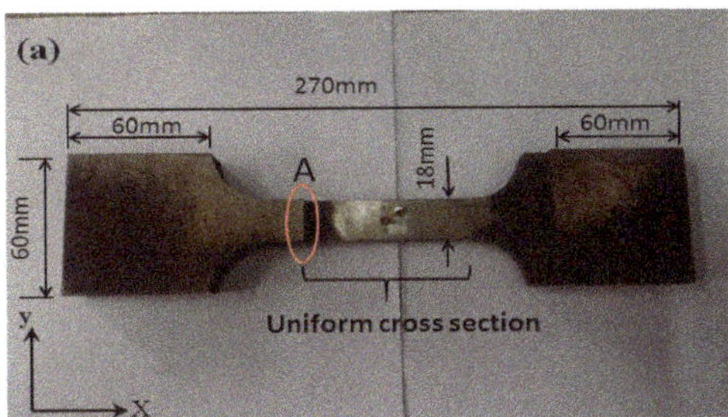

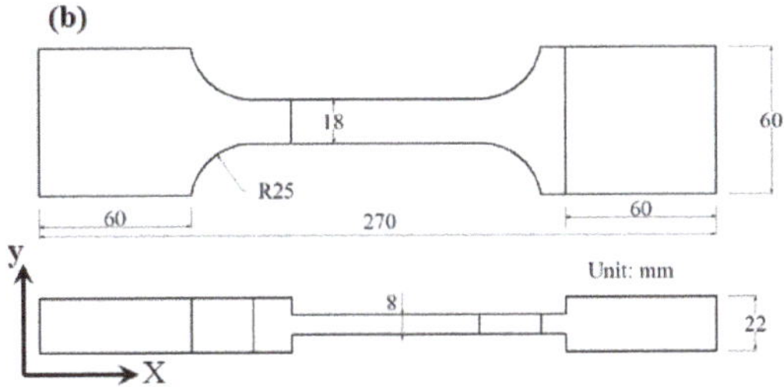

Figure 3. (**a**) The specimen and its dimensions after a fatigue test; (**b**) the dimensions of the specimen before a fatigue test.

2.3. Strain Measurement

The strain gauges were placed along the step on the surface of uniform cross section on both sides (shown in Figure 4) to measure the strain value variation and to monitor the crack initiation [41]. The stress concentration factor (SCF) at the T-type step ("A" in Figure 3a) was 1.9 according to the method proposed by Dong [42]. The value of the radius at the crack starting zone was 0.4 mm.

Figure 4. The arrangement of strain gauge on the specimen.

2.4. Fatigue Test

The test program involved both CA and VA loading tests on designed T-type specimens. Figure 5 shows the fatigue experiment setup. Figure 5 is the Microscopy System of KEYENCE which was used to observe the fatigue crack at the step ("A" in Figure 3a). All the fatigue tests were conducted with a frequency of 20 Hz in the force control mode. Loading of sinusoidal type was applied at room temperature. Fatigue tests stopped when the cracks on a specimen were found. Figure 6a shows the fracture surface of designed T-type specimen after fatigue failure. Figure 6b shows the crack initiation zone and the path of crack propagation. As shown in Figure 6b, the crack always initiated at one side of the specimen at the step ("A" in Figure 6b) and propagated in the Y direction on the surface. The crack changed its propagation direction into the Z direction after the crack reached one side of the cross section of the specimen. When the crack developed into a certain depth of the specimen then final instantaneous fracture occurred. Compared with the crack initiation life, the crack propagation life was very short. Therefore, the crack initiation life can be regarded as the whole fatigue life of the specimen.

Figure 5. Tensile tests setup-Microscopy System of KEYENCE(M).

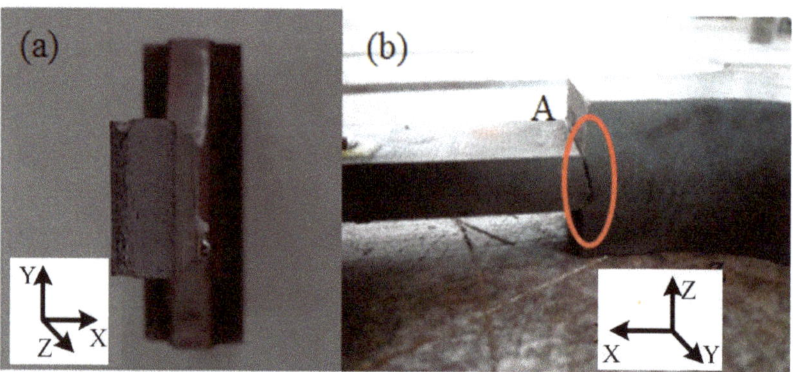

Figure 6. The fracture surface of designed T- type specimen (**a**), the crack growth path (**b**).

2.4.1. Constant Amplitude Fatigue Tests

Thirteen different stress amplitudes were chosen as CA loading conditions. Three repeated experiments were conducted for every stress amplitude condition. Table 3 shows the specific loading conditions. The stress ratio (R) of all loading conditions is 0.1. According to the recommendation of BS7608 [43], all the fatigue tests were stopped when the cyclic number was larger than 10^7 if fatigue failure didn't appear.

Table 3. Loading conditions of constant amplitude (CA) loading tests.

No.	1	2	3	4	5	6	7	8	9	10	11	12	13
σ_a	105	100	95	90	85	80	75	70	55	53	50	45	40
R	0.1	0.1	0.1	0.1	0.1	0.1	0.1	0.1	0.1	0.1	0.1	0.1	0.1

2.4.2. Variable Amplitude Loading Tests

Two-step repeating amplitude fatigue load tests were carried out after CA loading fatigue tests. Figure 7 shows an abridged general view which depicts the loading spectrum of two-step repeating amplitude fatigue load. In the two-step repeating amplitude fatigue load tests, the high cyclic stress and low cyclic stress levels σ_{a1} and σ_{a2} were set to values above and below the CAFL, respectively. The stress ratio of each stress amplitude was R = 0.1. n_1 and n_2 were the number of cycles of each stress levels σ_{a1} and σ_{a2} in a load block. In the experiments, n_1=20,000, and it is the same for all the experiments in this article. n_2 was different for different loading cycle ratios. For example, when the loading cycle ratio is 1:1, n_2=20,000. When the loading cycle ratio is 1:2, n_2=40,000. Before the failure, n_1 and n_2 kept the same in each loading block for a certain loading cycle ratio. When the failure occurred in the experiments, the failure might appear when the high or the low stress was applied since it is cyclic loading. n_1 was the instantaneous number at the failure moment and recorded in the experiment, not 20,000 any longer, if the failure appeared with the high stress applied. Correspondingly, n_2 was the instantaneous number at the moment when the failure was detected with the low stress applied, not the set number any longer.

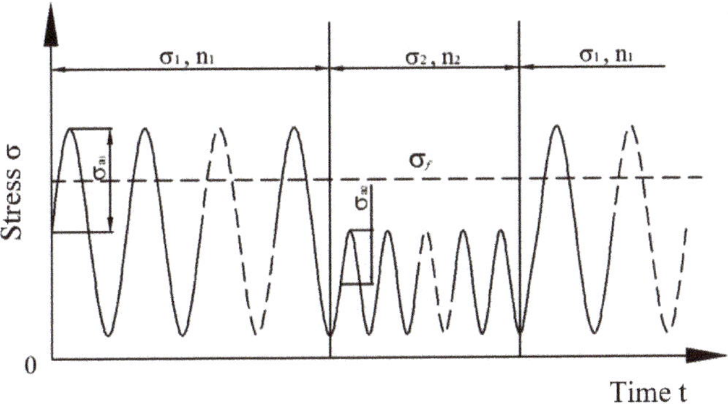

Figure 7. The loading sequence of two-step repeating test.

3. Results and Discussion

3.1. Constant Amplitude Test Results

The fatigue lives of different CA conditions were obtained with averaging experimental data from three repeated fatigue tests. Table 4 shows the average experimental results of every CA condition. Figure 8 shows the fitting S-N curve of experimental results. From the experiment results, it can be seen that the experimental scatter is small and the linearity agreement is good in the log—log scale.

Table 4. CA loading tests of designed T-type specimens.

Specimen No.	σ_a [MPa]	R	Average Cycles to Failure	Three Repeated Fatigue Experiment Results		
C1	105	0.1	273,187	267,358	270,591	281,612
C2	100	0.1	413,863	395,732	419,934	425,923
C3	95	0.1	706,895	575,216	719,683	825,786
C4	90	0.1	896,437	804,140	908,442	976,729
C5	85	0.1	925,700	1,035,762	898,867	842,471
C6	80	0.1	975,620	841,147	914,587	1,171,126
C7	75	0.1	1,156,460	1,400,569	1,075,784	993,027
C8	70	0.1	1,933,942	1,829,748	1,656,775	2,315,303
C9	55	0.1	10^7+	10^7+	10^7+	10^7+
C10	53	0.1	10^7+	10^7+	10^7+	10^7+
C11	50	0.1	10^7+	10^7+	10^7+	10^7+
C12	45	0.1	10^7+	10^7+	10^7+	10^7+
C13	40	0.1	10^7+	10^7+	10^7+	10^7+

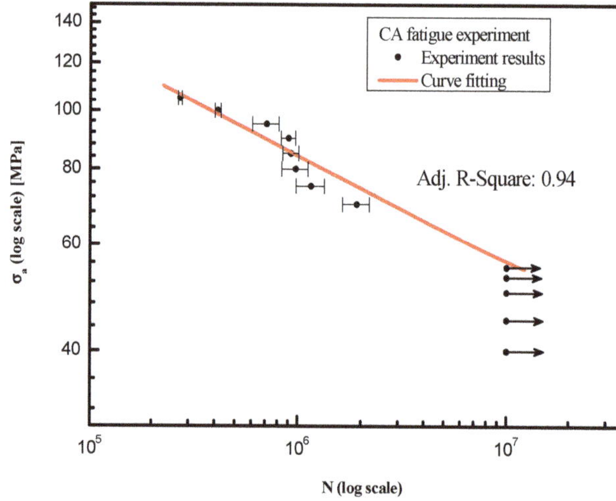

Figure 8. Results of experimental data and fitting curve.

3.2. Variable Amplitude Fatigue Experiment Results

The loading conditions of designed T-type specimens under variable amplitude test are mainly divided into two types, variable high cyclic stress (stress ratio) and high–low load cycle ratio. Table 5 lists the loading conditions of the test. The load type was a two-step repeating load block, and the low cyclic stress was 50 MPa which was below the CAFL according to the CA fatigue experiment results. Table 6 shows the experimental results of every load condition under a two-step repeating amplitude fatigue load block. The item $\sum n_i$ stands for the cyclic number contributed by the corresponding cyclic stress amplitude σ_{ai} and the item $\sum(n_1 + n_2)$ stands for the total fatigue life of designed T-type specimen when fatigue failure occurs. The item N_{f1} and N_{f2} is the cyclic number of fatigue failure under the cyclic stress of σ_{a1} and σ_{a2}, respectively. N_{f2} is constant and equals 16,000,000 corresponding to σ_{a2} = 50 MPa, which is calculated by extrapolating the fitting S-N curve in the log-log plot in Figure 8. σ_{a1} is different in different stress ratio cases and N_{f1} can also be calculated by extrapolating the fitting S-N curve in the log-log plot in Figure 8. The calculated values of N_{f1} for different high stresses are also shown in Table 6.

Table 5. Test loading conditions of the two-step repeating amplitude.

Stress (MPa) σ_{a1}/σ_{a2}	Specimen No. under Different Load Block (n_1/n_2)			
	20,000/20,000 (1:1)	20,000/40,000 (1:2)	20,000/60,000 (1:3)	20,000/80,000 (1:4)
105/50	R1	R8	R13	R19
100/50	R2	R9	R14	R20
95/50	R3	R10	R15	R21
93/50	-	-	-	R22
92.5/50	-	-	R16	R23
91/50	-	-	R17	-
90/50	R4	R11	R18	-
87.5/50	R5	R12	-	-
85/50	R6	-	-	-
84/50	R7	-	-	-

Table 6. The test results of designed T-type specimens under two-step repeating amplitude loading sequence.

Specimen No.	First Step N_{f1} (Cycle)	Second Step N_{f2} (Cycle)	$\sum n_1$	$\sum n_2$	$\sum(n_1+n_2)$
R1	293,018	16,000,000	147,790	145,000	292,790
R2	383,800	16,000,000	200,000	204,431	404,431
R3	509,075	16,000,000	310,000	315,434	625,434
R4	689,256	16,000,000	547,533	540,000	1,087,500
R5	804,793	16,000,000	695,000	702,573	1,397,573
R6	945,578	16,000,000	1,128,637	1,120,000	2,248,637
R7	1,007,230	16,000,000	1,544,489	1,530,000	3,174,489
R8	293,018	16,000,000	173,321	320,000	493,321
R9	383,800	16,000,000	245,315	480,000	725,315
R10	509,075	16,000,000	377,466	720,000	1,097,466
R11	689,256	16,000,000	640,000	1,276,359	1,916,359
R12	804,793	16,000,000	1,126,710	2,240,000	3,366,710
R13	293,018	16,000,000	167,548	480,000	657,548
R14	383,800	16,000,000	260,000	723,971	983,971
R15	509,075	16,000,000	400,000	1,109,637	1,509,637
R16	591,961	16,000,000	580,000	1,743,492	2,323,492
R17	647,055	16,000,000	780,000	2,326,483	3,106,483
R18	689,256	16,000,000	1,031,126	3,060,000	4,091,126
R19	293,018	16,000,000	180,000	677,563	837,563
R20	383,800	16,000,000	260,000	1,042,741	1,302,741
R21	509,075	16,000,000	460,000	1,811,567	2,271,567
R22	573,166	16,000,000	680,000	2,705,437	3,385,437
R23	591,961	16,000,000	906,292	3,600,000	4,506,292

4. Damage Accumulation of Variable Fatigue Experiment Analysis

The VA experimental results were analyzed in terms of Miner's rule to check its validity and to examine the effect of stress below the CAFL under the two-step repeating load block. The calculation method is as follows:

$$D = \frac{\sum n_1}{N_{f1}} + \frac{\sum n_2}{N_{f2}} \tag{3}$$

Figure 9 shows the damage accumulation results of designed T-type specimens under the two-step repeating load block by the Miner's rule. Figure 9 shows the damage accumulation value of every condition with a bar graph, and the different colors represent the fatigue damage contributed by each cyclic stress.

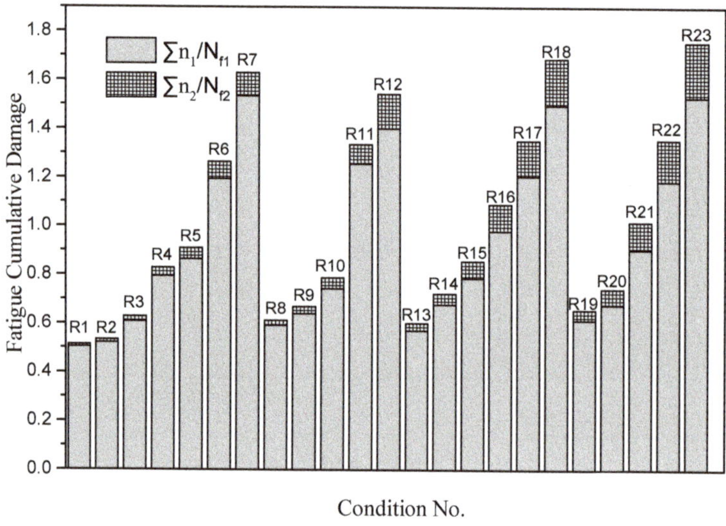

Figure 9. Damage accumulation of every condition.

It can be seen that the fatigue damage caused by low cyclic stress below the CAFL is relatively smaller than that caused by high cyclic stress above the fatigue limit. In order to better compare the fatigue damage evolution law of designed T-type specimens under two-step repeating VA loads, the results of the above-mentioned conditions are sorted out, and the variation law of cumulative damage with the high stress under different load cycle ratios is obtained.

As can be seen from Figure 10, the fatigue cumulative damage of designed T-type specimen shows the same change rule under different load cycle ratios. With the increase of high cyclic stress, the cumulative damage value decreases, and the rate slows down gradually. By comparing the cumulative damage of different load cycle ratios under the same high stress, it can be indicated that the greater the cycle ratio, the greater the fatigue cumulative damage of designed T-type specimens. Besides, the rule of fatigue cumulative damage varying with the cycle ratio under different high stresses is obtained. k_1, k_2, k_3 and k_4 represent the linear slope of fatigue cumulative damage with the cycle ratio under high-low stress of 105/50, 100/50, 95/50 and 90/50 MPa, respectively.

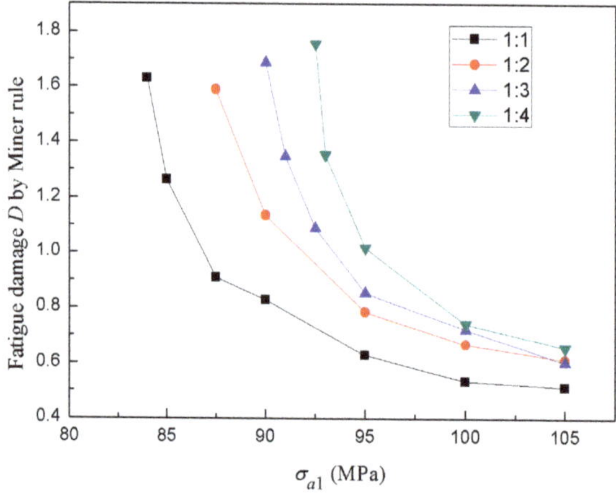

Figure 10. Fatigue cumulative damage of designed T-type specimens under different cycle ratios.

Under different high-low stress ratios, the fatigue cumulative damage calculated with the change of the cycle ratio based on Miner rule changes linearly, and the slope k of the curve is different. By fitting the relationship between the high stresses in different loading blocks and the slopes k of these curves in Figure 11, as shown in Figure 12, it can be found that the double logarithmic curve satisfies the data linearly. Finally, the logarithmic relationship between the high stress and the slope k of fatigue cumulative damage varying with the cycle ratio is obtained.

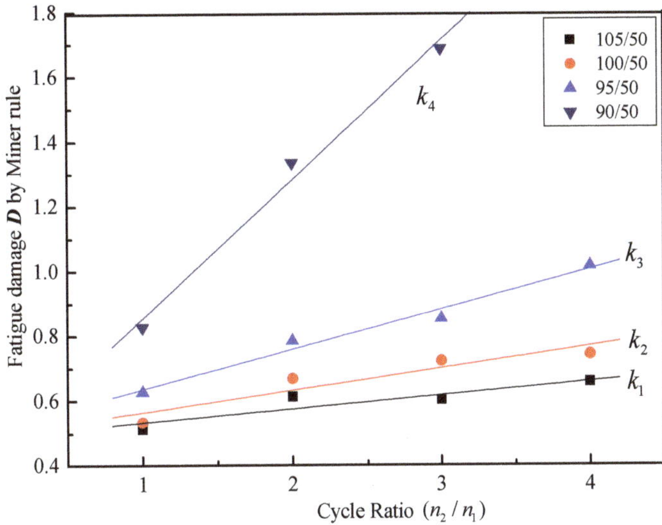

Figure 11. Fatigue damage of designed T-type specimens under different high cyclic stresses.

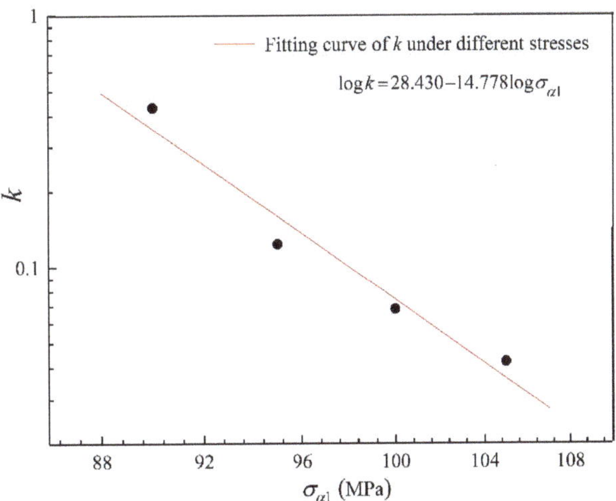

Figure 12. Double logarithmic curve between the high stress and slope k.

5. Non-Linear Damage Evaluation

As the load interaction caused the non-linear damage accumulation in the fatigue test with a two-step repeating amplitude loading block, therefore, the linear damage accumulation did not equal one. The parameter D_{nl} which defined the damage value of load interaction effect, was introduced into the Miner's law to evaluate the influence by the load interaction. The value of D_{nl} was proposed by Kim et al. [32] and could be calculated as the Formula (4).

$$D_{nl} = 1 - D = 1 - \sum_{i=1}^{n} \frac{n_i}{N_i} \tag{4}$$

When D_{nl} is positive, it indicates that the fatigue performance is damaged by load interaction effect, and when D_{nl} is negative, it indicates that the fatigue performance is strengthened by load interaction effect. The calculating results by the equation (4) were shown in the Figure 13.

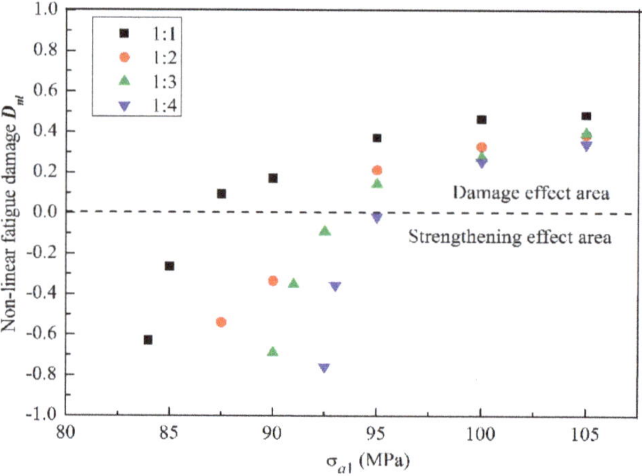

Figure 13. The calculation results of the damage value of load interaction effect D_{nl}.

As can be seen from Figure 13, the load interaction effect of high-low cyclic stress above and below CAFL shows different damage or strengthening effects with the change of cyclic ratio and high stress. Under the same cycle ratio, with the increase of high stress, the load interaction effect gradually transforms from strengthening effect to damage effect. Under the same stress, with the decrease of cycle ratio n_1/n_2, the damage effect decreases gradually at high stress area, and with the decrease of cycle ratio n_1/n_2 at low stress area, the effect of strengthening effect also increases. It can be seen that the load interaction effects of high-low cyclic stress have different damage or strengthening effects on fatigue performance with the change of stress ratio and cycle ratio.

The procedure for fatigue analysis also applies to other ships with variable loading. The linear damage rule must also be updated by considering the nonlinear damage. The form of the nonlinear damage can be developed further according to the actual loading condition.

6. Conclusions

This paper investigated the fatigue characteristics of designed T-type specimen under the two-step repeating load block. The designed T-type joint was common in the ship structure and the fatigue failure always occurred at this type of joint. The two-step repeating load block was introduced to simulate the load environment of voyage line of river-sea going ship and this type of load block was seldom studied systematically.

CA fatigue experiments were conducted first and then VA fatigue experiments. Every CA fatigue experiment was repeated three times. Twenty-three load conditions of VA fatigue experiment were conducted, Miner's rule was used to calculate the damage accumulation, and the load interaction effect of two-step repeating load block was also evaluated through the formula (4). From the analysis by the Sections 4 and 5, the following conclusions could be drawn:

(1) For the fatigue performance of designed T-type specimens under two-step VA load, the fatigue cumulative damage values calculated by linear Miner rule are not 1. Compared with the cumulative damage caused by high cyclic stress, the cumulative damage caused by low stress below CAFL is relatively small. It is found that the cumulative damage decreases with the increase of the high stress under the same cycle ratio. While under the same high-low load stress ratio, the cumulative fatigue damage increases with the linear increase of the cycle ratio of the low stress to the high stress, and logarithmic curve between the slope k and high stress above CAFL is linear.

(2) Nonlinear cumulative damage D_{nl} is used to evaluate the load interaction between low and high cyclic stress. It is found that the load interaction shows different damage or strengthening effects with the change of low-high load cycle ratio and high cyclic stress. The area of strengthening effect occurs at high cyclic stress and low load cycle ratio.

(3) From Figure 11, it can be seen that the linear slope of fatigue cumulative damage with the cycle ratio n_2/n_1 increases with the decrease of high stress in the loading block. The slopes k_3 and k_4 represent the linear slope of fatigue cumulative damage with the cycle ratio n_2/n_1 under high-low stress of 95/50 and 90/50 MPa, respectively. The reason is that the high stress in the loading block is closer to the low stress 50 MPa, the fatigue damage caused by both the high and the low stresses is much smaller. When the cycles of low stress take a larger percentage, i.e., a larger ratio of n_2/n_1, the strengthening effects are more obvious. As shown in Figure 13, two of the three points at the high stress of 90 MPa are in the strengthening effect area, and only one of the four points at the high stress of 95 MPa is in the strengthening effect area. All of the four points at the high stress of 100 MPa or 105 MPa are in the damage effect area.

The method also applies to other ship types, like the ship only voyaging on the sea or ship only voyaging on the inland river. The ranges of the high and low loading levels should be set according to the actual loading on the ship. For the river-sea-going ship in this study, it mostly voyages in the inland river and then the load below the CAFL takes the most part in the loading. That's why the chosen high cyclic loads were larger than the CAFL and the chosen low cyclic loads were below the CAFL. When the ranges of the high and low loading levels in the variable cyclic loading spectra are different, the experiments may show some different results. More investigation will be necessary.

Author Contributions: Conceptualization, J.G., D.S. and Z.W.; data curation, J.G., D.S. and H.D.; formal analysis, J.G., D.S., Z.W. and X.W.; funding acquisition, J.G., Z.W., L.Y. and W.W.; investigation, H.D.; methodology, J.G. and D.S.; project administration, J.G., Z.W. and W.W.; resources, J.G., Z.W. and W.W.; supervision, J.G. and X.W.; visualization, D.S., H.D. and X.W.; writing—original draft, J.G., D.S., Z.W. and X.W.; writing—review & editing, J.G., Z.W., X.W. and L.Y. All authors have read and agreed to the published version of the manuscript.

Funding: This research was funded by the National Natural Science Foundation of China (NSFC) (grant number 51879208), the National Scientific Research Project of the High-tech Ship of China (grant number 2014[493]) and the Fundamental Research Funds for the Central Universities (grant number 205207019).

Institutional Review Board Statement: Not applicable.

Informed Consent Statement: Not applicable.

Data Availability Statement: The data presented in this study are available in Tables 3–6.

Acknowledgments: All the fatigue experiments were carried out via an MTS 322 250 kN Dynamic Fatigue Testing System at the Ship Structure Laboratory of Wuhan University of Technology (Wuhan, China). The authors are grateful to all the staff of School of Transportation at Wuhan University of Technology for supporting this work.

Conflicts of Interest: The authors declare no conflict of interest. The funders had no role in the design of the study; in the collection, analyses, or interpretation of data; in the writing of the manuscript, or in the decision to publish the results.

References

1. Gan, J.; Sun, D.; Wang, Z.; Luo, P.; Wu, W. The effect of shot peening on fatigue life of Q345D T-welded joint. *J. Constr. Steel Res.* **2016**, *126*, 74–82. [CrossRef]
2. Li, Z.; Mao, W.; Ringsberg, J.W.; Johnson, E.; Storhaug, G. A comparative study of fatigue assessments of container ship structures using various direct calculation approaches. *Ocean. Eng.* **2014**, *82*, 65–74. [CrossRef]
3. Beretta, S.; Carboni, M. Variable amplitude fatigue crack growth in a mild steel for railway axles: Experiments and predictive models. *Eng. Fract. Mech.* **2011**, *78*, 848–862. [CrossRef]
4. Warhadpande, A.; Jalalahmadi, B.; Slack, T.S.; Sadeghi, F. A new finite element fatigue modeling approach for life scatter in tensile steel specimens. *Int. J. Fatigue* **2010**, *32*, 685–697. [CrossRef]
5. Sun, D.; Gan, J.; Wang, Z.; Luo, P.; Wu, W. Experimental and analytical investigation of fatigue crack propagation of T-welded joints considering the effect of boundary condition. *Fatigue Fract. Eng. Mater. Struct.* **2016**, *40*, 894–908. [CrossRef]
6. Miner, M.A. Cumulative damage in fatigue. *J. Appl. Mech.* **1945**, *67*, A159–A164.
7. Zhang, Y.-H.; Maddox, S. Investigation of fatigue damage to welded joints under variable amplitude loading spectra. *Int. J. Fatigue* **2009**, *31*, 138–152. [CrossRef]
8. Stäcker, C.; Sander, M. Experimental, analytical and numerical analyses of constant and variable amplitude loadings in the very high cycle fatigue regime. *Appl. Fract. Mech.* **2017**, *92*, 394–409. [CrossRef]
9. Tilly, G. Fatigue of land-based structures. *Int. J. Fatigue* **1985**, *7*, 67–78. [CrossRef]
10. Dahle, T. Spectrum fatigue life of welded specimen in relation to the linear damage rule. In *Fatigue Under Spectrum Loading and in Corrosive Environments*; Blom, A.F., Ed.; EMAS Publishing: Birchwood Park Warrington, UK, 1993; pp. 133–147.
11. Tubby, P.J.; Razmjoo, G.R.; Gurney, T.R.; Priddle, E.K. *Fatigue of Welded Joints Under Variable Amplitude Loading, In: OTO 94 804*; HSE Books: Sudbury, ON, Canada, 1996.
12. Gurney, T.R. *Exploratory Investigation of the Significance of the Low Stresses in a Fatigue Loading Spectrum*; TWI Member Report 718/2000; TWI Ltd: Cambridge, UK, December 2000.
13. Berger, C.; Eulitz, K.G.; Heuler, P.; Kotte, K.L.; Naundorf, H.; Schuetz, W.; Sonsino, C.M.; Wimmer, A.; Zenner, H. Betriebsfestigkeit in Germany—An overview. *Int. J. Fatigue* **2002**, *24*, 603–625. [CrossRef]
14. Dattoma, V.; Giancane, S.; Nobile, R.; Panella, F. Fatigue life prediction under variable loading based on a new non-linear continuum damage mechanics model. *Int. J. Fatigue* **2006**, *28*, 89–95. [CrossRef]
15. Fatemi, A.; Yang, L. Cumulative fatigue damage and life prediction theories: A survey of the state of the art for homogeneous materials. *Int. J. Fatigue* **1998**, *20*, 9–34. [CrossRef]
16. Bolchoun, A.; Baumgartner, J.; Kaufmann, H. A new method for fatigue life evaluation under out-of-phase variable amplitude loadings and its application to thin-walled magnesium welds. *Int. J. Fatigue* **2017**, *101*, 159–168. [CrossRef]
17. Marco, S.M.; Starkey, W.L. A concept of fatigue damage. *Trans. ASME* **1954**, *76*, 627–632.
18. Hell, M.; Wagener, R.; Kaufmann, H.; Melz, T. Fatigue life design of components under variable amplitude loading with respect to cyclic material behaviour. *Procedia Eng.* **2015**, *101*, 194–202. [CrossRef]
19. Gatts, R.R. Cumulative fatigue damage with random loading. *ASME J. Basic Eng.* **1962**, *84*, 403–409. [CrossRef]
20. He, L.; Akebono, H.; Kato, M.; Sugeta, A. Fatigue life prediction method for AISI 316 stainless steel under variable-amplitude loading considering low-amplitude loading below the endurance limit in the ultrahigh cycle regime. *Int. J. Fatigue* **2017**, *101*, 18–26. [CrossRef]
21. Gan, J.; Zhao, K.; Wang, Z.; Wang, X.; Wu, W. Fatigue damage of designed T-type specimen under different proportion re-peating Two-Step variable amplitude loads. *Eng. Fract. Mech.* **2019**, *221*, 106684. [CrossRef]
22. Xi, L.; Zheng, S. Strengthening of transmission gear under low-amplitude loads. *Mat. Sci. Eng. A-Struct.* **2008**, *488*, 55–63. [CrossRef]
23. Xi, L.; Songlin, Z. Changes in mechanical properties of vehicle components after strengthening under low-amplitude loads below the fatigue limit. *Fatigue Fract. Eng. Mater. Struct.* **2009**, *32*, 847–855. [CrossRef]
24. Lee, C.-H.; Chang, K.-H.; Van Do, V.N. Modeling the high cycle fatigue behavior of T-joint fillet welds considering weld-induced residual stresses based on continuum damage mechanics. *Eng. Struct.* **2016**, *125*, 205–216. [CrossRef]
25. Ciavarella, M.; D'Antuono, P.; Demelio, G. A simple finding on variable amplitude (Gassner) fatigue SN curves obtained using Miner's rule for unnotched or notched specimen. *Eng. Fract. Mech.* **2017**, *176*, 178–185. [CrossRef]
26. Keating, P.B.; Fisher, J.W. *Evaluation of Fatigue Tests and Design Criteria on Welded Details*; NCHRP Report 286; Transportation Research Board: Washington, DC, WA, USA, 1986; pp. 163–172.
27. Keating, P.B.; Fisher, J.W. Full-scale welded details under random variable amplitude loading. In *Fatigue of Welded Constructions*; Maddox, S.J., Ed.; Elsevier: Brighton, UK, 1987.
28. Dahle, T. Long-life spectrum fatigue tests of welded joints. *Int. J. Fatigue* **1994**, *16*, 392–396. [CrossRef]
29. Marquis, G. Long Life Spectrum Fatigue Of Carbon And Stainless Steel Welds. *Fatigue Fract. Eng. Mater. Struct.* **1996**, *19*, 739–753. [CrossRef]

30. Hobbacher, A.F. Recommendations for fatigue design of welded joints and components. In *Recommendations for Fatigue Design of Welded Joints and Components*; Springer Nature: London, UK, 2016.
31. Lv, Z.; Huang, H.-Z.; Zhu, S.-P.; Gao, H.; Zuo, F. A modified nonlinear fatigue damage accumulation model. *Int. J. Damage Mech.* **2014**, *24*, 168–181. [CrossRef]
32. Kim, M.H.; Kim, S.M.; Kim, Y.N.; Kim, S.G.; Lee, K.E.; Kim, G.R. A comparative study for the fatigue assessment of a ship structure by use of hot spot stress and structural stress approaches. *Ocean. Eng.* **2009**, *36*, 1067–1072. [CrossRef]
33. Erny, C.; Thévenet, D.; Cognard, J.-Y.; Körner, M. Fatigue life prediction of welded ship details. *Mar. Struct.* **2012**, *25*, 13–32. [CrossRef]
34. Li, Z.; Ringsberg, J.W.; Storhaug, G. Time-domain fatigue assessment of ship side-shell structures. *Int. J. Fatigue* **2013**, *55*, 276–290. [CrossRef]
35. Xu, X.; Augello, R.; Yang, H. The generation and validation of a CUF-based FEA model with laser-based experiments. *Mech. Adv. Mater. Struct.* **2019**, 1–8. [CrossRef]
36. Xu, X.; Yang, H.; Augello, R.; Carrera, E. Optimized free-form surface modeling of point clouds from laser-based measurement. *Mech. Adv. Mater. Struct.* **2019**, 1–9. [CrossRef]
37. Carrera, E.; Pagani, A.; Augello, R. On the role of large cross-sectional deformations in the nonlinear analysis of composite thin-walled structures. *Arch. Appl. Mech.* **2020**, 1–17. [CrossRef]
38. Carrera, E.; Pagani, A.; Augello, R. Evaluation of geometrically nonlinear effects due to large cross-sectional deformations of compact and shell-like structures. *Mech. Adv. Mater. Struct.* **2018**, *27*, 1269–1277. [CrossRef]
39. Sun, D.; Gan, J.; Wang, Z.; Wu, W.-G. Experimental study on fatigue characteristics of T-welded joint under repeating two-step load. In Proceedings of the International Offshore and Polar Engineering Conference, ISOPE-1-17-236, 27th International Ocean and Polar Engineering Conference, San Francisco, CA, USA, 25–30 June 2017.
40. CCS. *Metallic Materials—Tensile Testing. China Classification of Society*; Standards Press of China: Beijing, China, 2016.
41. Gao, X.; Gan, J.; Sun, D.; Wu, W. Experimental study on fatigue characteristics of typical welded joints of river-sea-going ship. In Proceedings of the International Offshore and Polar Engineering Conference, ISOPE-1-17-358, 27th International Ocean and Polar Engineering Conference, San Francisco, CA, USA, 25–30 June 2017.
42. Dong, P. A structural stress definition and numerical implementation for fatigue analysis of welded joints. *Int. J. Fatigue* **2001**, *23*, 865–876. [CrossRef]
43. BSI. *BS 7608: 2014+A1: 2015 Guide to Fatigue Design and Assessment of Steel Products*; British Standards Institution: London, UK, 2015.

Journal of
Marine Science
and Engineering

MDPI

Article

Small–Scale Experimental Investigation of Fatigue Performance Improvement of Ship Hatch Corner with Shot Peening Treatments by Considering Residual Stress Relaxation

Jin Gan [1,2], Zi'ang Gao [1,2,3], Yiwen Wang [1,2,3,*], Zhou Wang [4] and Weiguo Wu [1,3]

[1] Key Laboratory of High Performance Ship Technology (Wuhan University of Technology), Ministry of Education, Wuhan 430063, China; ganjinwut@163.com (J.G.); gaoziang_wut@163.com (Z.G.); mailjt@163.com (W.W.)

[2] Department of Naval Architecture, Ocean and Structural Engineering, School of Transportation, Wuhan University of Technology, Wuhan 430063, China

[3] Green & Smart River–Sea–Going Ship, Cruise and Yacht Research Center, Wuhan University of Technology, Wuhan 430063, China

[4] Department of Automotive Engineering, School of Automotive Engineering, Wuhan University of Technology, Wuhan 430070, China; wangzhou@whut.edu.cn

* Correspondence: yiwenwang90@whut.edu.cn

Citation: Gan, J.; Gao, Z.; Wang, Y.; Wang, Z.; Wu, W. Small–Scale Experimental Investigation of Fatigue Performance Improvement of Ship Hatch Corner with Shot Peening Treatments by Considering Residual Stress Relaxation. *J. Mar. Sci. Eng.* **2021**, *9*, 419. https://doi.org/10.3390/jmse9040419

Academic Editor: Joško Parunov

Received: 15 March 2021
Accepted: 7 April 2021
Published: 13 April 2021

Abstract: Ship hatch corner is a common structure in a ship and its fatigue problem has always been one of the focuses in ship engineering due to the long–term high–stress concentration state during the ship's life. For investigating the fatigue life improvement of the ship hatch corner under different shot peening (SP) treatments, a series of fatigue tests, residual stress and surface topography measurements were conducted for SP specimens. Furthermore, the distributions of the surface residual stress are measured with varying numbers of cyclic loads, investigating the residual stress relaxation during cyclic loading. The results show that no matter which SP process parameters are used, the fatigue lives of the shot–peened ship hatch corner specimens are longer than those at unpeened specimens. The relaxation rate of the residual stress mainly depends on the maximum compressive residual stress (σ^{RS}_{max}) and the depth of the maximum compressive residual stress (δ_{max}). The larger the values of σ^{RS}_{max} and δ_{max}, the slower the relaxation rates of the residual stress field. The results imply that the effect of residual stress field and surface roughness should be considered comprehensively to improve the fatigue life of the ship hatch corner with SP treatment. The increase in peening intensity (PI) within a certain range can increase the depth of the compressive residual stress field (CRSF), so the fatigue performance of the ship hatch corner is improved. Once the PI exceeds a certain value, the surface damage caused by the increase in surface roughness will not be offset by the CRSF and the fatigue life cannot be improved optimally. This research provides an approach of fatigue performance enhancement for ship hatch corners in engineering application.

Keywords: shot peening; ship hatch corner; fatigue life; surface roughness; residual stress relaxation

1. Introduction

As a large–scale structure integrating safety, economy and practicability, the ship's fatigue problem is the focus of attention. The International Association of Classification Society (IACS) Regulations for Bulk Carriers [1] and China Classification Society (CCS)'s "Guidelines for Fatigue Strength Assessment of Hull Structures" [2] both provide relevant regulations for ship hull structures that are prone to fatigue problems. During the service life, the hull structure is affected by alternating stress loads such as wave–induced load, which have a great impact on its fatigue properties. Figure 1 shows a fatigue crack in the hatch corner of a certain ship. Guoqing et al. [3] proposed to use the equivalent wave method to evaluate the fatigue life of hatch corners. The results of the equivalent wave method agree well with those from the spectral fatigue analysis. Selle et al. [4] used the finite element method with a fatigue assessment technique developed by Germanischer Lloyd (GL) to analyze the effect of radii of ship hatch corners on fatigue performance. Yong et al. [5] conducted stress analysis on ship hatch corner structures based on the S–N curve method and studied the effects of different structure details on fatigue performance of ship hatch corners. Xu et al. [6,7] did a detailed study on the method of establishing a finite element analysis (FEA) model by laser measurement, proposed some optimized model generation methods, and used the CUF (Carrera unified formula) method developed by Carrera. [8] for numerical analysis. Jiancheng et al. [9] conducted research on stress concentration at ship hatch corners with large openings, and analyzed the effect of stress releasing hole and strengthened plates for reducing the stress concentration. However, researches on improving the fatigue performance of ship hatch corners has mainly been conducted to optimize the structure to reduce the stress concentration, and there are few analyses on improving fatigue performance of ship hatch corner directly through a certain treatment.

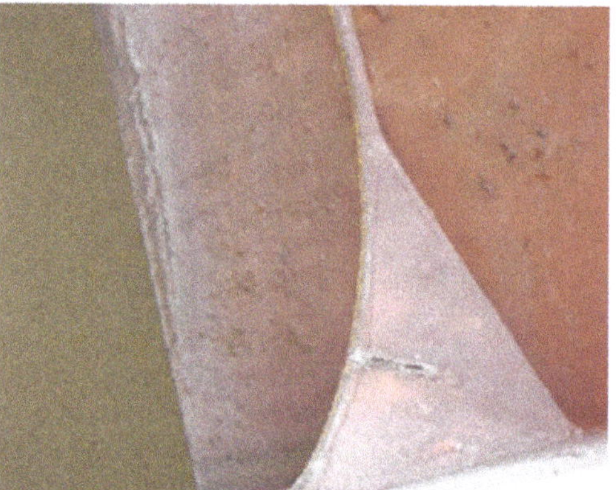

Figure 1. A fatigue crack in the hatch corner of a certain ship.

Compressive residual stresses on the surface layer help to inhibit crack initiation and expansion, which helps to improve the fatigue life and reliability of structures. For introducing compressive residual stress in the surface layer to prolong the fatigue life of the structure, some methods of surface treatment have been applied such as shot peening (SP), laser shock peening [10–13], high–frequency mechanical impact treatment [14–16], and ultrasonic impact treatment [17–20]. Among these kinds of treatments, SP is the most used treatment for its flexibility, low cost, and dispensing with pretreatment. SP is an effective cold working surface treatment in improving fatigue properties of structures [21,22] and is widely used in the aerospace industry, automobile industry, and other fields. However, there are few applications of SP in the shipbuilding industry. It can introduce compressive residual stresses to the surface layer and enhance the surface hardness and roughness by impacting on the surface with a mass of metallic or ceramic particles into the surface region. The compressive residual stress layers in the surface and subsurface of the shot–peened workpiece can partially offset the tensile stress generated by the alternating external load. Therefore, the compressive residual stress is recognized as the main strengthening factor to improve the fatigue performance of the specimen surface [23,24]. Gan et al. [25] studied the effect of SP on the fatigue life of T–welded joints via fatigue tests and residual stress measurements, proving that SP has significant benefits on specimens' fatigue performance. Wohlfahrt [26] studied the influence of residual stress change by different peening conditions. For shot peening, it is indispensable to evaluate the residual stress field. However, these studies only consider the effect of the initial residual stress field and ignore the relaxation of the residual stress field during the cyclic loading process.

Generally, SP treatment can cause an increase in the surface roughness of the peened structure while excessive surface roughness has a negative impact on fatigue performance. Surface roughness is one of the important parameters for evaluating the surface integrity of materials, which has a significant effect on fatigue performance. The larger the roughness, the easier it is to cause local stress concentration and induce fatigue crack initiation [27]. Novovic et al. [28] studied the correlation between the surface roughness (R_a) and fatigue life, showing that when R_a is less than 0.1 μm, its effect is insignificant. If R_a is between 2 μm and 5 μm, it will have a positive effect on the fatigue life. Ruihong et al. [29] investigated the effect of SP on the surface roughness and fatigue properties of 300M steel by tests, proving that the effect of SP is not proportional to the peening intensity (PI).

There are also some investigations related to the relaxation of the residual stress field of the shot peening treatment. Residual stress relaxation in fatigue loading is one of the main factors that affect the improvement of fatigue strength by shot peening, as well as the surface conditions created by shot peening and the possibility of the relaxation of compressive residual stress field (CRSF) to push the crack source beneath the surface [30]. In terms of the effect of the stability of the residual stress field on the fatigue life, the shot–peened specimens will exhibit longer life times than the unpeened ones as long as the CRSFs of the shot–peened specimens remain stable [31]. Dalaei et.al. [32,33] considered the main relaxation of the residual stresses that take place during the first sub–block which is due to the large relaxation during the first loading cycle as well as the maximum strain amplitude present in the first sub–block. The effect of shot peening conditions on the stability of the residual stress field is also significant, Huang et al. [34] investigated fatigue performance improvement of different modified SP treatments, considered that CRSF stability has a great influence on fatigue strength of shot–peened specimens and suitable peening temperature can lead CRSF to be more stable. However, the actual structure in practical engineering has not been paid enough attention until now. All of the objects in related research are simple standard specimens and the locations of interest are not well described which makes the results not suitable for complex structures and makes it difficult to reflect the evolution law of the residual stress field on the actual structure in practical applications. In the study of the numerical method of the relaxation of CRSF, Ruiz et al. [35] analyzed the relaxation of high frequency mechanical impact treatment induced CRSF of a single–sided out–of–plane gusset welded joint by numerical analysis. This analysis method is also applicable to SP–induced CRSF.

As for engineering applications, even though applying suitable SP for a specific structure and mastering the law of relaxation of its residual stress field will improve the fatigue strength and life of the structure, which cannot be solved by structure design or optimization effectively, there are still few applications of SP treatment in improving fatigue performance of structures in ships, especially for a structure like the ship hatch corner ,where fatigue damage often occurs. For this specific structure, it is not enough to consider the introduced initial residual compressive stress field alone because the residual stress field will relax during cyclic fatigue loading, which leads to the decline of the SP–induced fatigue strengthening effect. If the SP treatment is to be used in the fatigue performance improvement of ship structures, the law of residual stress relaxation and how it is affected by the load and SP parameters should be investigated in detail. The research on SP and residual stress relaxation in ship hatch corners is worthy of being focused on since it can provide an approach to improving the fatigue life of the structures in ships that suffer from serious fatigue problems. Beyond that, a procedure of fatigue life assessment after SP enhancement can provide some reference for researchers and designers who focus on ship fatigue problems.

In order to provide a practical assessment of actual ship hatch corner structures in engineering application, the effects of CRSF and surface topographies on the fatigue performances of ship hatch corners were analyzed by experimental and numerical methods. Experimental research based on fatigue tests, surface topography measurements, and residual stress measurements were conducted to investigate the effect of SP on the fatigue life of ship hatch corner structures. Variations of residual stress fields at a specific area on the structure was derived by measurement during cyclic loading. The variation law of the residual stress field in ship hatch corners was analyzed. The effect of shot peening treatment on the surface topography of the structure and the relationship between the residual stress relaxation and fatigue life were investigated.

2. Experimental Setup

2.1. Specimen Design and Processing

The small–scale experimental specimen and the geometry of the simplified ship hatch corner are shown in Figures 2 and 3a, respectively. In order to control the positions of crack initiation, two arc notches with smooth transition were set on the free edges at specific locations individually. The geometry of the arc notch is shown in Figure 3c while the corresponding location of the arc origination point is demonstrated in Figure 3b. These notches can be regarded as prefabricated defects, which could cause stress concentration. Therefore, the initiation area of the crack can be observed conveniently as can the measurement of the residual stress field's variation at a specific location. The model is composed of two parts: one is the foundation support of the T section, an opening made at the web to prevent the web bulking during the fatigue tests, and the other is the ship hatch corner specimen which is welded to the panel and welded with the U–shaped chuck symmetrically on both sides of the upper end.

The material of the specimen in this test was Q235B steel. For ascertaining the related parameters of material performance accurately, a series of static tensile tests were carried out as well. The model of the static tensile specimen is shown in Figure 4, and the test procedures were conducted according to GB/T 228.1(2010) [36]. The related parameters of material properties measured in the static tensile test are listed in Table 1. All the specimens were cut from the corresponding 6 mm thick plate by laser cutting directly.

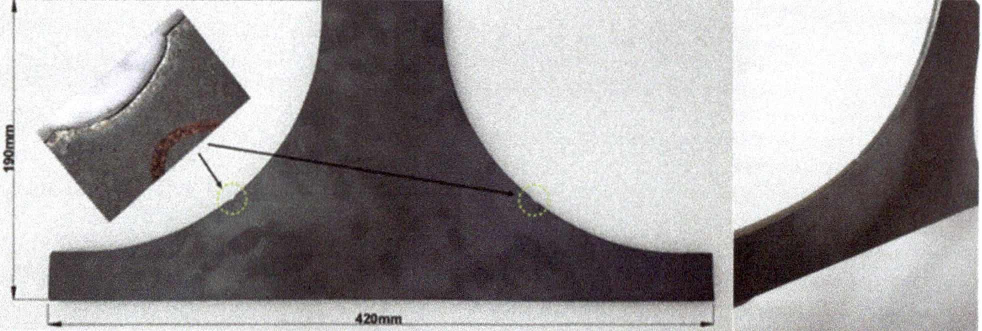

Figure 2. The experimental ship hatch corner specimen.

Table 1. Mechanical properties of Q235B steel.

Mechanical Properties	Value
Ultimate tensile stress R_m (MPa)	473
Yield stress σ_s (MPa)	294
Young's modulus E (GPa)	206
Poisson's ratio ν	0.26
Elongation δ (%)	26

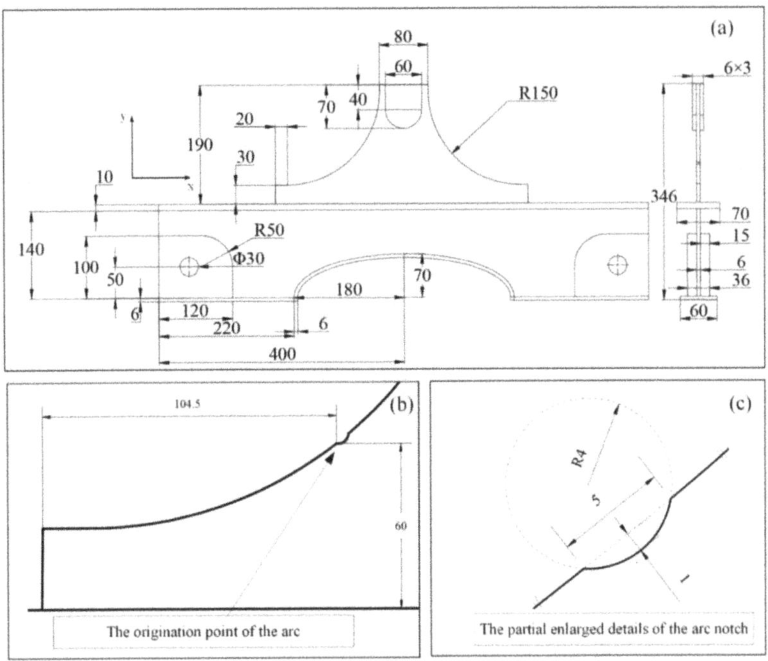

Figure 3. The geometry of the hatch corner specimen (**a**) and the arc notch (**c**) with the location of its origination point (**b**).

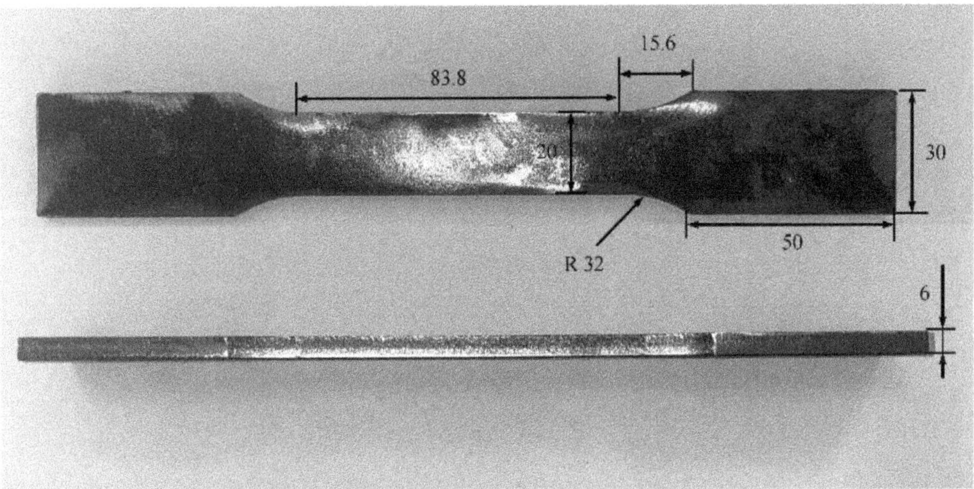

Figure 4. The static tensile specimen.

2.2. SP Treatment

SP treatments were carried out on ship hatch corner specimens by air blast SP equipment. Cast steel shots S110, S230, S280, and S330 were used individually for different specimens with the same peening pressures of 3 bar and a constant media flow rate of 25 kg/min. The mean diameters of shots (MDS) were 0.3 mm (S110), 0.6 mm (S230), 0.8 mm (S280), and 1.0 mm (S330), respectively. The diameter of the peening nozzle was 20 mm, while the distance from the peening nozzle to the specimen surface was about 50 mm. The nozzle was kept perpendicular to the surface of the specimen. The peening area included the specific area around both the left and right arc notches on the front and back surfaces, as shown in Figure 5. The peening time was 90 s and the peening coverage was close to 100% in all specimens. Both the left and right peening areas on the front and back surfaces of the specimen were treated with the same PI. Table 2 shows the different SP parameters for hatch corner specimens. Three specimens were processed corresponding to each SP treatment group. All five groups of specimens add up to a total of 15 specimens. All the specimens were conducted with cyclic fatigue tests, and the first specimen of each group was conducted with surface roughness measurement and residual stress relaxation measurement.

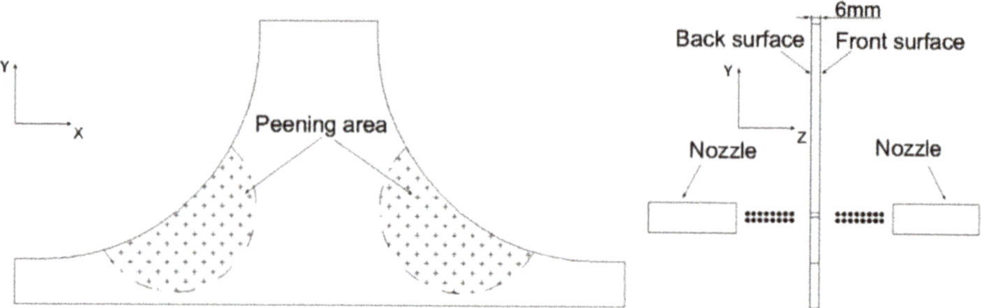

Figure 5. Schematic diagram of shot peening (SP) treatment for a hatch corner specimen.

Table 2. The different SP parameters for T–welded joint specimens.

Specimen	Mass Flow Rate (kg/min)	Air Pressure (bar)	Mean Diameter of Shots (MDS) d (mm)	Coverage	Shot Media Material
P1	–	–	–	–	–
P2	25	3	0.3 (S110)	100%	Cast steel
P3	25	3	0.6 (S230)	100%	Cast steel
P4	25	3	0.8 (S280)	100%	Cast steel
P5	25	3	1.0 (S330)	100%	Cast steel

2.3. Surface Roughness Measurement

In order to investigate the effect of SP on specimens, surface topography measurements were performed by using a Keyence 3D microscope. The first specimen of each group is conducted with surface roughness measurement. Thus, the heights of fluctuations in the horizontal and vertical directions of specimens' surfaces could be measured. Each direction generated a contour curve by data averaging from 30 paths, as shown in Figure 6. Since R_a could be measured directly by the microscope, a program was written based on ISO 4287–2010 [37], which is commonly used in the calculation of R_a. The calculation method of R_a is briefly introduced as follows: The baseline is a line with the smallest sum of squares of the distance from each point on the contour curve to the corresponding point on the contour curve, which is generated by using the least square method. R_a is extracted as follows

$$R_a = \frac{\sum_1^N |Y_{xi} - m_{xi}|}{N} = \frac{\sum_1^N |Z_{xi}|}{N} \tag{1}$$

where Y_{xi} is the measured value on the contour curve, m_{xi} is a value of the corresponding point on the baseline, Z_{xi} is the absolute value of the distance between Y_{xi} and m_{xi}, and N is the total number of points on the contour curve.

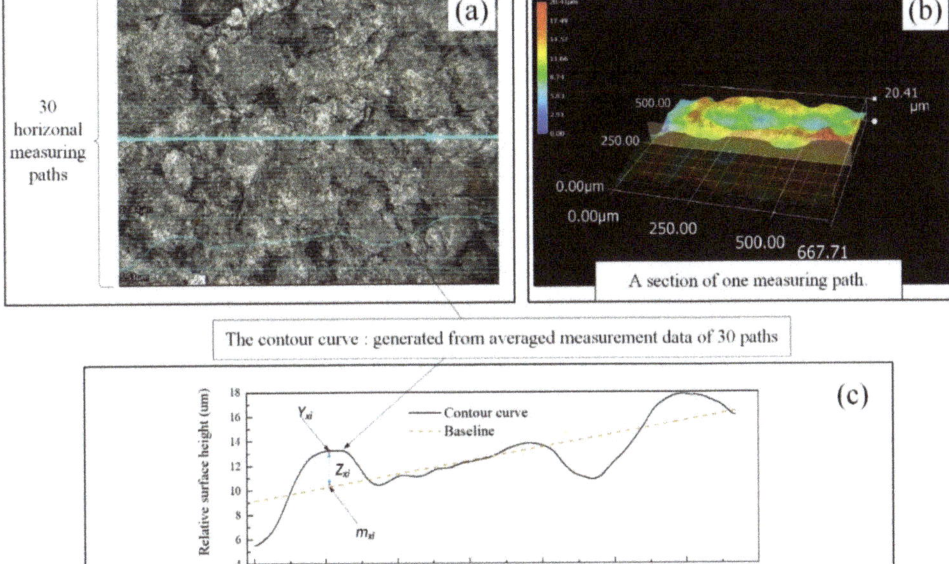

Figure 6. The schematic diagram of R_a measurement and calculation: (**a**) the measuring paths along the specimen surface, (**b**) a section of surface measured by one path, (**c**) the contour curve and the baseline for R_a calculation.

2.4. Finite Element Method (FEM) Analysis

FEM analysis was performed in order to verify the effectiveness of the arc notch design. The finite element model contained 973,456 elements including two element types, C3D8R and C3D4, as shown in Figure 7a. Moreover, the gradually refined transition method was adopted for the arc notch region, and the final element size near the arc notch was not more than 0.5 mm, as shown in Figure 7b. The material properties of the corresponding Q235B were ascertained by the static tensile test and shown in Table 1. It is an ideal elastoplastic material without considering the hardening effect of the material.

The boundary conditions of the finite element analysis are as follows: MPC constraints are used on the surface nodes of the fixed holes on both sides of the foundation support, and the MPC control point is the geometric center of the fixed hole, as shown in Figure 7c; the constraint condition is defined as $U_x = U_y = U_z = 0$, $R_x = R_y = 0$ (U represents the displacement, R represents the rotation, x, y, and z represent the directions of U and R). MPC constraints were performed on the U–shaped chuck surface nodes on both sides of the loading end. The MPC control point was 30 mm above the centerline of the U–shaped chuck surface, and the load is applied to the MPC control points symmetrically, as shown in Figure 7d.

The loads were increased from 5 kN to 80 kN gradually and the stress distributions of the hull hatch corner model were analyzed as well. Results show that the regions near the two arc notches were always the most serious regions of stress concentration in the whole model no matter how the tensile loads changed. Figure 8 shows the stress distribution of the hatch corner FE model which verifies that the design of the arc notch was effective.

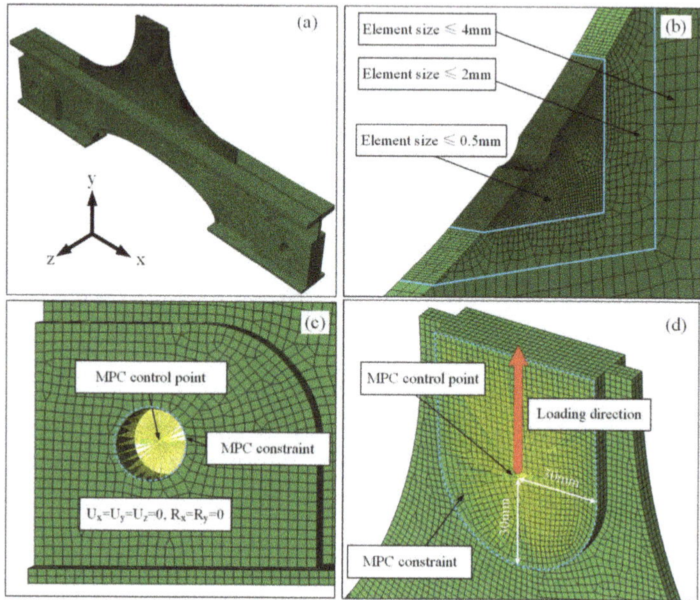

Figure 7. (**a**) The FE model of the hatch corner, (**b**) the local mesh refinement around arc notch, (**c**) the boundary conditions of the FE model, (**d**) setting of the loading side.

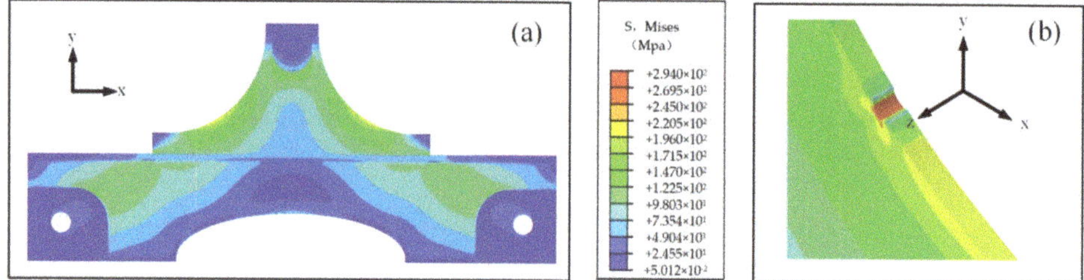

Figure 8. The (**a**) overall stress distribution and (**b**) local stress distribution of FE model with 70 kN loading.

2.5. Residual Stress Measurement

In order to investigate the effect of SP on surface residual stress, a portable X–ray two–dimensional residual stress analyzer μ–X360s (Pulstec® μ–X360s, Hamamatsu–City, Janpan) was used to measure the residual stress at every shot–peened region on each specimen applying the measuring principle of the cosα method [38] (also referred as the single incident angle method). This residual stress analyzer only requires one measurement operation at a specific angle ψ_0 to collect the diffraction angle transition of a certain surface, which is used to analyze and obtain the residual stress of the specimen. The schematic diagram of the measuring principle of u–X360s is shown in Figure 9. The angle α is the diffraction angle and the angle η is the complementary angle of α. The optical path of the X–ray and the obtained complete Debye ring are shown in Figure 9b,c, respectively. The angle selected for the Debye ring is presented in Figure 9d. The definition of the parameter a_1 can be given as follows:

$$a_1 = \frac{1}{2}\left[\left(\varepsilon_\alpha - \varepsilon_{\pi-\alpha}\right) + \left(\varepsilon_{-\alpha} - \varepsilon_{-\pi-\alpha}\right)\right] \tag{2}$$

where ε_α is the strain in the Debye ring, $\varepsilon_{\pi-\alpha}$ is the complementary strain, $\varepsilon_{-\alpha}$ is the contrary strain.

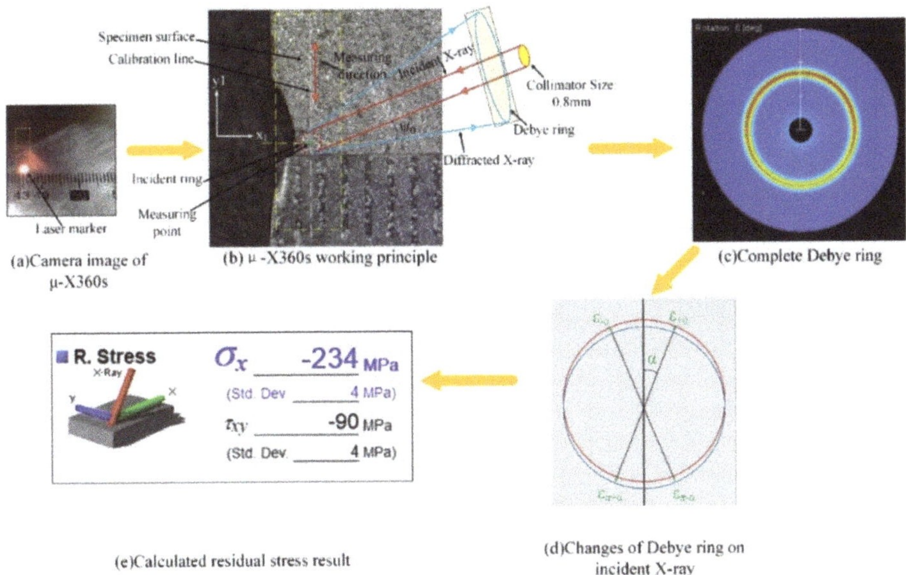

Figure 9. The schematic diagram of measuring principle of μ–X360s: (**a**)the camera image of μ–X360s, (**b**) the working principle, (**c**) the detected Debye ring, (**d**) the changes of Debye ring, and (**e**) the calculated residual stress result.

With the strain ε_α on the Debye ring, the residual stress can be calculated by the formula given as follows:

$$\sigma = -\frac{E}{1+v}\frac{1}{sin2\psi_0}\frac{1}{sin2\eta}\left(\frac{\partial a_1}{\partial cos\alpha}\right) \tag{3}$$

where σ represents the residual stress, E represents the Young's modulus, and v denotes the Poisson's ratio. In this work, the Young's modulus E and the Poisson's ratio v of Q235B steel specimen were determined as 206 GPa and 0.26 according to Section 2.1, respectively.

The first specimen of each group was conducted with residual stress relaxation measurement. In order to show the variation of the residual stress field near the arc notch in detail, some measuring points were defined near the arc notch to measure the dynamic change of the residual stress field. The location of the first measuring point was defined in detail as schematically shown in Figure 10 and the residual stress measuring procedures were as follows: (1) Draw the vertical line of the tangent to the deepest point of the arc notch as the measuring path. (2) Attach a ruler sticker parallel and below at least 0.5 mm to the vertical line to avoid any influence on measurement and align a tick mark on the ruler with the tangent line of the arc notch. (3) Adjust the analyzer to align the green cross in its camera image with the laser mark irradiated on the surface of the specimen to calibrate the measurement distance. (4) Adjust the analyzer to align line1 in the middle of the first two tick marks and overlap line2 in the camera image, the first measuring point is defined as Point1, and then start the measuring program of the analyzer. (5) Move the analyzer 1 mm along the measuring path to the next measuring point and repeat the measuring operation, as shown in Figure 11.

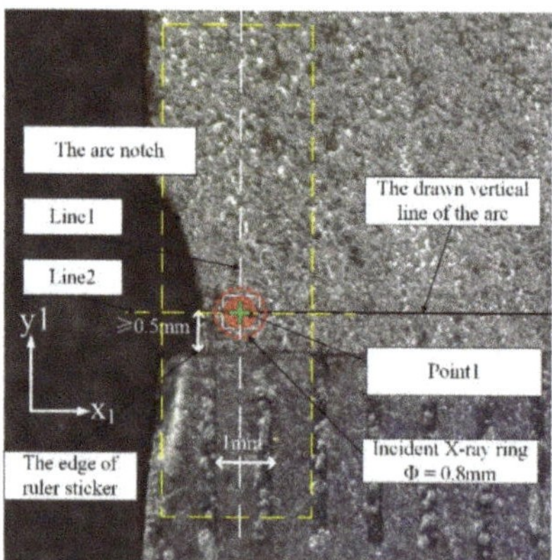

Figure 10. The location of Point1 and residual stress measurement setup.

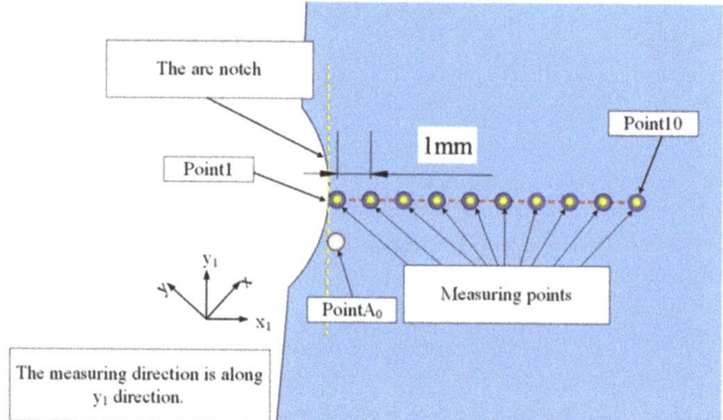

Figure 11. The schematic diagram of the measuring points at one shot–peened region.

There were a total of ten measurement points, and the residual stresses in the y_1 direction (the vertical line of the tangent point) was measured where x and y were the original coordinate system axes. The measuring points were named as Point1–Point10 from the distance to the arc notch in sequence. There was another measuring point named as $PointA_0$ which was 1 mm below Point1 along y_1 direction near the arc notch of each specimen. In order to obtain a value of residual stresses in depth before fatigue tests, residual stress measurements in $PointA_0$ were also carried out by iterative electrolytic removal of thin surface layer. Although the measurement path was not necessarily the same as the crack propagation path, it was enough to reflect the changing law of the residual stress field near the arc notch.

2.6. Static Loading Test and Fatigue Tests

The static loading test and fatigue tests were carried out by the MTS322 250 kN Dynamic Fatigue Testing System at the Ship Structure Laboratory in Wuhan University of Technology (Wuhan, China) under room temperature. Figure 12 shows the installation of the hatch corner test model and the loading direction of tests. Both ends of the hatch corner model were fixed on two ear seats (1) (2) by bolts, and the ear seats were fixed on the test base (3) by bolts; the U–shaped chuck (4) at the upper end of the model was clamped by the fixture (5) of the MTS testing system, and the loading direction is shown in Figure 12. The maximum and minimum stresses near the arc notch were about 218 and 21 MPa with

70 and 7 kN loading measured by strain gauges, respectively. The hot spot stresses at the left and right arc notches are shown in Figure 13.

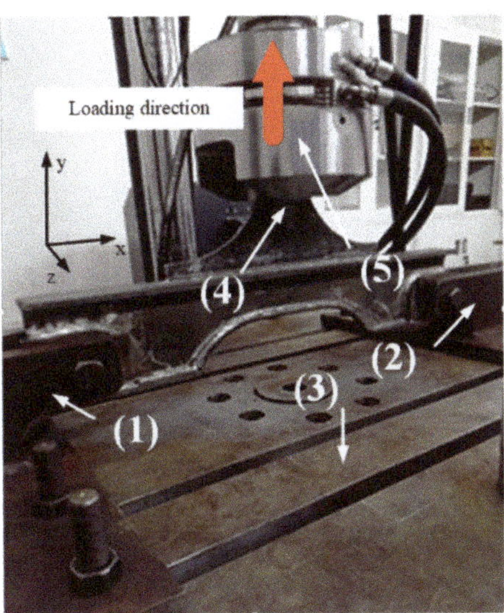

Figure 12. The installation of hatch corner specimen: (1), (2) the ear seats, (3) the test base, (4) the U–shaped chuck, and (5) the fixture of the MTS testing system.

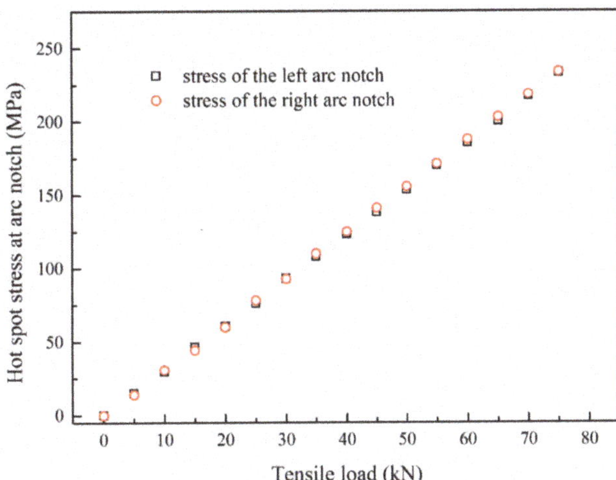

Figure 13. The relationship between tensile load and hot spot stresses at the left and the right arc notches.

Fatigue tests were carried out after the static loading test. All the specimens were conducted with constant amplitude cyclic loading of 31.5 kN and the stress ratio was 0.1. The loading frequency was 5 Hz and the loading direction was the same as static loading, as shown in Figure 13. The sinusoidal loading curve is shown in Figure 14. The criterion for judging crack initiation was when the crack was beyond 0.1 mm, and the criterion for judging the fatigue life was the number of fatigue cycle loading when the crack was beyond 1 mm.

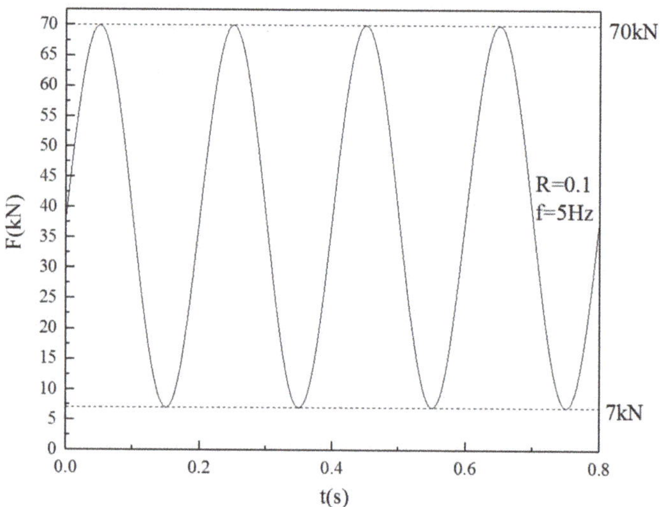

Figure 14. The schematic diagram of the sinusoidal cyclic loading curve for fatigue tests.

3. Results and Discussion

3.1. Surface Roughness and Topography

The changes in surface roughness of specimens caused by the different diameters of shots can be calculated by the method mentioned in Section 2.3. The calculated surface roughness data of specimens are shown in Table 3, which show that SP treatment can increase the surface roughness of peened specimens. The R_a value of unpeened specimen P1 was about 0.07 μm, which was much smaller than the minimum R_a value of peened specimen that appears in P2. The surface roughness increased with a value of MDS. The maximum R_a value appeared in P5, which was peened by the largest shots with diameters of 1 mm.

Table 3. The calculation results of surface roughness of specimens.

Specimen	MDS d (mm)	R_a (μm)		Average R_a (μm)
		Horizontal (x)	Vertical (y)	
P1	–	0.08	0.06	0.07
P2	0.3	1.41	1.26	1.34
P3	0.6	2.27	2.44	2.36
P4	0.8	3.14	3.28	3.21
P5	1.0	3.38	3.92	3.65

Figure 15 illustrates the differences in surface topography between each specimen. Compared with the unpeened specimen P1, which had only a few tiny bumps on the surface, all the shot–peened specimens became rough after SP. It can be concluded that the diameter and the depth of a single dimple increase with a value of MDS. The maximum dimple depth appeared in P5 with a value of 16.30 μm, the minimum dimple depth appeared in P2 with a value of 12.02 μm. It was complicated to define the diameter of a dimple due to the difficulties in finding clear boundaries, but it could be estimated based on the color of the cloud in Figure 15. The case of P5 had the greatest diameter of dimple caused by a single shot for SP with the largest MDS.

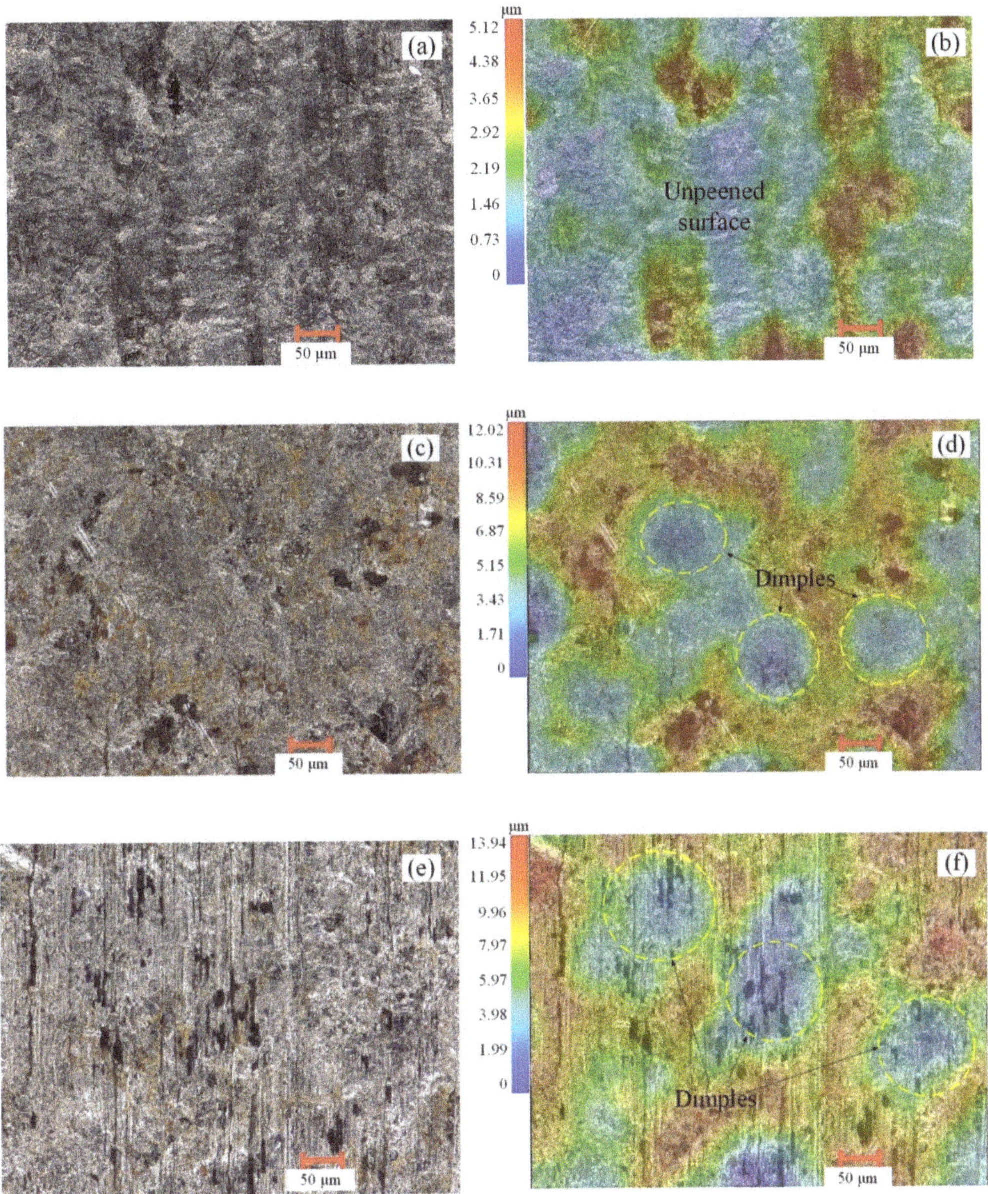

Figure 15. *Cont.*

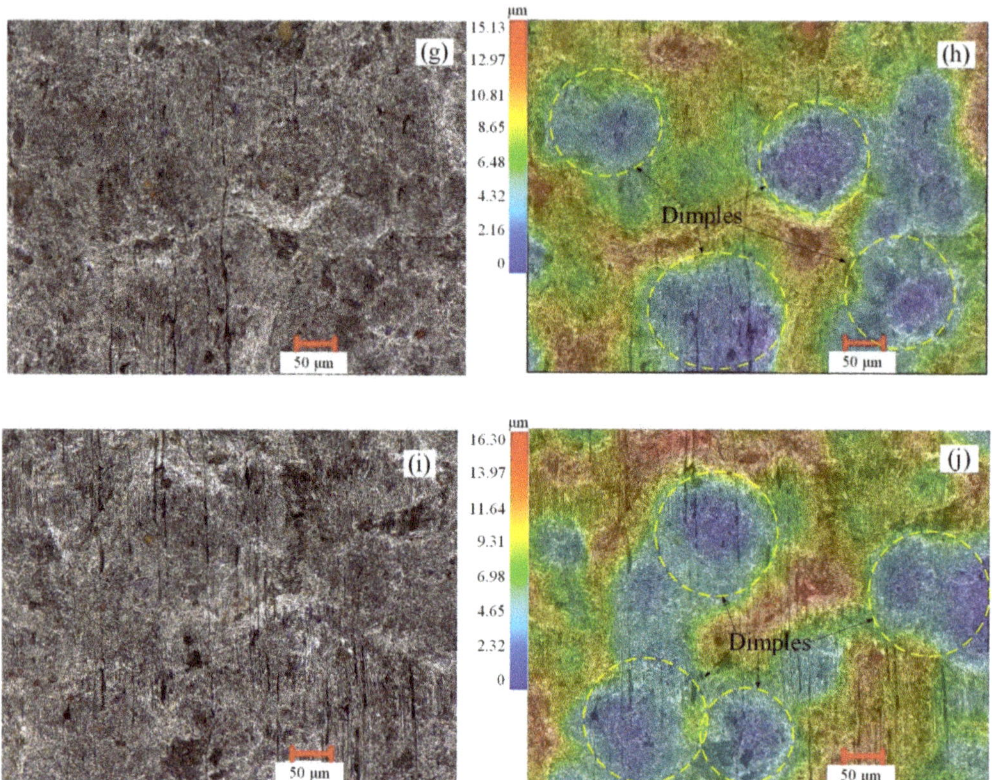

Figure 15. The specimen surface graph of (**a**) P1, (**c**) P2, (**e**) P3, (**g**) P4 and (**i**) P5 and the processed 2D topography of (**b**) P1, (**d**) P2, (**f**) P3, (**h**) P4 and (**j**) P5.

3.2. CRSF and Residual Stress Relaxation

Before residual stress measuring, the peened surface was electrically polished to avoid test error due to an uneven surface. In order to describe the CRSF quantitatively, three characteristic parameters including the surface residual stresses (σ^{RS}_{surf}), the maximum compressive residual stresses (σ^{RS}_{max}), and the depth of σ^{RS}_{max} (δ_{max}), were defined. The specimens were measured for CRSF at PointA$_0$. Figure 16 shows the residual stress depth profiles of PointA$_0$ in the y_1 direction. The maximum value of σ^{RS}_{surf} appeared in P3 with a value of -376 MPa, the minimum value of σ^{RS}_{surf} appeared in P2 with a value of -364 MPa, and the values of σ^{RS}_{surf} in the other two specimens were both around 370 MPa. The maximum value of σ^{RS}_{max} appeared in P5 with a value of -463 MPa, and the values of σ^{RS}_{max} in other three specimens increased with MDS from -392 MPa ($d = 0.3$ mm) to -447 MPa ($d = 0.8$ mm). It can be seen from Figure 16 that values of δ_{max} in all specimens also increased with MDS from 0.063 mm ($d = 0.3$ mm) to 0.144 mm ($d = 1.0$ mm). This phenomenon shows that the values of σ^{RS}_{surf} will not increase with the values of MDS remarkably. Addtionally, the values of σ^{RS}_{max} and δ_{max} in all specimens increased greatly with the values of MDS, which means an increase of PI.

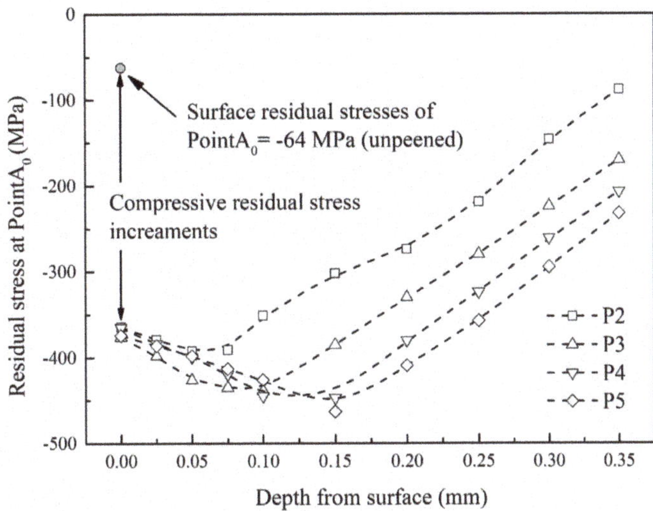

Figure 16. The compressive residual stress field in depth at PointA$_0$ of different specimens.

In the process of fatigue loading, the residual stresses were measured at 10 measuring points every 50,000 times of fatigue loading until crack initiation. Figure 17 shows the variety of residual stresses with fatigue loading of five ship hatch corner specimens. For the unpeened specimen P1, it can be seen from Figure 17a that the residual stress at each point on the surface was not uniform compared with other specimens. The residual stress fields began to relax just after the fatigue loading began.

It can be seen from Figure 17 that relaxation rates of the residual stress field were very fast at the first 50,000 cycles of fatigue loading, and then they were reduced to varying degrees. The relaxation rates of the residual stress field gradually decreased from the measuring point from the arc notch. The closer the measuring point to arc notch was, the greater was the relaxation of the residual stress field at this measuring point. During the fatigue loading process, the relaxation rate of the residual stress field decreased gradually.

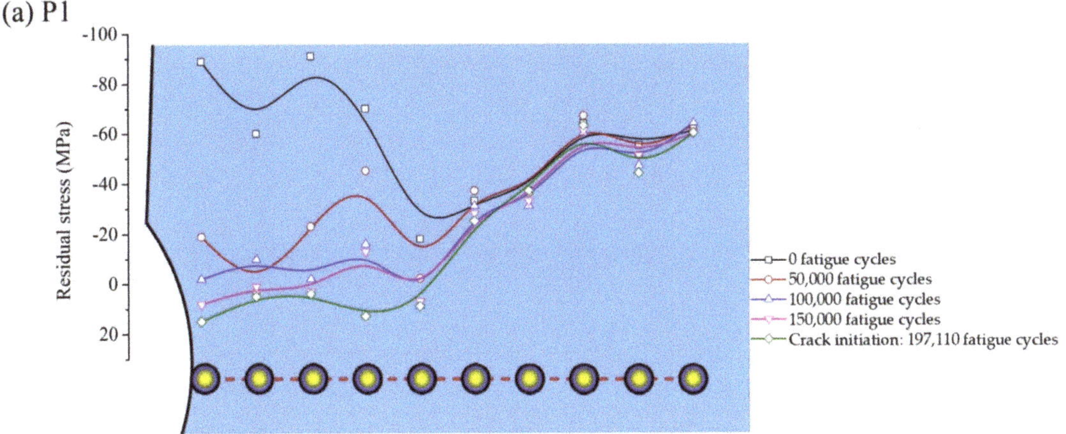

Figure 17. *Cont.*

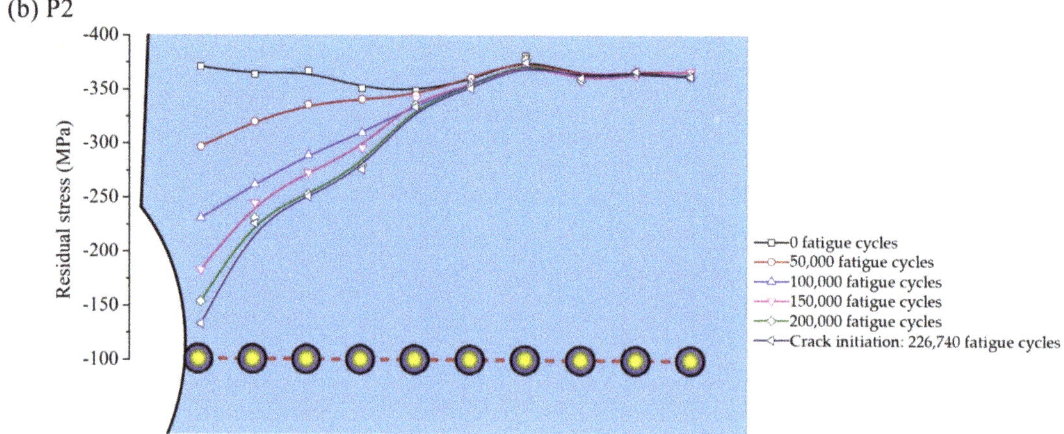

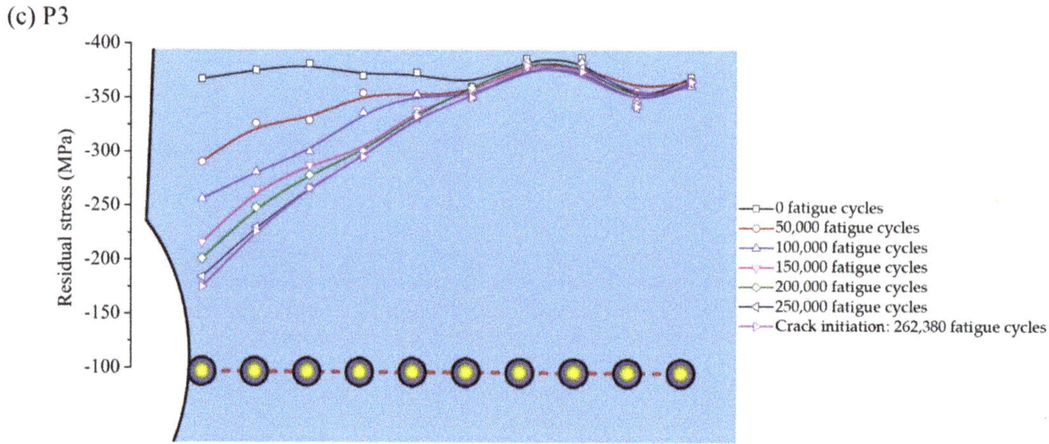

Figure 17. *Cont.*

(d) P4

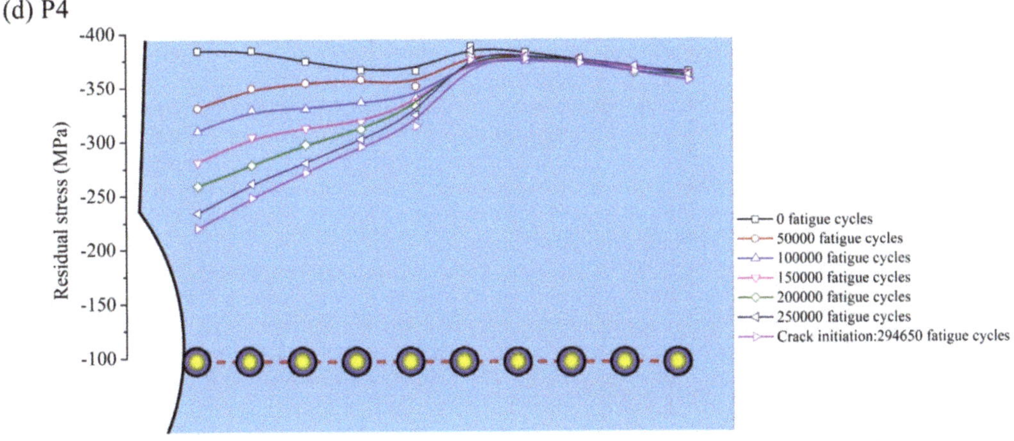

(e) P5

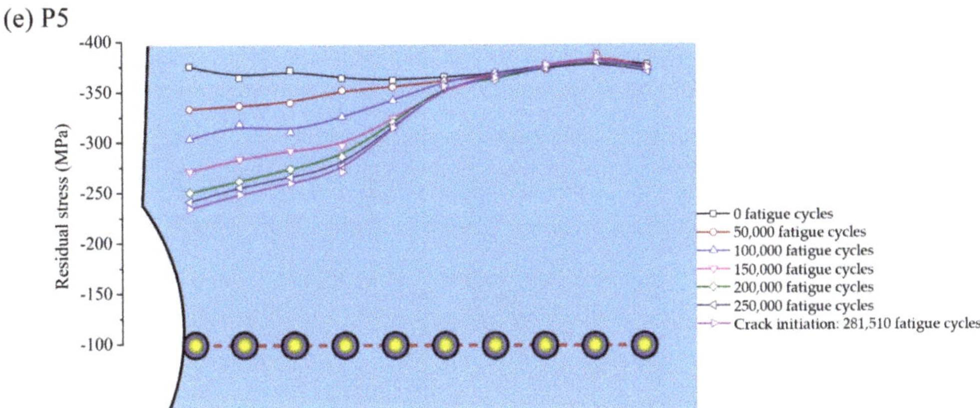

Figure 17. The variety of residual stresses with fatigue loading at each measuring point of (**a**) P1, (**b**) P2, (**c**) P3, (**d**) P4 and (**e**) P5.

Although all the specimens were observed to relax the residual stress field, their relaxation rates were not the same under the same fatigue load, which can be reflected by the curve distance in Figure 17. As seen in Figure 17a, the compressive residual stresses of the unpeened specimen P1 reduced by 78.7% (−89 MPa to −19 MPa) after the 1st 50,000 fatigue cycles at Point1 and turns to tensile residual stresses with a value of 15 MPa after crack initiation. As for Figure 17b, the compressive residual stresses of P2 reduced by 19.9% (−371 MPa to −297 MPa) after the first 50,000 fatigue cycles and by 64% (−371 MPa to −133 MPa) at Point1 after the crack initiation. As for Figure 17c, the compressive residual stresses of P3 reduced by 20.9% (−367 MPa to −290 MPa) after the first 50,000 fatigue cycles and by 52.3% (−367 MPa to −175 MPa) at Point1 after the crack initiation. As for Figure 17d, the compressive residual stresses of P4 reduced by 13.8% (−384 MPa to −331 MPa) after the first 50,000 fatigue cycles and by 42.7% (−384 MPa to −220 MPa) at Point1 after the crack initiation. In Figure 17e, the compressive residual stresses of P5 reduced by 11.2% (−375 MPa to −333 MPa) after the first 50,000 fatigue cycles and by 37.6% (−375 MPa to −234 MPa) at Point1 after the crack initiation. After 50,000 fatigue loadings, the residual stress relaxation rate at each point also decreased, but the trend was still similar to before.

Combined with Figures 16 and 17, it can be found that the specimens with greater values of $\sigma^{RS}{}_{max}$ and δ_{max} had lower residual stress relaxation rates. Although their surface residual stresses were similar before fatigue loading, they show different "retention capabilities" of residual stress or stability of CRSF during fatigue loading. P5 had the greatest value of $\sigma^{RS}{}_{max}$ and δ_{max}, and the remaining residual stresses were the greatest when the crack was initiated and the remaining residual stresses of other specimens decrease with the decrease of MDS. In other words, the specimen with deeper CRSF had better retention capability of surface residual stress.

Figure 18 shows the local stress distribution of numerical analysis results at 70 kN loading and the Von Mises stress of each measuring point. The Von Mises stresses could be alternatively regarded as the amplitude stresses in fatigue loading here. Combined with the residual stress data obtained in X–ray diffraction measurement, the relaxation ratio of the residual stress at each measuring point of shot–peened specimens P2 to P5 could be compared. Comparing the residual stress after every 50,000 fatigue cycles with that before fatigue cycles, the results shown in Figure 19 are the calculated relaxation ratios of the residual stress per 50,000 fatigue cycles. It can be seen that for all the shot–peened specimens, the relaxation of the compressive residual stress was also greater where the stress is greater. As the stress decreased linearly, the relaxation of the compressive residual stress also decreased, but it was nonlinear. Although the specimens P2 to P5 were processed by SP with different MDS, they all showed insensitivity to residual stress relaxation when the stress was below 174 MPa (Point6 and subsequent measurement points). It can be noticed that there was a threshold value of the stress amplitude that allowed the residual stress to relax. When the stress amplitude was lower than the threshold value, regardless of the size of the MDS as well as the number of fatigue cycles, the compressive residual stress barely changed. Combining the experimental data and the FEA results, the threshold value of the stress amplitude should be between 171 MPa and 174 MPa.

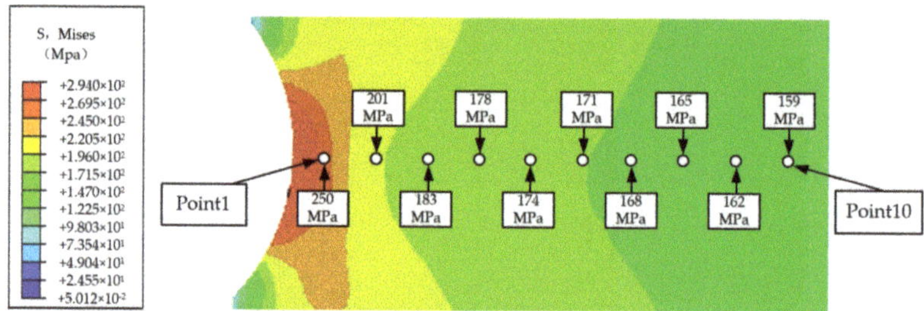

Figure 18. The finite element analysis (FEA) stress distribution near the arc notch and the stresses at measuring points.

The residual stress relaxation ratio of each specimen at each fatigue loading stage were also different. For example, P2 and P3 had large residual stress relaxation ratios in the first and second 50,000 fatigue cycles; the relaxation ratios of P2 after the first and second 50,000 fatigue cycles were about 20% and 22% at Poin1, 12% and 18% at Point2, and 9% and 14% at Point3, respectively; the relaxation ratios of P3 after the first and second 50,000 fatigue cycles were about 21% and 12% at Poin1, 13% and 14% at Point2, and 13% and 9% at Point3 respectively. These two specimens had a great stress relaxation in the early stage (100,000 fatigue cycles) of fatigue loading. By contrast, the relaxation ratios of P4 after the first and second 50,000 fatigue cycles were about 14% and 6% at Poin1, 9% and 6% at Point2, 5% and 7% at Point3, respectively; the relaxation ratios of P5 after the first and second 50,000 fatigue cycles are about 11% and 9% at Poin1, 8% and 8% at Point2, and 3% and 7% at Point3 respectively. Compared with P2 and P3, P4 and P5, which had deeper CRSFs, had better performances in resisting compressive residual stress relaxation. It also can be seen from Figure 19 that nearly all the residual stress relaxation ratios of P4 and P5 were below 10% at all measuring points in every fatigue loading stage.

For the actual design and construction of River–Sea–Going ships, the stress concentration at the corners of the hatch is difficult to avoid, but the stress amplitude should be controlled. It can be found that if the fatigue performances of the hatch corners are expected to be improved by SP treatments, the residual stress relaxation should be considered when local stress is beyond the threshold value. If the stress amplitude at the hatch corner is lower than the stress threshold value of residual stress relaxation, the residual stress field will barely change. It is also a feasible method to change the parameters of SP treatment to introduce a deeper residual stress field into the surface of the hatch corner structure to reduce the residual stress relaxation ratio and keep the CRSF more stable in fatigue loading.

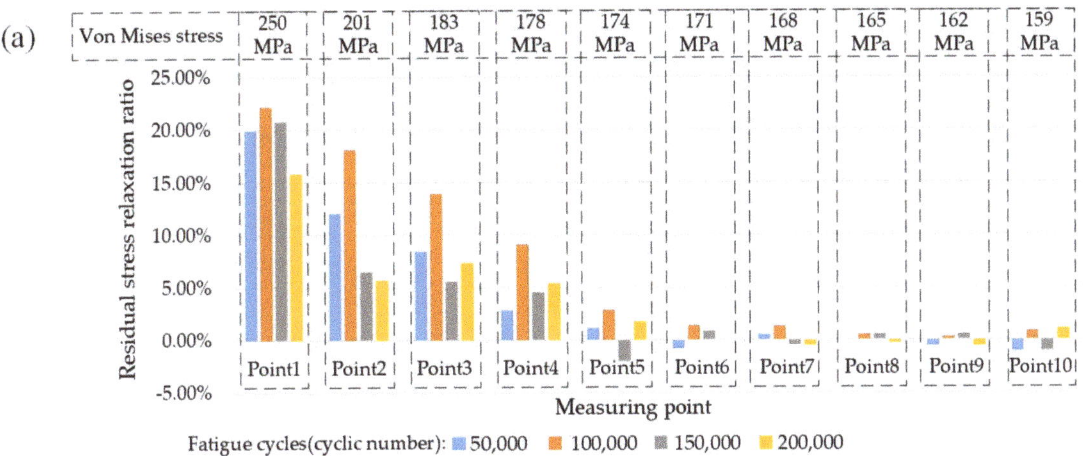

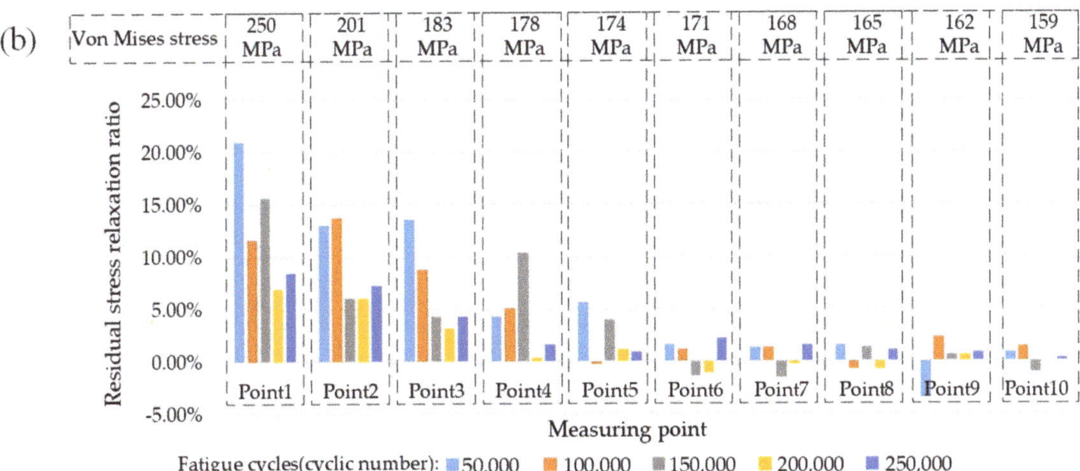

Figure 19. *Cont.*

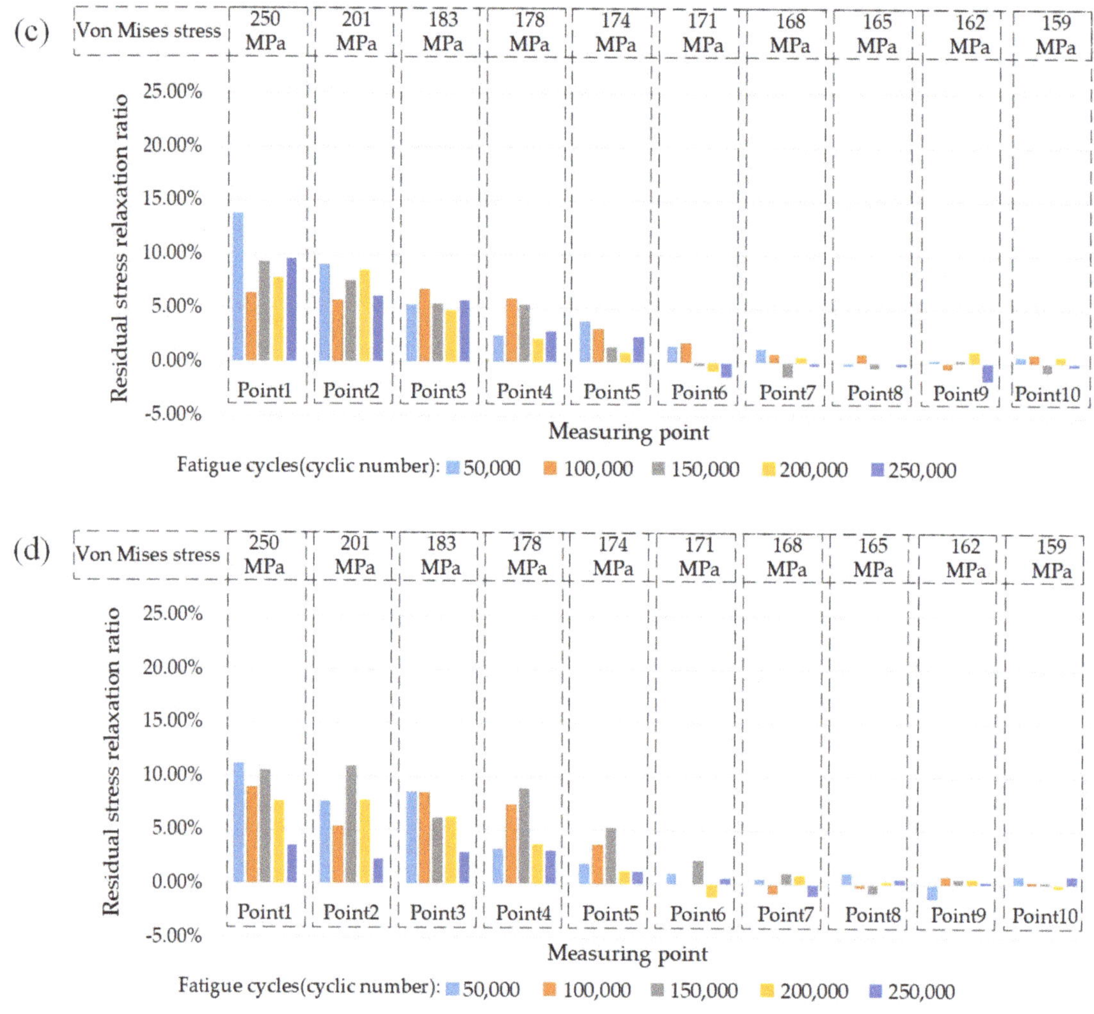

Figure 19. The residual stress relaxation ratio of (**a**) P2, (**b**) P3, (**c**) P4, and (**d**) P5.

3.3. Fatigue Life

Table 4 shows the fatigue life and other characteristic parameters of all specimens, including not only the specimens for residual stress relaxation research but also the other two specimens in the same SP treatment group that were conducted with fatigue tests. In Table 4, N_{in} represents the crack initiation life (crack from 0.0 mm to 0.1 mm), ΔN_{in} represents the ratio of crack initiation life increment, N_{pr} represents the crack propagation life (0.1 mm to 1 mm), ΔN_{pr} represents the ratio of crack propagation life increment, N_f represents fatigue life (crack from 0.0 mm to 1 mm), and ΔN_f represents the ratio of fatigue life increment, where all the increments are based on the comparison with P1. It can be seen that the fatigue life of P1 without SP was about 214,550 in the first fatigue test. The increment comparisons of subsequent data were based on the first fatigue test data of the P1 specimen. The maximum fatigue life appeared at the P4 specimen in the third fatigue test with a value of 347,710 when MDS was 0.8 mm. It can be concluded through Table 4 that all the SP treatments can improve fatigue lives of ship hatch corner specimens greatly.

Table 4. Fatigue lives of ship hatch corner specimens.

Specimen	Number of Test	MDS d (mm)	R_a (μm)	N_{in} (Cyclic Number)	ΔN_{in}	N_{pr} (Cyclic Number)	ΔN_{pr}	N_f (Cyclic Number)	ΔN_f
P1	1st	/	0.07	197,110	/	17,440	/	214,550	/
	2nd	/	/	173,530	−12.0%	17,210	−1.3%	190,740	−11.1%
	3rd	/	/	186,220	−5.5%	21,460	23.1%	207,680	−3.2%
P2	1st	0.3	1.34	226,740	15.0%	20,420	17.1%	247,160	15.2%
	2nd	0.3	/	237,270	20.4%	19,850	13.8%	257,120	19.8%
	3rd	0.3	/	206,160	4.6%	21,584	23.8%	227,744	6.1%
P3	1st	0.6	2.36	262,380	33.1%	25,160	44.3%	287,540	34.0%
	2nd	0.6	/	248,220	25.9%	21,240	21.8%	269,460	25.6%
	3rd	0.6	/	278,600	41.3%	24,770	42.0%	303,370	41.4%
P4	1st	0.8	3.21	294,650	49.5%	27,610	58.3%	322,260	50.2%
	2nd	0.8	/	315,510	60.1%	28,120	61.2%	343,630	60.2%
	3rd	0.8	/	319,380	62.0%	28,330	62.4%	347,710	62.1%
P5	1st	1.0	3.65	281,510	42.8%	27,120	55.5%	308,630	43.8%
	2nd	1.0	/	291,940	48.1%	26,430	51.5%	318,370	48.4%
	3rd	1.0	/	303,770	54.1%	28,020	60.7%	331,790	54.6%

Specifically, the average of ΔN_{in}s within each group were 13.3% (P2), 33.5% (P3), 57.2% (P4), and 48.3% (P5); the averages of ΔN_{pr}s within each group were 18.2% (P2), 36.0% (P3), 60.7% (P4), and 55.9% (P5); the average of ΔN_fs within each group were 13.7% (P2), 33.7% (P3), 57.5% (P4), and 48.9% (P5). Although the fatigue life extension of each group showed differences for a single specimen, it can be found through the comparison between the groups that when the PI increased, the differences in the ΔN_fs within each group were gradually reduced. It can be analyzed by the comparison of ΔN_{in} of each group that in the crack initiation stage, shot peening had a very good effect on the suppression of the crack initiation. SP also suppressed the crack growth rate in the crack propagation stage, which can be analyzed by the comparison of ΔN_{pr} of each group. ΔN_{in}, ΔN_{pr}, and ΔN_f all increased initially with the increase of MDS (when d= 0.3 mm, 0.6 mm and 0.8 mm), but when d = 1.0 mm, the increments of ΔN_{in}, ΔN_{pr}, and ΔN_f were not obvious compared with those when d = 0.8 mm, and some of the data even decreased. This means that prolonging the fatigue life of the hatch corner by only increasing the MDS has a limited effect. This can be caused by the excessive surface roughness caused by the increase in MDS. P4 and P5 both had better residual stress stability, and thus they had larger ΔN_{in} values than P2 and P3. With the increase of MDS, the crack initiation life and fatigue life both showed an increasing trend. It can be seen from this that the fatigue lives of P2, P3, and P4 all increased with the increase of σ^{RS}_{max} and δ_{max} values. At the same time, the effective residual stress at the arc notch at the crack initiation time also increased with the increase of σ^{RS}_{max} and δ_{max} values. On the other hand, P5 had larger σ^{RS}_{max} and δ_{max} values than P4 and its remaining effective residual stress when the crack initiated was also greater, but its fatigue life was not improved; however, it was reduced. The reason for this may be that P5 has a larger R_a than P4. Although the remaining effective residual stress on the surface is slightly larger than that of P4, it is not enough to offset the effect of surface defects caused by the larger R_a value, so the fatigue life of P5 is slightly smaller than that of P4. For P5, which has larger σ^{RS}_{max} and δ_{max}, the influence of surface roughness was even more severe. When increasing the PI of SP treatments, a value of R_a would also increase. If R_a is beyond a certain value, the limited improvement of CRSF cannot offset the negative effect of increased surface roughness on surface integrity. This is manifested in the local defects caused by the increase of surface micro–damage, and the micro–stress concentration, which makes it easier for cracks to initiate and propagate at the defects.

It is generally believed that the surface compressive residual stress has an important influence on the fatigue performance of the structure. In fact, the surface residual stress changes during the fatigue load, especially in the initial stage of the fatigue loading, it has a large relaxation rate. Under the fatigue loading, the relaxation performance of the residual stress field of different specimens was different and closely related to the characteristic parameters of CRSF. The surface residual stress relaxation rate of the specimen with deeper CRSF was slower, and the surface compressive residual stress before the crack initiation was larger, the crack initiation suppression effect was better, and the crack growth rate could also be slowed down.

In the aim of improving fatigue lives of hatch corners in the River–Sea–Going ship by SP treatment, both residual stress relaxation and surface roughness should be considered. It can be concluded that when the parameters of SP treatment are correctly chosen, the fatigue life of the hatch corner can be improved greatly. In terms of optimizing shot peening parameters, the approach to reduce the increase in roughness induced by shot peening intensity should be investigated. On the other hand, the assessment of the residual stress stability of the hatch corners should be combined with the actual load of the River–Sea–Going ship.

4. Conclusions

Given the fact that there is almost no application of fatigue improvement by SP treatment in the ship engineering field, this paper illustrates the relationship between the stress relaxation and fatigue life of the actual ship hatch corner structure in the shot–peened region under a constant amplitude fatigue load, and shows the sensitivity of the residual stress relaxation at different positions during the cyclic fatigue loading. Moreover, it provides an approach to improve the fatigue strength and life of ship hatch corner structures by SP treatment and a procedure for evaluating the fatigue life of the actual ship hatch corner structure after SP. The effect of SP on the fatigue life of ship hatch corner specimens made of Q235B steel was investigated at a stress ratio of 0.1 and cyclic loading amplitude of 31.5 kN. SP intensity was changed by changing the mean diameter of shots of SP. The surface roughness, residual stress, and fatigue life tests were carried out on ship hatch corner specimens. The surface residual stresses were measured with the aim of analyzing the residual stress variation at different locations in each specimen during cyclic loading. The residual stress relaxation ratios of different specimens at different locations were compared and analyzed. The following conclusions have been reached:

1. Introducing compressive residual stress into the surface layer by SP is an effective method to prolong the fatigue life of the ship hatch corner. The larger the values of σ^{RS}_{max} and δ_{max}, the slower the residual stress field relaxation rates are under cyclic loading.

2. It was found that the compressive residual stress near surface layer is beneficial to both the initiation life and the propagation life of fatigue crack. Compared with the unpeened specimen, the increments of crack initiation life are about 13.3%, 33.5%, 57.2%, and 48.3% on average, and the increments of crack propagation lives, are about 18.2%, 36.0%, 60.7%, and 55.9% with MDS of 0.3 mm, 0.6 mm, 0.8 mm, and 1.0 mm, respectively.

3. The magnitude of residual stress relaxation does not decrease linearly with the linear decrease of stress under the same number of cyclic loading. When the stress is less than the threshold value for the relaxation of the residual stress, the residual stress on the surface of the structure will not relax. Specimens P4 and P5 with deeper CRSFs have better residual stress stabilities and better fatigue performances compared to P2 and P3.

4. In practical engineering applications, increasing SP intensity can increase the values of σ^{RS}_{max} and δ_{max} in the compressive residual stress field, as well as the residual stress stability. Moreover, it also increases the surface roughness, which has adverse effects on the fatigue life of the hatch corner specimen. Therefore, the effect of residual stress field and surface roughness should be considered comprehensively in the SP process.

Results of this research prove that shot peening is feasible in improving the fatigue performance of ship hatch corner structures. In order to achieve better fatigue property improvement, residual stress relaxation and surface roughness should be taken into consideration comprehensively. Residual stress relaxation is highly concerned with the depth of CRSF and the stress amplitude. Increasing the depth and value of CRSF via increasing peening intensity can reduce the residual stress relaxation at the same stress amplitude in practical application.

Author Contributions: Conceptualization, J.G., W.W., and Z.G.; Methodology, J.G., Z.G., and Z.W.; writing–original manuscript, Z.G. and Y.W.; writing–review and editing, Y.W. and Z.W. All authors have read and agreed to the published version of the manuscript.

Funding: This research was funded by the National Natural Science Foundation of China (No. 51879208) and the National Scientific Research Project of the High–tech Ship of China (No.2014[493]). The authors are grateful to all the staff of Green & Smart River–Sea–Going Ship, Cruise and Yacht Research Center for supporting this work.

Conflicts of Interest: The authors declare no conflict of interest.

Abbreviations

SP	Shot peening
MDS	Mean diameter of shots
d	Value of MDS
PI	Peening intensity
CRSF	Compressive residual stress field
FEM	Finite element method
FEA	Finite element analysis
U_x	Displacement in the x direction
U_y	Displacement in the y direction
U_z	Displacement in the z direction
R_x	Rotation in the x direction
R_y	Rotation in the y direction
$\sigma^{RS}{}_{max}$	Maximum residual stress in depth
$\sigma^{RS}{}_{surf}$	Surface residual stress
δ_{max}	Depth of $\sigma^{RS}{}_{max}$
R_a	Surface roughness
N	Numbers of points on the contour curve
Y_{xi}	Value of contour curve
m_{xi}	Value of baseline
Z_{xi}	The absolute value of the distance between Y_{xi} and m_{xi}
E	Young's modulus
R_m	Ultimate tensile stress
σ_s	Yield stress
v	Poisson's ratio
δ	Elongation
α	Diffraction angle
η	The complementary angle of α
a_1	Intermediate parameters
ε_α	Strain in the Debye ring
$\varepsilon_{\pi-\alpha}$	Complementary strain
$\varepsilon_{-\alpha}$	Contrary strain
ψ_0	Operating angle of the X–ray analyzer
N_{in}	Crack initiation life
ΔN_{in}	The ratio of crack initiation life increment
N_{pr}	Crack propagation life
ΔN_{pr}	The ratio of crack propagation life increment
N_f	Fatigue life
ΔN_f	The ratio of fatigue life increment

References

1. IACS. *Common Structural Rules for Bulk Carriers*; IACS: London, UK, 2006.
2. CCS. *Guidelines for Fatigue Strength Assessment of Hull Structures*; CCS: Edmond, OK, USA, 2015.
3. Guoqing, F.; Hao, S.; Dongping, L.; Hui, L. The stress combination method for the fatigue assessment of the hatch corner of a bulk carrier based on equivalent waves. *J. Mar. Sci. Appl.* **2012**, *11*, 68–73.
4. Selle, H.; Doerk, O.; Scharrer, M. Global strength analysis of ships with special focus on fatigue of hatch corners. In *MARSTRUCT Book Series*; Taylor Francis Group: Abingdon, UK, 2009; pp. 255–260.
5. Yong, H.X.; Wei, W.; Heng, Z. Fatigue strength study on different structure type of hatch corner. *Adv. Mat. Res.* **2011**, *233*, 2580–2583.
6. Xu, X.; Augello, R.; Yang, H. The generation and validation of a CUF–based FEA model with laser–based experiments. *Mech. Adv. Mater. Struct.* **2019**, 1–8. [CrossRef]
7. Xu, X.; Yang, H.; Augello, R.; Carrera, E. Optimized free–form surface modeling of point clouds from laser–based measurement. *Mech. Adv. Mater. Struct.* **2019**, 1–9. [CrossRef]
8. Carrera, E.; Pagani, A.; Augello, R. Evaluation of geometrically nonlinear effects due to large cross–sectional deformations of compact and shell–like structures. *Mech. Adv. Mater. Struc.* **2020**, *27*, 1269–1277. [CrossRef]
9. Jiancheng, L.; Yongning, G. A Study on Stress Concentration at Hatch Corner for Ship with Large Openings. *Ship Eng.* **2000**, *6*, 9–12.
10. Petronic, S.; Colic, K.; Dordevic, B.; Milovanovic, D.; Burzic, M.; Vucetic, F. Effect of laser shock peening with and without protective coating on the microstructure and mechanical properties of Ti–alloy. *Opt. Laser. Eng.* **2020**, *129*, 106052. [CrossRef]
11. Hackel, L.; Dane, C. Reliable laser technology for laser peening applications. In Proceedings of the 2012 Conference on Lasers and Electro–Optics (CLEO), San Jose, CA, USA, 6–11 May 2012; pp. 1–2.
12. Leap, M.J.; Rankin, J.; Harrison, J.; Hackel, L.; Nemeth, J.; Candela, J. Effects of laser peening on fatigue life in an arrestment hook shank application for Naval aircraft. *Int. J. Fatigue* **2011**, *33*, 788–799. [CrossRef]
13. Rajan, S.S.; Swaroop, S.; Manivasagam, G.; Rao, M.N. Fatigue life enhancement of titanium alloy by the development of nano/micron surface layer using laser peening. *J. Nanosci. Nanotechnol.* **2019**, *19*, 7064–7073. [CrossRef] [PubMed]
14. Kasra, G.; Rakesh, R.; Scott, W.; Ayhan, I. Fatigue strength improvement of aluminum and high strength steel welded structures using high frequency mechanical impact treatment. *Proc. Eng.* **2015**, *133*, 465–476.
15. Harati, E.; Svensson, L.; Karlsson, L.; Widmark, M. Effect of high frequency mechanical impact treatment on fatigue strength of welded 1300MPa yield strength steel. *Int. J. Fatigue* **2016**, *92*, 96–106. [CrossRef]
16. Jan, F.; Volker, H.; Majid, F. High frequency mechanical impact treatment (HFMI) for the fatigue improvement: Numerical and experimental investigations to describe the condition in the surface layer. *Weld. World* **2016**, *60*, 749–755.
17. Zhenwen, Y.; Qi, L.; Jiahui, W.; Zongqing, M.; Ying, W.; Dongpo, W. Effect of ultrasonic impact treatment on the microstructure and mechanical properties of diffusion–bonded TC11 alloy joints. *Arch. Civ. Mech. Eng.* **2019**, *19*, 1341–1441.
18. Kahraman, F. Surface layer properties of ultrasonic impact–treated AA7075 aluminum alloy. *Proc. Inst. Mech. Eng. Part B J. Eng.* **2018**, *232*, 2218–2225. [CrossRef]
19. Bolin, H.; Haipeng, D.; Mingming, J.; Kang, W.; Li, L. Effect of ultrasonic impact treatment on the ultra high cycle fatigue properties of SMA490BW steel welded joints. *Int. J. Adv. Manuf. Technol.* **2018**, *96*, 1571–1577.
20. Chuan, L.; Yi, Y.; Xiaohua, C.; ChunJing, W.; Yong, Z. Residual stress in a restrained specimen processed by post–weld ultrasonic impact treatment. *Sci. Technol. Weld. JOI* **2019**, *24*, 1–7.
21. Mohammad, A.O.; Hamzah, M.M.; Faris, M.A.; Mohammad, A. Enhancing the surface hardness and roughness of engine blades using the shot peening process. *Int. J. Min. Met. Mater.* **2019**, *26*, 999–1004.
22. Lukáš, F.; Werner, D.; Werner, E.; Thomas, K.; Michael, T.; Christoph, C. Effect of shot peening on residual stresses and crack closure in CVD coated hard metal cutting inserts. *Int. J. Refract. Met. H* **2019**, *82*, 174–182.
23. Bag, A.; Delbergue, D.; Ajaja, J.; Bocher, P.; Levesque, M.; Brochu, M. Effect of different shot peening conditions on the fatigue life of 300 M steel submitted to high stress amplitudes. *Int. J. Fatigue* **2020**, *130*, 105271–105274. [CrossRef]
24. Martín, V.; Vázquez, J.; Navarro, C.; Domínguez, J. Effect of shot peening residual stresses and surface roughness on fretting fatigue strength of Al 7075–T651. *Tribol. Int.* **2020**, *142*, 106004. [CrossRef]
25. Gan, J.; Sun, D.; Wang, Z.; Luo, P.; Wu, W. The effect of shot peening on fatigue life of Q345D T–welded joint. *J. Constr. Steel. Res.* **2016**, *126*, 74–82. [CrossRef]
26. Wohlfahrt, H. The influence of peening conditions on the resulting distribution of residual stress. In Proceedings of the Second International Conference on Shot Peening, Chicago, IL, USA, 14–17 May 1984; pp. 316–331.
27. Maiya, P.S. Geometrical characterization of surface roughness and its application to fatigue crack initiation. *Mater. Sci. Eng.* **1975**, *21*, 57–62. [CrossRef]
28. Novovic, D.; Dewes, R.C.; Aspinwall, D.K.; Voice, W. The effect of machined topography and integrity on fatigue life. *Int. J. Mach. Tool. Manuf.* **2004**, *44*, 125–134. [CrossRef]
29. Ruihong, L.; Daoxin, L.; Wei, Z.; Xuan, L.; Mingjie, Q.; Mingli, X. Influence of Shot Peening and Surface Integrity on the Fatigue Properties of 300M Steel. *Mech. Sci. Tech. Aerosp. Eng.* **2011**, *9*, 1418–1423.
30. Torres, M.A.S.; Voorwald, H.J.C. An evaluation of shot peening, residual stress and stress relaxation on the fatigue life of AISI 4340 steel. *Int. J. Fatigue* **2002**, *24*, 877–886. [CrossRef]

31. Dalaei, K.; Karlsson, B.; Svensson, L.E. Stability of residual stresses created by shot peening of pearlitic steel and their influence on fatigue behaviour. *Proc. Eng.* **2010**, *2*, 613–622. [CrossRef]
32. Dalaei, K.; Karlsson, B.; Svensson, L.E. Stability of shot peening induced residual stresses and their influence on fatigue lifetime. *Mater. Sci. Eng. A* **2011**, *528*, 1008–1015. [CrossRef]
33. Dalaei, K.; Karlsson, B. Influence of shot peening on fatigue durability of normalized steel subjected to variable amplitude loading. *Int. J. Fatigue* **2012**, *38*, 75–83. [CrossRef]
34. Huang, J.; Wang, Z.; Gan, J.; Yang, Y.; Wu, G.; Meng, Q. Investigation of fatigue performance improvement in SiCw/Al composites with different modified shot peening treatments by considering surface mechanical properties. *J. Alloys Compd.* **2017**, *728*, 169–178. [CrossRef]
35. Ruiz, H.; Osawa, N.; Rashed, S. Study on the stability of compressive residual stress induced by high–frequency mechanical impact under cyclic loadings with spike loads. *Weld. World* **2020**, *64*, 1855–1865. [CrossRef]
36. Gb/T 228.1. In *Metallic Materials—Tensile Testing, Part I: Method of Test at Room Temperature*; Standards Press of China: Beijing, China, 2010. (In Chinese)
37. ISO 4287. *Geometrical Product Specifications (GPS), Surface Texture: Profile Method—Terms, Definitions and Surface Texture Parameters*; The Spanish Association for Standardization and Certification: Madrid, Spain, 2010.
38. Lee, S.; Ling, J.; Wang, S.; Ramirez–Rico, J. Precision and accuracy of stress measurement with a portable X–ray machine using an area detector. *J. Appl. Crystallogr.* **2017**, *50*, 131–144. [CrossRef]

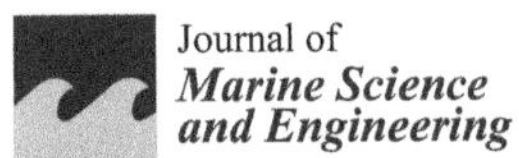

Journal of
Marine Science and Engineering

Article

Methods for Fitting the Limit State Function of the Residual Strength of Damaged Ships

Zhiyao Zhu [1,2], Huilong Ren [1,2], Xiuhuan Wang [3], Nan Zhao [4] and Chenfeng Li [1,2,*]

[1] College of Shipbuilding Engineering, Harbin Engineering University, Harbin 150001, China; zhuzhiyao@hrbeu.edu.cn (Z.Z.); renhuilong@263.net (H.R.)
[2] International Joint Laboratory of Naval Architecture and Offshore Technology between Harbin Engineering University and Lisbon University, Harbin 150001, China
[3] College of Liberal Education, Chongqing Vocational and Technical University of Mechatronics, Chongqing 402760, China; haibei0115@163.com
[4] China Ship Scientific Research Center, Wuxi 214082, China; zhaonan702@126.com
* Correspondence: lichenfeng@hrbeu.edu.cn; Tel.: +86-451-8251-9902

Abstract: The limit state function is important for the assessment of the longitudinal strength of damaged ships under combined bending moments in severe waves. As the limit state function cannot be obtained directly, the common approach is to calculate the results for the residual strength and approximate the limit state function by fitting, for which various methods have been proposed. In this study, four commonly used fitting methods are investigated: namely, the least-squares method, the moving least-squares method, the radial basis function neural network method, and the weighted piecewise fitting method. These fitting methods are adopted to fit the limit state functions of four typically sample distribution models as well as a damaged tanker and damaged bulk carrier. The residual strength of a damaged ship is obtained by an improved Smith method that accounts for the rotation of the neutral axis. Analysis of the results shows the accuracy of the linear least-squares method and nonlinear least-squares method, which are most commonly used by researchers, is relatively poor, while the weighted piecewise fitting method is the better choice for all investigated combined-bending conditions.

Keywords: limit state function; longitudinal strength; least-squares method; moving least-squares method; radial basis function neural network method; weighted piecewise fitting method

Citation: Zhu, Z.; Ren, H.; Wang, X.; Zhao, N.; Li, C. Methods for Fitting the Limit State Function of the Residual Strength of Damaged Ships. *J. Mar. Sci. Eng.* **2022**, *10*, 102. https://doi.org/10.3390/jmse10010102

Academic Editors: Joško Parunov and Yordan Garbatov

Received: 13 December 2021
Accepted: 10 January 2022
Published: 13 January 2022

Publisher's Note: MDPI stays neutral with regard to jurisdictional claims in published maps and institutional affiliations.

1. Introduction

Ship safety is a major concern to researchers, and the number of damaged ships in accidents has been decreasing with advances in technology. According to the statistics of the International Association of Dry Cargo Shipowners, the loss of bulk carriers over 10,000 DWT has decreased from more than 500 in 1994–2003 to 202 in 2008–2017, and the number of casualties has also decreased from more than 200 to 53. In order to further reduce the casualties and property losses, researchers have continued their effort on the improvement of ship safety. When the ship is subjected to collision and grounding, which are the major types of accidents [1], the longitudinal strength will decrease and the wave load will change as well. Owing to the damage-induced change in floating state of the damaged ship, the wave load behavior is more complicated than the intact ship. Chen et al. [2] investigated the wave load of a damaged RO-RO ship and found the horizontal load is as large as 1.73 times the vertical load in the oblique wave. The ultimate-strength assessment method is well established for the vertical load on intact ships [3] but not for damaged ships. Therefore, it is necessary to devise a method for the damaged ship under combined bending moments. For this purpose, the limit state function is key to the method.

The longitudinal-strength assessment method for intact ships can be calculated directly by the Smith method and the Finite Element Method (FEM). There are more studies about the ultimate strength based on these methods [4–8]. However, it is difficult to directly obtain the longitudinal strength of the ship under combined bending moments, and one has to rely on the limit state function for the assessment of ship safety. The accuracy of the limit state function depends on the ultimate-strength calculation method and the fitting method. To obtain accurate samples for the fitting of the limit state function, a variety of methods have been proposed. Yao et al. [9] applied a simplified progressive collapse method to study the longitudinal strength of the bulk carrier under bi-axis bending moments. Parunov et al. [10] investigated the longitudinal strength of a damaged tanker under combined bending moments with the FEM, and the relationship between the damage size and the longitudinal strength was discussed. Paik et al. [11,12] compared the longitudinal strength of unstiffened and stiffened plates under combined loads. Dow et al. [13] explored the longitudinal strength of a stiffened box girder under combined bending moments with the Smith method and the FEM. Paik et al. [14] studied corroded stiffened plates under combined compression loads with the FEM. Fujikubo et al. [15] investigated the longitudinal strength of the stiffened plate under the combined shear and thrust force with the FEM. Dow et al. [16] studied the longitudinal strength of an alloy plate under combined shear and compression/tension with the FEM. Among the methods widely used in the community, the FEM can account for the initial implication, the material nonlinearity and the geometric nonlinearity. The finite element models provide more details about the structures, and the relationship between adjacent parts is taken into account. However, the method costs much more time for modeling and computation. The Smith method only needs to discretize a cross section into stiffened plate elements, plate elements, and hard-corner elements. As the curvature increases, the stress and the bending moment of each element are calculated according to the strain–stress relationship curves of different element types, and the sum of the element bending moment is the bending moment of the cross section. This method is much more efficient, but the accuracy is poor when applied to ships with asymmetric cross sections. To improve the accuracy of the method in nonsymmetric applications, improved Smith methods were devised by Fujikubo et al. [17] and Joonmo et al. [18]. Fujikubo et al. [17] applied an improved Smith method to damaged ships, and proposed an equation to describe the relationship between the increments of vertical and horizontal bending moments and the increments of the curvatures. Joonmo et al. [19] proposed the force vector equilibrium criterion to track the rotation of the neutral axis so as to obtain the bending moments of the asymmetric cross section.

Once samples of longitudinal strength are obtained, the limit state function can be approximated by means of fitting. Gordo et al. [19,20] applied a nonlinear fitting method to obtain the limit state function of the longitudinal strength of a tanker under combined bending moments. Monsour et al. [21] applied a nonlinear fitting method to obtain the limit state functions of different ships. Luis et al. [22] adopted a nonlinear fitting method to fit the limit state functions of two damaged tankers. Khan et al. [23] also used a nonlinear fitting method to fit the limit state functions and explored the reliabilities of a damaged tanker and a bulk carrier. Shahid [24] used the response surface method and the artificial neural network to fit the limit state function. Zhu et al. [25] proposed the weighted piecewise fitting method for the limit state function. Paik et al. [26] investigated the longitudinal strength of the as-built ultra-large containership under combined vertical bending and torsion, then fitted the limit state function with a nonlinear method and obtained the design load area. Kim et al. [27] studied the longitudinal strength of the hull girder under combined bending and torsion and obtained a $\frac{1}{4}$ circular form of the limit state function by fitting. Shi et al. [28] compared the longitudinal strength of open box girders with cracked damage under pure vertical bending load and combined loads, and a circular form of the limit state function was also obtained by fitting. Hu et al. [29] studied the longitudinal strength of a large opening box girder with a crack under torsion and bending loads and presented the interaction between the two loads in a circular form. The same form of the limit state function was also used by Li et al. [30], who investigated the pipe under combined bending and torsion moment. Though the fitting methods are used in many research studies, the accuracy and the applicability of the fitting methods remain unclear and need to be studied.

As the ship structure and damage condition can be different from one another, the limit state curves may also be different from one another, but all of them are closed curves in the coordinate system. In the existing studies, four typical closed curves are adopted to approximate the curve: (1) circle, (2) transverse ellipse, (3) vertical ellipse, and (4) oblique ellipse. Four fitting methods are proposed and investigated in this study: namely, (1) the least-squares method, (2) the moving least-squares method (MLS), (3) the radial basis artificial neural network method (RBFNN), and (4) the weighted piecewise fitting method (WP). Finally, samples of longitudinal strength for a damaged tanker and damaged bulk carrier calculated by Fujikubo et al. [12] were obtained for the fitting method study.

2. Fitting Methods for the Limit State Function

2.1. The Least-Squares Method

The least-squares method adopts the linear or nonlinear regression to establish a polynomial function. The general form of the linear polynomial function is $a + b = c$:

$$\overline{Y} = C_0 + \sum_{i=1}^{n} C_i X_i + \sum_{i=1}^{n} \sum_{j=1}^{n} C_{ij} X_i X_j + \varepsilon, \tag{1}$$

where $\overline{Y}$ is the regression result of the fitting function for the n random variables X_i, C is the regression coefficient, and ε is the error between the regression result and the actual response.

Quadratic functions and cubic functions are commonly used linear least-squares fitting functions. They take the following forms, respectively:

$$y = a + bx + cx^2, \tag{2}$$

$$y = a + bx + cx^2 + dx^3 \tag{3}$$

where $x = M_H / M_{UH}$ and $y = M_V / M_{UV}$; where M_V and M_H are the vertical and horizontal bending moment, respectively, and M_{UV} and M_{UH} are the ultimate bending moments resulted from pure vertical and horizontal bending, respectively. In the case of multiple-valued functions, the following functions are used instead:

$$x = a + by + cy^2, \tag{4}$$

$$x = a + by + cy^2 + dy^3, \tag{5}$$

Nonlinear functions are also widely used in least-squares fitting, where an often-used form is

$$x^{\alpha_1} + y^{\alpha_2} = 1, \tag{6}$$

2.2. The Moving Least-Squares Method

Different from the classic polynomial function, the coefficient vectors of the moving least-squares-based fitting function and the basis functions are determined by the fitting results. The function usually takes the form

$$y = \sum_{i=1}^{m} \alpha_i(x) p_i(x) = \mathbf{p}^{\mathrm{T}}(x) \boldsymbol{\alpha}(x), \tag{7}$$

where $\mathbf{p}^{\mathrm{T}}(x)$ is the basis function, $\boldsymbol{\alpha}(x)$ is the coefficient vector, m is the number of terms, and

$$\mathbf{p}^{\mathrm{T}}(x) = [p_1(x), p_2(x), \cdots, p_m(x)], \tag{8}$$

$$\boldsymbol{\alpha}^{\mathrm{T}}(x) = [\alpha_1(x), \alpha_2(x), \cdots, \alpha_m(x)], \tag{9}$$

To extend the fitting range, the samples are described in polar coordinates, and the fitting function is

$$r = \mathbf{p}^{\mathrm{T}}(\theta)\boldsymbol{\alpha}(\theta), \tag{10}$$

where r is the distance between the sample and the origin and θ is the angle between the point and the positive direction of the x-axis at the origin.

2.3. The Radial Basis Function Neural Network Method

The radial basis function is a function that relies only on the distance between the point x and the origin (or the calculation point c), which takes the form

$$\Phi(x) = \Phi(\|x\|), \tag{11}$$

or

$$\Phi(x,c) = \Phi(\|x - c\|), \tag{12}$$

where $\|x - c\|$ is the Euclidean distance between point x and point c.

The radial basis function neural network (RBFNN) is a neural network comprised of three layers: the input layer, the hidden layer, and the output layer, as shown in Figure 1. The RBF is the basis of the hidden layer. The transformation between the input layer and the hidden layer is nonlinear, while the transformation between the hidden and output layers is linear.

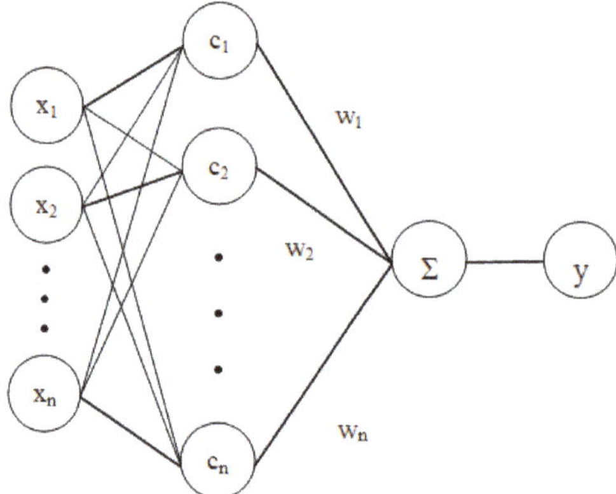

Figure 1. Typical form of the RBFNN.

The activation function is

$$\varphi(x_k - c_i) = \exp\left(-\frac{1}{2\sigma^2}\|x_k - c_i\|^2\right) \quad (k = 1,2,\ldots,l; i = 1,2,\ldots,m), \tag{13}$$

where x_k is the k-th input data, c_i is the center of the i-th neuron, and σ is the standard deviation

$$\sigma = \frac{c_{\max}}{\sqrt{2h}}, \tag{14}$$

where h is the number of the centers that determines the K-means clustering and $c_{\max}$ is the maximum distance among the chosen centers.

The output of the neural network is

$$y_i = \sum_{i=1}^{h} w_{ij}\exp\left(-\frac{1}{2\sigma^2}\|x_k - c_i\|^2\right) \quad (j = 1,2,\ldots,n), \tag{15}$$

where w_{ij} is the weight of the i-th neuron for the output

$$w = \exp\left(\frac{h}{c_{max}^2}\|x_k - c_i\|^2\right)(j = 1, 2, \ldots, n), \tag{16}$$

Once the RBFNN-based calculation is carried out, the input vector $X_{p\times1}$ can be transformed into the output vector $Y_{n\times1}$.

When the method is used to fit the samples in the polar coordinate system, the output of the neural network takes the form

$$r_i = \sum_{i=1}^{h} w_{ij}\exp(-\tfrac{1}{2\sigma^2}\|\theta_k - c_i\|^2) \quad (j = 1, 2, \ldots, n), \tag{17}$$

2.4. The Weighted Piecewise Fitting Method

The weighted piecewise fitting method provides a series of functions to describe the response relationship [20]. It adopts a piecewise regression method to fit the Function (3), and the weights matrix **w** is introduced to improve the accuracy. With the sample matrix **X** and **Y**, the coefficient matrix **N** is

$$\mathbf{N} = \left(\mathbf{X}^T\mathbf{w}\mathbf{X}\right)^{-1}\mathbf{X}^T\mathbf{w}\mathbf{Y}, \tag{18}$$

where

$$\mathbf{N} = \{a, b, c, d\}^T, \tag{19}$$

$$\mathbf{X} = \begin{bmatrix} 1\ x_1\ x_1{}^2\ x_1{}^3 \\ 1\ x_2\ x_2{}^2\ x_2{}^3 \\ \vdots\ \vdots\ \vdots\ \vdots \\ 1\ x_m\ x_m{}^2\ x_m{}^3 \end{bmatrix}, \tag{20}$$

$$\mathbf{Y} = \{y_1, y_2, \ldots, y_m\}^T, \tag{21}$$

where a, b, c, and d are the coefficients of the function. x_i and y_i is the i-th sample for the fitting.

Once the fitting functions for all samples are obtained, the values at both ends of each piece, y_{i1} and y_{i2}, and the slopes y'_{i1} and y'_{i2} can be found. As the two adjacent piece functions must be smooth at the joint, the following boundary conditions are applied:

$$y_{i2} = y_{(i+1)1} = y_i, \tag{22}$$

$$\begin{cases} y'_{i1} = (y'_{(i-1)2} + y'_{i1})/2 \\ y'_{i2} = (y'_{i2} + y'_{(i+1)1})/2 \end{cases}, \tag{23}$$

In order to obtain a closed curve, the following boundary conditions are used:

$$\begin{cases} y_{(n+1)1} = y_1 \\ y'_{(n+1)1} = y'_{11} \end{cases}, \tag{24}$$

The coefficients of the fitting function of each piece need be recalculated with y and y' at the curve ends as follows:

$$\begin{cases} y_{i1} = a_i + b_i x_{i1} + c_i x_{i1}^2 + d_i x_{i1}^3 \\ y'_{i1} = b_i + 2c_i x_{i1} + 3d_i x_{i1}^2 \\ y_{i2} = a_i + b_i x_{i2} + c_i x_{i2}^2 + d_i x_{i2}^3 \\ y'_{i2} = b_i + 2c_i x_{i2} + 3d_i x_{i2}^2 \end{cases}, \tag{25}$$

3. Calculation and Analysis of Typical Fitting Sample Distributions

In order to compare the difference between the fitting methods, six fitting methods are adopted to fit the function using typical fitting sample distributions: namely, the least-squares method with the quadratic function (LS-Q), the cubic function (LS-C), and the nonlinear function (LS-N); the moving least-squares method; the radial basis function neural network method; and the weighted piecewise fitting method.

3.1. Typical Fitting Sample Distribution

The distribution pattern of the sample of the longitudinal strength of the ship structures under combined bending moments is usually a closed curve, and the shape depends on the type and the damage of the structures. To present the representative conclusion, four typical distributions (TD1–TD4) are obtained for research after comparing the fitting sample distributions in the literature [15–24], as shown in Figure 2, where x and y represent the input data and output data of the fitting sample.

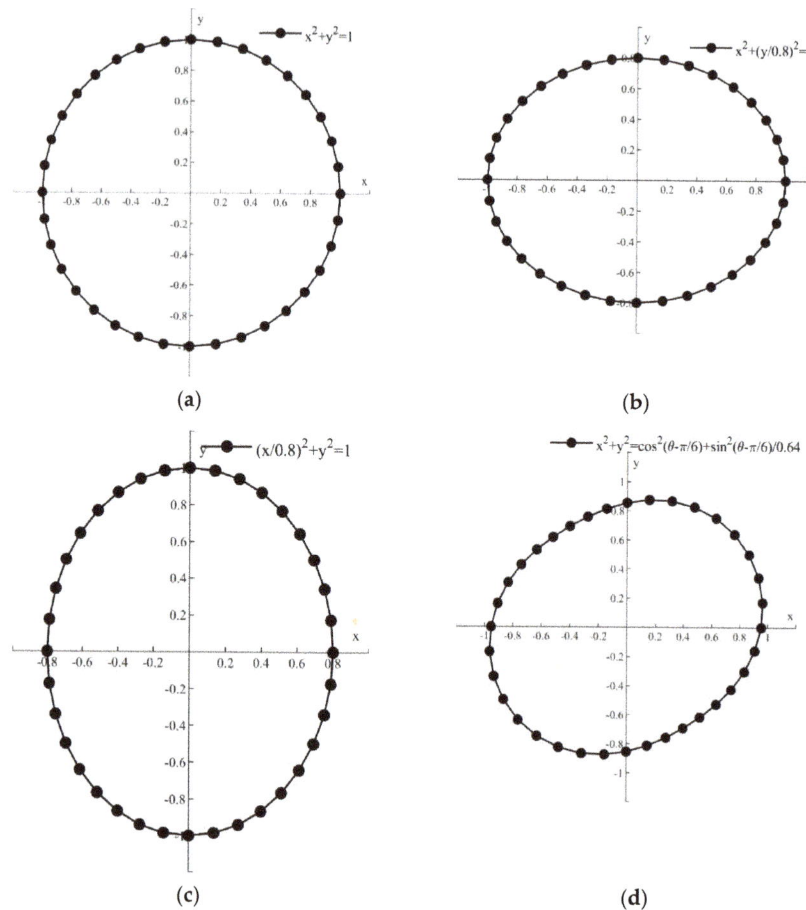

Figure 2. Fitting sample distributions for (**a**) TD1, (**b**) TD2, (**c**) TD3, and (**d**) TD4.

3.2. Fitting and Analysis of the Typical Distribution

The results obtained with the fitting methods for the typical distributions are shown in Figure 3.

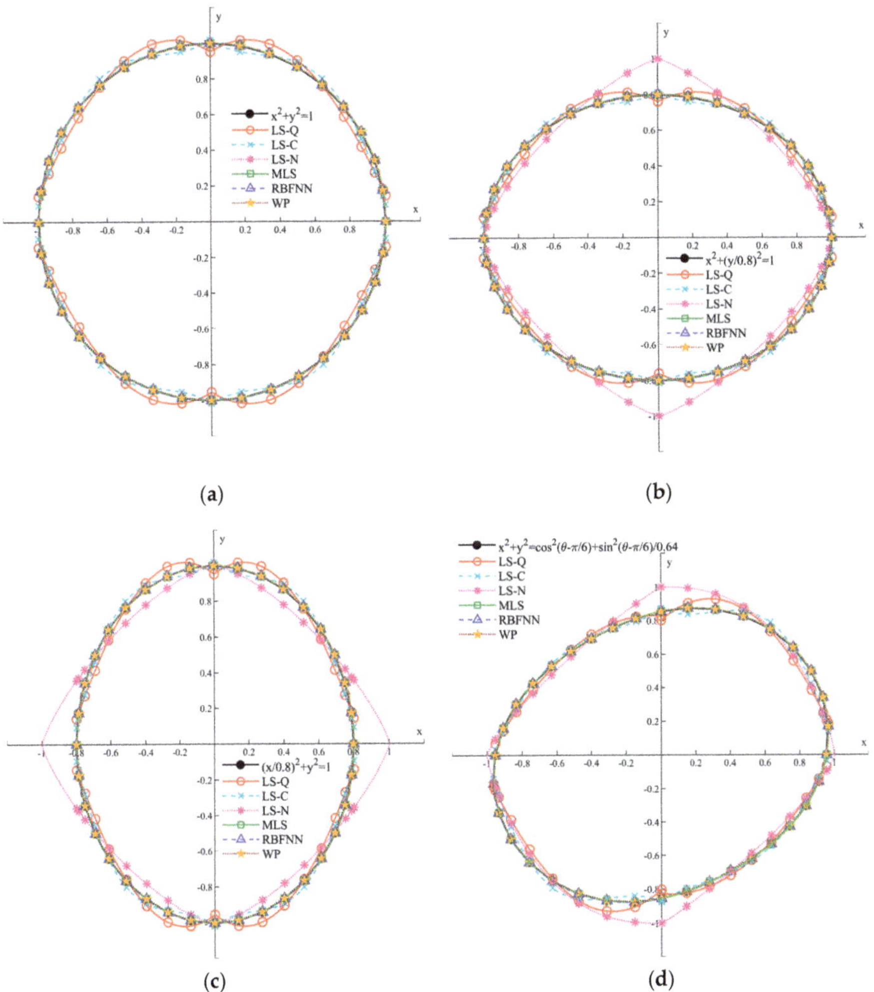

Figure 3. Fitting results for (**a**) TD1, (**b**) TD2, (**c**) TD3, and (**d**) TD4.

In order to compare the accuracy of the fitting methods, the y of the sample and the fitting results $y_{fitting}$ are shown in Figure 3, where the abscissa represents the y of the sample and the ordinate denotes the fitted results $y_{fitting}$. The vertical distance between the fitting results for y and the line $y_{fitting} = y$ is the error of the fitting, and when the point is located on the line $y_{fitting} = y$, it means that the fitting result is accurate. Two types of fitting are performed: (1) Case 1, fitting for the sample data and (2) Case 2, fitting for the removed-sample data.

The maximum error and the mean square errors (*MSE*) of the fitting results of the sample data and the sample-removed data are calculated. The errors are shown in Tables 1–4, and the *MSE* is

$$MSE = \frac{\sum\limits_{i=1}^{n} \left(y_{fitting} - y \right)^2}{n},$$

(26)

As shown in Figure 3, some fitting curves are not coincident with the sample curves, and the reason is that the shape and accuracy of the fitting curves depend on the sample distribution and the fitting method used. In Figure 4, it is found that the results calculated by the least-squares method have larger errors than others. The errors of LS-Q are larger than LS-C and LS-N when the sample distributions are TD1 and TD4. The errors of LS-N are larger than LS-Q and LS-C in TD2 and TD3. It is also observed in Tables 1–4 that the maximum errors and *MSE* of LS-Q, LS-C, and LS-N are larger than that of MLS, RBFNN, and WP, and the error comparison between LS-Q, LS-C, and LS-N is the same as shown in Figure 4. The comparison between LS-Q and LS-C shows that increasing of the function order barely improves the fitting accuracy. The comparison of results obtained with LS-N shows the errors are larger when the sample is not −1 or 1 on the coordinate axis. It is also shown in Tables 1–4 that the errors of LS-Q, LS-C, and LS-N in Case 1 are larger than in Case 2. The reason is that the least-squares method requires all data to obtain the minimum sum-of-squares errors. Therefore, the new sample for the fitting may change the fitting function and increase the fitting error. MLS, RBFNN, and WP need the sample near the fitting points, and the reduction of the sample may have a large influence on the fitting accuracy.

Table 1. Maximum error and *MSE* for TD1.

Project	Case	LS-Q	LS-C	LS-N	MLS	RBFNN	WP
maximum error	Case 1	0.1427	0.0930	$<1 \times 10^{-4}$	$<1 \times 10^{-4}$	$<1 \times 10^{-4}$	0
	Case 2	0.0996	0.0823	$<1 \times 10^{-4}$	$<1 \times 10^{-4}$	$<1 \times 10^{-4}$	0.0081
MSE	Case 1	0.0043	0.0019	$<1 \times 10^{-8}$	$<1 \times 10^{-8}$	$<1 \times 10^{-8}$	0
	Case 2	0.0041	0.0024	$<1 \times 10^{-8}$	$<1 \times 10^{-8}$	$<1 \times 10^{-8}$	1.09×10^{-5}

Table 2. Maximum error and *MSE* for TD2.

Project	Case	LS-Q	LS-C	LS-N	MLS	RBFNN	WP
maximum error	Case 1	0.1142	0.0744	0.2000	0.0008	$<1 \times 10^{-4}$	0
	Case 2	0.0796	0.0658	0.1582	0.0108	$<1 \times 10^{-4}$	0.0073
MSE	Case 1	0.0028	0.0012	0.0105	2.80×10^{-7}	$<1 \times 10^{-8}$	0
	Case 2	0.0026	0.0016	0.0134	1.20×10^{-5}	$<1 \times 10^{-8}$	7.71×10^{-6}

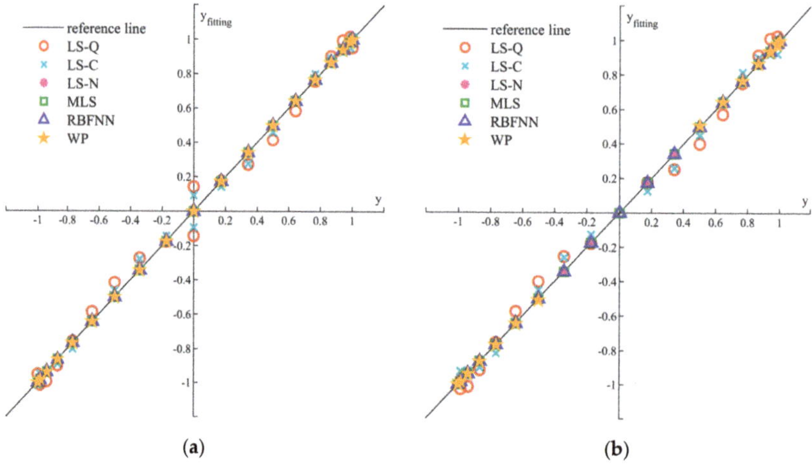

Figure 4. *Cont.*

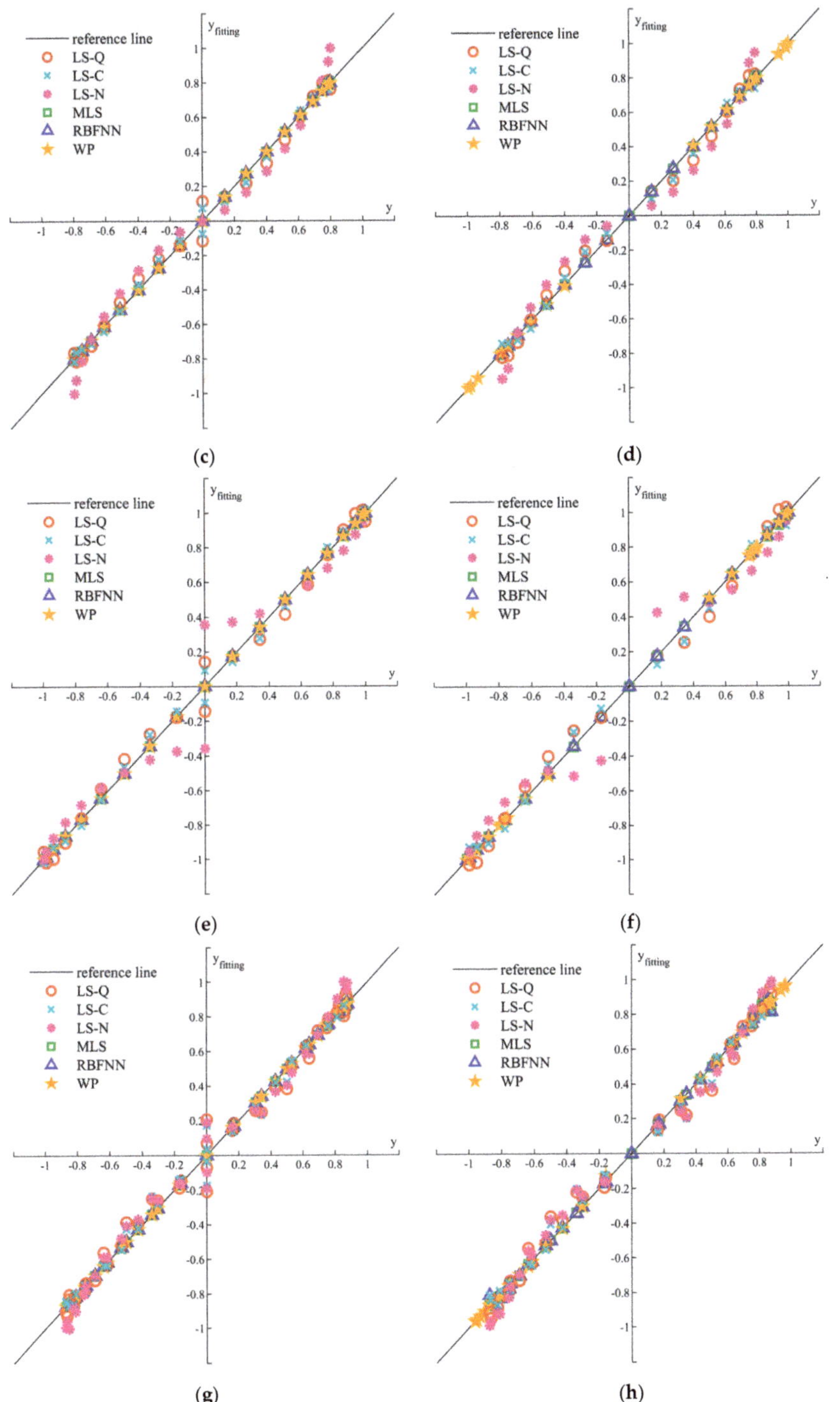

Figure 4. Accuracy comparison of fitting results for (**a**) TD1 in Case 1, (**b**) TD1 in Case 2, (**c**) TD2 in Case 1, (**d**) TD2 in Case 2, (**e**) TD3 in Case 1, (**f**) TD3 in Case 2, (**g**) TD4 in Case 1, and (**h**) TD4 in Case 2.

Table 3. Maximum error and *MSE* for TD3.

Project	Case	LS-Q	LS-C	LS-N	MLS	RBFNN	WP
maximum error	Case 1	0.1427	0.0930	0.3566	0.0024	$<1 \times 10^{-4}$	01
	Case 2	0.0996	0.0823	0.2511	0.0097	0.0003	0.00181
MSE	Case 1	0.0043	0.0019	0.0197	1.23×10^{-6}	$<1 \times 10^{-8}$	0
	Case 2	0.0041	0.0024	0.0161	1.34×10^{-5}	$<1 \times 10^{-8}$	1.02×10^{-5}

Table 4. Maximum error and *MSE* for TD4.

Project	Case	LS-Q	LS-C	LS-N	MLS	RBFNN	WP
maximum error	Case 1	0.2071	0.1732	0.1891	0.0016	$<1 \times 10^{-4}$	0
	Case 2	0.1389	0.1386	0.1330	0.0593	0.0611	0.0334
MSE	Case 1	0.0046	0.0028	0.0076	6.84×10^{-7}	$<1 \times 10^{-8}$	0
	Case 2	0.0040	0.0029	0.0063	0.0002	0.0003	3.78×10^{-5}

4. Calculation and Fitting for the Sample of the Damaged Ships

4.1. The Improved Smith Method

In the improved Smith method [17], the rotation of the neutral axis can be taken into account, which makes it applicable to asymmetric hull girder, for instance, damaged hull like the cross-section in Figure 5 under combined bending moments.

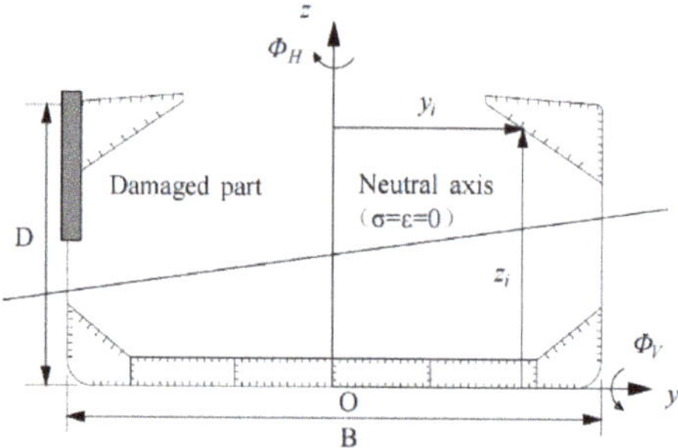

Figure 5. Cross section of a damaged ship.

The process for calculating the residual strength consists of 10 steps [3]:

(1) The cross section of concern is divided into different types of stiffened plate units, plating units, and hard-corner units;

(2) For a given curvature, the stress–strain relationships for all types of units are defined as shown in Figure 6. Then, the strain of the i-th unit is

$$\varepsilon_i(y_i, z_i) = \varepsilon_0 + y_i\Phi_H + z_i\Phi_V, \tag{27}$$

where ε_0 is the strain at the origin O and Φ_H and Φ_V are the horizontal curvature and vertical curvature, respectively. Then, the tangential stiffness D_i can be calculated as

$$D_i = \frac{df_i(\sigma)}{d\varepsilon} = \frac{\Delta\sigma_i}{\Delta\varepsilon}, \tag{28}$$

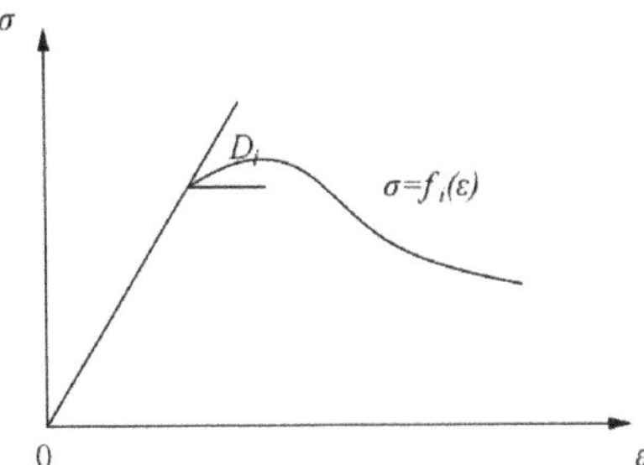

Figure 6. Stress–strain curve.

(3)   The axial force P should satisfy

$$P = \sum_{i=1}^{N} \sigma_i A_i \equiv 0,\tag{29}$$

The vertical bending moment M_V and the horizontal bending moment M_H can be calculated as

$$\begin{cases} M_H = \sum\limits_{i=1}^{N} \sigma_i y_i A_i \\ M_V = \sum\limits_{i=1}^{N} \sigma_i z_i A_i \end{cases},\tag{30}$$

where A_i is the cross-sectional area of the i-th unit;

(4)   The position of point G is shown in Figure 7 and can be obtained by

$$\begin{cases} y_G = \left(\sum\limits_{i=1}^{n} y_i D_i A_i \right) \Big/ \left(\sum\limits_{i=1}^{n} D_i A_i \right) \\ z_G = \left(\sum\limits_{i=1}^{n} z_i D_i A_i \right) \Big/ \left(\sum\limits_{i=1}^{n} D_i A_i \right) \end{cases},\tag{31}$$

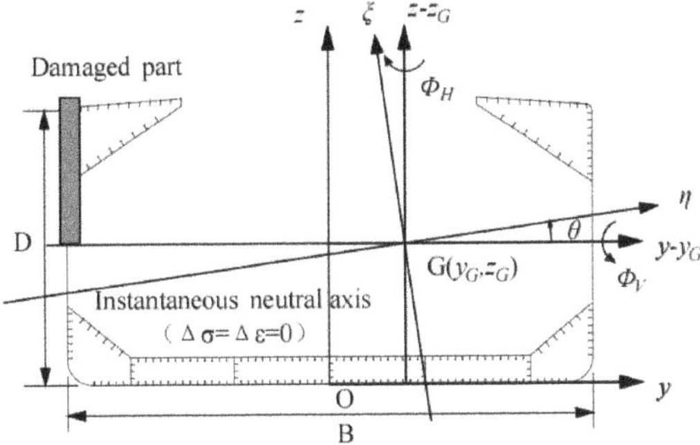

Figure 7. Schemes follow the same formatting.

(5)   The flexural stiffness should satisfy the function

$$\begin{Bmatrix} \Delta M_H \\ \Delta M_V \end{Bmatrix} = \begin{bmatrix} D_{HH} & D_{HV} \\ D_{VH} & D_{VV} \end{bmatrix} \begin{Bmatrix} \Delta \Phi_H \\ \Delta \Phi_V \end{Bmatrix},\tag{32}$$

where

$$\begin{cases} D_{HH} = \sum_{i=1}^{N} D_i(y_i - y_G)^2 A_i \\ D_{VV} = \sum_{i=1}^{N} D_i(z_i - z_G)^2 A_i \\ D_{HV} = D_{VH} = \sum_{i=1}^{N} D_i(y_i - y_G)(z_i - z_G)A_i \end{cases} , \tag{33}$$

(6) The increments of the next curvature and/or bending moment can be calculated by

$$\left\{ \begin{array}{c} \alpha\Delta M_H \\ \Delta M_V \end{array} \right\} = \left[\begin{array}{cc} D_{HH} & D_{HV} \\ D_{VH} & D_{VV} \end{array} \right] \left\{ \begin{array}{c} \Delta\Phi_H \\ \Delta\Phi_V^0 \end{array} \right\}, \tag{34}$$

(7) Increase the curvature, calculate the increment in strain and stress according to the stress–strain curve, and then the cumulative results of bending moment, strain, and stress of each unit can be obtained;

(8) The position of the neutral axis can be calculated with the stress and strain

$$\varepsilon_0 + y\Phi_H + z\Phi_V = 0 \tag{35}$$

(9) When the ultimate strength is reached, the calculation is stopped.

4.2. Distribution of the Sample of Damaged Ships

In order to compare the accuracy of these fitting methods when applied to the limit state function, the sample of the longitudinal strength of a damaged single-hull bulk carrier (DB) and a damaged double-hull oil tanker (DT) calculated by Fujikubo et al. [17] are adopted. The main dimensions of the ships are shown in Table 5. The diagrammatic sketch of the cross sections and the damage conditions are shown in Figure 8.

Table 5. The main dimensions of the ships.

Ship Parameter	DB	DT
L (mm)	217,000	219,000
B (mm)	32,236	32,240
D (mm)	18,300	19,900

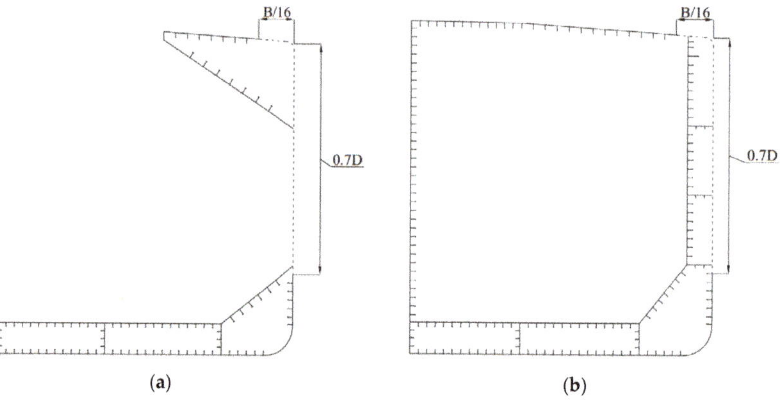

Figure 8. Cross sections of (**a**) DB and (**b**) DT.

The distribution of the results for the residual strength of DB and DT are shown in Figure 9.

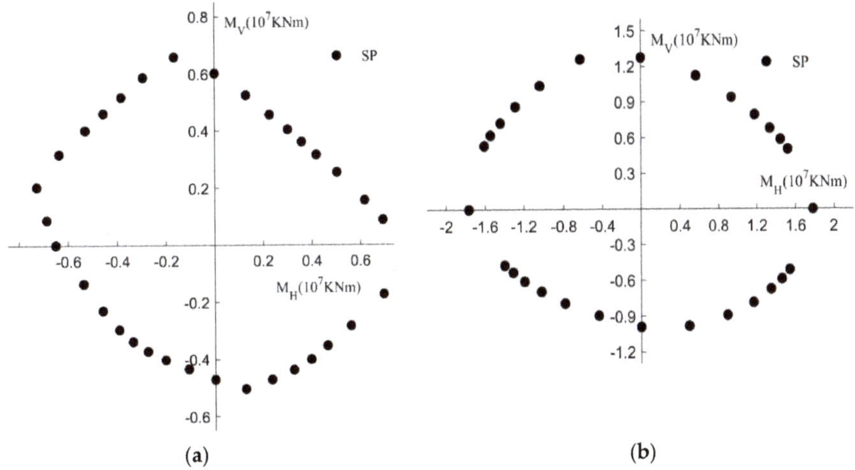

Figure 9. Distribution of sample for (**a**) DB and (**b**) DT.

4.3. Residual Strength Fitting Results for the Damaged Ships

The envelopes of bending moments of different ships or damaged cases are not different. In order to facilitate the comparison, the vertical bending moments and horizontal bending moments are nondimensionalized separately by the maximum vertical bending moment and the maximum vertical bending moment, respectively. The fitting results are shown in Figure 10. The fitting results of Case 1 and Case 2 are shown in Figure 11. The maximum errors and *MSEs* are shown in Tables 6 and 7.

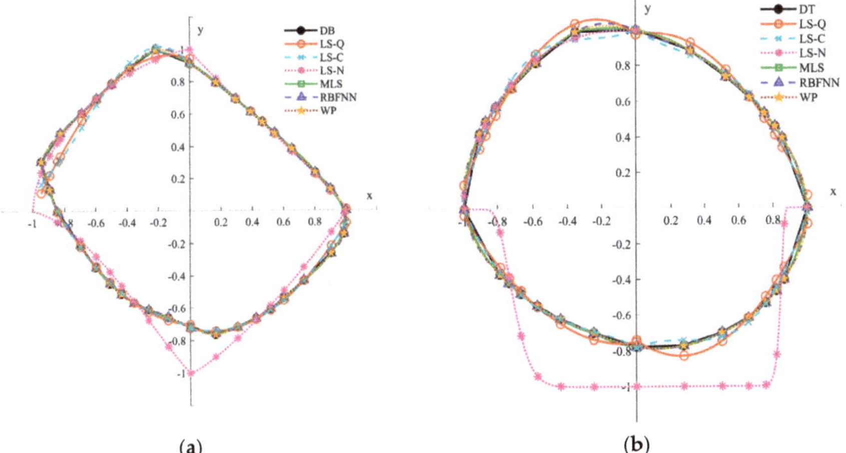

Figure 10. Fitting results for (**a**) DB and (**b**) DT.

Table 6. Maximum error and *MSE* for TD1.

Project	Case	LS-Q	LS-C	LS-N	MLS	RBFNN	WP
maximum error	Case 1	0.3121	0.2944	0.4264	0.0129	0.0087	0.0000
	Case 2	0.3267	0.3640	0.2209	0.0471	0.0650	0.0404
MSE	Case 1	0.0046	0.0043	0.0138	8.53×10^{-6}	8.21×10^{-6}	0.0000
	Case 2	0.0054	0.0075	0.0072	0.0001	0.0005	0.0001

Table 7. Maximum error and *MSE* for TD2.

Project	Case	LS-Q	LS-C	LS-N	MLS	RBFNN	WP
maximum error	Case 1	0.1326	0.0740	0.4658	0.0024	0.0008	0.0000
	Case 2	0.1069	0.2451	0.4682	0.0700	0.0378	0.0750
MSE	Case 1	0.0026	0.0008	0.0419	1.30×10^{-6}	1.66×10^{-7}	0.0000
	Case 2	0.0035	0.0047	0.0576	0.0008	0.0001	0.0003

It is shown in Figure 10 that the deviation between fitting curves of LS-Q, LS-C, and LS-N and sample curves is larger than the curves of MLS, RBFNN, and WP. The fitting accuracy of LS-Q and LS-C is poor when the sample curves have more than one inflection point in a single quadrant, such as the curve in the second quadrant in Figure 10a, of which y increases and then decreases near $x = 0$ and of which x decreases and then increases near $x = -1$. The fitting accuracy of LS-N is poor when the sample curves do not pass the points (1,0), (0,1), (−1,0), and (0,−1), such as the curve in the fourth quadrant in Figure 10b, which is flat, vertical, and then flat with decreasing x. It depends on the function of LS-N, and influences the curve shape and accuracy. The fitting curves of MLS, RBFNN, and WP are near the sample curves. In Figure 11, it can be found that when y is near −1, 0 and 1, $y_{fitting}$ of LS-Q, LS-C, and LS-N is usually far away from the reference line, which is disadvantageous for the assessment of the ship-hull girder residual strength under vertical bending moments. Comparison of Case 1 and Case 2 shows that the fitting accuracy of LS-Q, LS-C, and LS-N is also poor for the nonsample, and the fitting errors obviously increment or reduce. Comparing Case 1 of MLS for DT, the fitting errors of Case 2 increase when y is near 1 or −1, and the reason is that the lack of inflection point decreases the accuracy. The fitting accuracy of RBFNN and WP is high. In Tables 6 and 7, it is shown that the maximum error and *MSE* of Case 1 and Case 2 of LS-Q, LS-C, and LS-N are large and close, but the maximum error and *MSE* of Case 2 of MLS, RBFNN, and WP are larger than in Case 1. The fitting results also show that LS-Q, LS-C, and LS-N can provide the explicit fitting functions in a single quadrant, and the fitting curves of LS-Q and LS-C are not continuous as well as LS-N not being smooth. The fitting curves of MLS, RBFNN, and WP are continuous and smooth, and a series of explicit fitting functions is obtained with WP, while the implicit fitting functions are obtained with MLS and RBFNN.

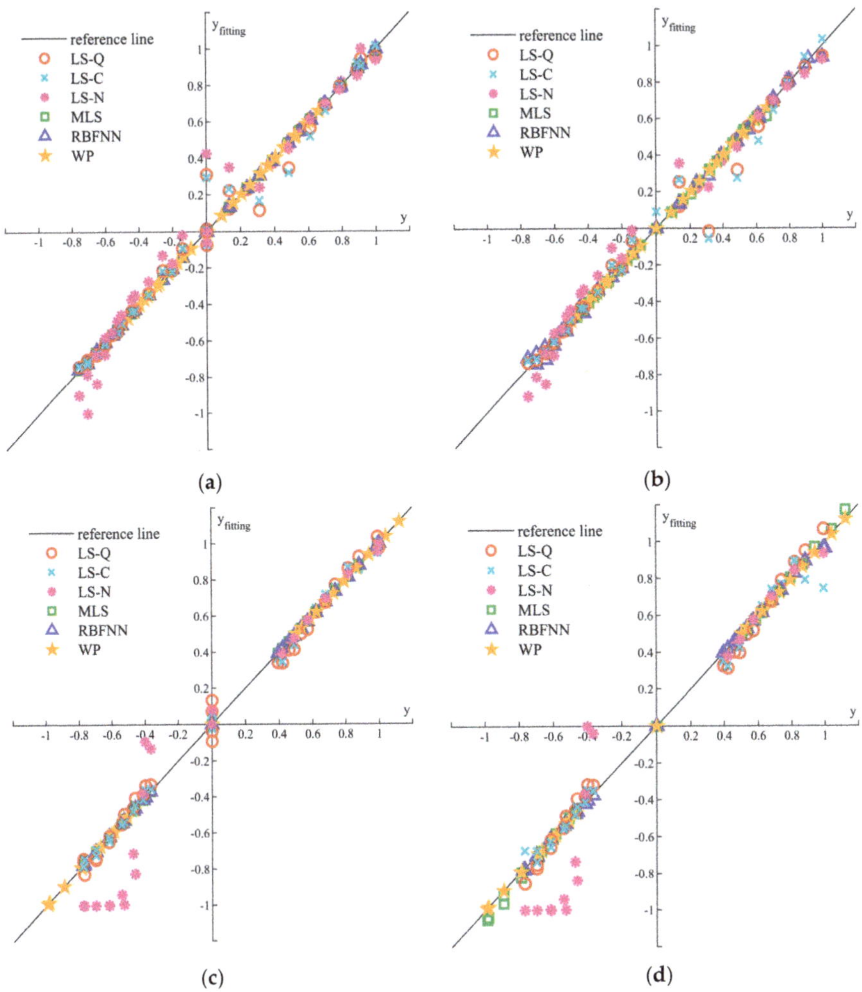

Figure 11. Accuracy comparison of fitting results for (**a**) DB in Case 1, (**b**) DB in Case 2, (**c**) DT in Case 1, and (**d**) DT in Case 2.

5. Conclusions

In this study, four fitting methods and six fitting functions are applied to calculate the fitting curves of four typical sample distributions and the longitudinal strength of two damaged ships under combined bending moments. Based on analysis of the results, the following conclusions are drawn:

(1) The distribution of the sample influences the fitting accuracy. When the sample curves have multiple inflection points in a single quadrant and the curves do not pass the points $(1,0)$, $(0,1)$, $(-1,0)$, and $(0,-1)$, the difference between the fitting curves and the sample curves is large.

(2) The least-squares method can fit the curves with different fitting functions and all the functions are explicit, but the fitting accuracies of the quadratic function, the cubic function, and the nonlinear function are not satisfactory. The fitting curves of the linear functions are not continuous, and the nonlinear functions are not smooth. The

fitting results also show that the increase in the sample and the order of the function has little contribution to the fitting accuracy.

(3) Application of MLS, RBFNN, and WP are more complex than the least-squares method, but the fitting accuracy is much better. All the fitting curves are continuous and smooth in the four quadrants, and they are able to improve the assessment accuracy of the residual strength under pure or combined bending moments. It can also be found that the increment of the sample has little contribution to the fitting accuracy.

(4) The implicit fitting function can be obtained with MLS and RBFNN, and a series of explicit fitting functions can be obtained with WP.

Author Contributions: Conceptualization, C.L.; Formal analysis, Z.Z., X.W., N.Z. and C.L.; Funding acquisition, C.L.; Methodology, Z.Z., H.R., X.W. and C.L.; Project administration, C.L.; Resources, H.R.; Supervision, H.R. and C.L.; Writing—original draft, Z.Z.; Writing—review and editing, H.R. All authors have read and agreed to the published version of the manuscript.

Funding: This research was supported by the Science Fund Project of Heilongjiang Province (LH2020E078) and the National Natural Science Foundation of China (52171305).

Institutional Review Board Statement: Not applicable.

Informed Consent Statement: Not applicable.

Data Availability Statement: Not applicable.

Acknowledgments: The authors thank all the reviewers for their valuable comments.

Conflicts of Interest: The authors declare no conflict of interest.

References

1. European Maritime Safety Agency (EMSA). *Annual Overview of Marine Casualties and Incidents 2019*; EMSA: Lisbon, Portugal, 2019.
2. Chan, H.S.; Incecik, A.; Atlar, M. Structural Integrity of a Damaged Ro-Ro Vessel. In Proceedings of the Second International Conference on Collision and Grounding of Ships, Copenhagen, Denmark, 1–3 July 2001; Technical University of Denmark: Lyngby, Denmark, 2001; pp. 253–258.
3. International Association of Classification Societies (IACS). *Harmonized Common Structural Rules for Oil Tankers and Bulk Carriers*; IACS: London, UK, 2014.
4. Vu, V.T.; Yang, P.; Doan, V.T. Effect of uncertain factors on the hull girder ultimate vertical bending moment of bulk carriers. *Ocean Eng.* **2018**, *148*, 161–168.
5. Vu, V.T.; Yang, P. Effect of corrosion on the ship hull of a double hull very large crude oil carrier. *J. Mar. Sci. Appl.* **2017**, *16*, 334–343.
6. Vu, V.T.; Dong, D.D. Hull girder ultimate strength assessment considering local corrosion. *J. Mar. Sci. Appl.* **2020**, *19*, 693–704. [CrossRef]
7. Gordo, J.M.; Teixeira, A.P.; Guedes Soares, C. Ultimate strength of ship structures. *Mar. Technol. Eng.* **2011**, *2*, 889–900.
8. Estefen, S.F.; Chujutalli, J.H.; Guedes Soares, C. Influence of geometric imperfections on the ultimate strength of the double bottom of a Suezmax tanker. *Eng. Struct.* **2016**, *127*, 287–303. [CrossRef]
9. Yao, T.; Nikolov, P.I. Progressive Collapse Analysis of a Ship's Hull under Longitudinal Bending (2nd Report). *J. Soc. Nav. Archit. Jpn.* **1992**, *172*, 437–446. [CrossRef]
10. Parunov, J.; Rudan, S.; Gledić, I.; Bužančić Primorac, B. Finite Element Study of Residual Longitudinal Strength of a Double Hull Oil Tanker with Simplified Collision Damage and Subjected to Bi-axial Bending. *Ships Offshore Struct.* **2018**, *13*, 25–36. [CrossRef]
11. Paik, J.K.; Kim, B.J.; Seo, J.K. Methods for Ultimate Limit State Assessment of Ships and Ship-shaped Offshore Structures: Part I Unstiffened Plates. *Ocean Eng.* **2008**, *35*, 261–270. [CrossRef]
12. Paik, J.K.; Kim, B.J.; Seo, J.K. Methods for Ultimate Limit State Assessment of Ships and Ship-shaped Offshore Structures: Part II Stiffened Panels. *Ocean Eng.* **2008**, *35*, 271–280. [CrossRef]
13. Benson, S.; Abubakar, A.; Dow, R.S. A Comparison of Computational Methods to Predict the Progressive Collapse Behaviour of a Damaged Box Girder. *Eng. Struct.* **2013**, *48*, 266–280. [CrossRef]
14. Kim, D.K.; Kim, S.J.; Kim, H.B.; Zhang, X.M.; Li, C.G.; Paik, J.K. Longitudinal Strength Performance of Bulk Carriers with Various Corrosion Additions. *Ships Offshore Struct.* **2015**, *10*, 59–78. [CrossRef]
15. Ogawa, H.; Takami, T.; Tatsumi, A.; Tanaka, Y.; Hirakawa, S.; Fujikubo, M. Buckling/Longitudinal Strength Evaluation For Continuous Stiffened Panel under Combined Shear and Thrust. In Proceedings of the ASME 35th International Conference on Ocean, Offshore and Arctic Engineering (OMAE), Busan, Korea, 19–24 June 2016.

16. Syrigou, M.S.; Dow, R.S. Strength of Steel and Aluminium Alloy Ship Plating under Combined Shear and Compression/Tension. *Eng. Struct.* **2018**, *166*, 128–141. [CrossRef]
17. Fujikubo, M.; Zubair Muis Alie, M.; Takemura, K.; Iijima, K.; Oka, S. Residual Hull Girder Strength of Asymmetrically Damaged Ships. *J. Jpn. Soc. Nav. Archit. Ocean. Eng.* **2012**, *16*, 131–140. [CrossRef]
18. Joonmo, C.; Nam, J.M.; Ha, T.B. Assessment of Residual Ultimate Strength of an Asymmetrically Damaged Tanker Considering Rotational and Translational Shifts of Neutral Axis Plane. *Mar. Struct.* **2012**, *25*, 71–84.
19. Gordo, J.M.; Guedes Soares, C. Collapse of Ship Hulls under Combined Vertical and Horizontal Bending Moments. In Proceedings of the 6th International Symposium on Practical Design of Ships and Mobile Units (PRADS'95), Seoul, Korea, 17–22 September 1995.
20. Gordo, J.M.; Guedes Soares, C. Interaction Equation for the Collapse of Tankers and Containerships under Combined Bending Moments. *J. Ship Res.* **1997**, *41*, 230–240. [CrossRef]
21. Mansour, A.E.; Lin, Y.H.; Paik, J.K. Ultimate Strength of Ships under Combined Vertical and Horizontal Moments. *J. Ship Ocean. Technol.* **1998**, *2*, 31–41.
22. Luís, R.M.; Hussein, A.W.; Guedes Soares, C. On the Effect of Damage to the Ultimate Longitudinal Strength of Double Hull Tankers. In Proceedings of the 10th International Symposium on Practical Design of Ships and Other Floating Structures (PRADS'07), Houston, TX, USA, 30 September–5 October 2007.
23. Khan, I.A.; Das, P.K. Random Design Variables and Sensitivity Factors Applicable to Ship Structures Considering Combined Bending Moments. *Proc. Inst. Mech. Eng. Part M J. Eng. Marit. Environ.* **2008**, *222*, 133–143. [CrossRef]
24. Shahid, M. Development of Structural Reliability Techniques and Their Application to Marine Structural Components and Systems. Ph.D. Thesis, Universities of Glasgow and Strathclyde, Glasgow, UK, 2008.
25. Zhu, Z.; Ren, H.; Li, C.; Zhou, X. Ultimate Limit State Function and Its Fitting Method of Damaged Ship under Combined Loads. *J. Mar. Sci. Eng.* **2020**, *8*, 117. [CrossRef]
26. Lee, D.H.; Paik, J.K. Ultimate Strength Characteristics of As-built Ultra-Large Containership Hull Structures under Combined Vertical Bending and Torsion. *Ships Offshore Struct.* **2020**, *15*, 143–160. [CrossRef]
27. Kim, K.; Yoo, C.H. Ultimate Strengths of Steel Rectangular Box Beams Subjected to Combined Action of Bending and Torsion. *Eng. Struct.* **2008**, *30*, 1677–1687. [CrossRef]
28. Shi, G.J.L.; Wang, D.Y. Residual Ultimate Strength of Open Box Girders with Cracked Damage. *Ocean Eng.* **2012**, *43*, 90–101. [CrossRef]
29. Hu, K.; Yang, P.; Xia, T.; Peng, Z. Residual Ultimate Strength of Large Opening Box Girder with Crack Damage under Torsion and Bending Loads. *Ocean Eng.* **2018**, *162*, 274–289. [CrossRef]
30. Li, J.; Zhou, C.Y.; Cui, P.; He, X.H. Plastic limit loads for pipe bends under combined bending and torsion moment. *Int. J. Mech. Sci.* **2015**, *92*, 133–145. [CrossRef]

Journal of
*Marine Science
and Engineering*

Article

Improvement of the Ship Emergency Response Procedure in Case of Collision Accident Considering Crack Propagation during Salvage Period

Ivana Gledić, Antonio Mikulić and Joško Parunov *

Faculty of Mechanical Engineering and Naval Architecture, University of Zagreb, 10000 Zagreb, Croatia;
ivana.gledic@fsb.hr (I.G.); antonio.mikulic@fsb.hr (A.M.)
* Correspondence: josko.parunov@fsb.hr

Abstract: Specialized procedures to help in the emergency response situations following ship accidents have been under development by the Classification Societies. Such procedures consider the hull-girder collapse as the most important failure mode, without the possibility of crack propagation caused by fluctuating wave loads. In the present study, the fatigue crack propagation in the main deck of the oil tanker damaged in collision during salvage is investigated. The shape and size of the damage are modelled using the realistic bow shape of the striking ship and historical data of ship accidents. The stress intensity factor (SIF) across the main deck of the struck ship is calculated numerically and by the method based on the available experimental results of the crack propagation in the stiffened panel. Fluctuating wave–induced stresses in short-term sea conditions during salvage are obtained by Monte Carlo simulation (MC) based on Rayleigh distribution. Cycle-by-cycle crack propagation is calculated using Paris law. Many salvage simulations are performed to cover different possible time-histories of the fatigue loading. Results of the analysis are presented as histogram of the crack increase during salvage. Parametric analysis is performed to investigate the influence of the sea state severity, initial crack size, and towing duration on the final crack size. The proposed procedure can be considered as a part of a software tool for emergency response action during salvage of damaged ship.

Keywords: oil tanker; collision; salvage; crack propagation; Monte Carlo simulation

Citation: Gledić, I.; Mikulić, A.; Parunov, J. Improvement of the Ship Emergency Response Procedure in Case of Collision Accident Considering Crack Propagation during Salvage Period. *J. Mar. Sci. Eng.* **2021**, *9*, 737. https://doi.org/10.3390/jmse9070737

Academic Editor: José A.F.O. Correia

Received: 10 June 2021
Accepted: 1 July 2021
Published: 3 July 2021

Publisher's Note: MDPI stays neutral with regard to jurisdictional claims in published maps and institutional affiliations.

1. Introduction

Research on accidental loads and the hull girder strength after collision and grounding is nowadays one of the priorities in the field of marine structures [1]. This is particularly the case for an oil tanker, as an accident involving this type of vessel can have disastrous economic and environmental repercussions. To limit the escalation of accident scenarios, the accidental limit state (ALS) is incorporated in the Harmonized Common Structural Rules [2]. Also, specialized software to help in the emergency response actions following marine accidents has been developing by the Classification Societies [3].

The main hypothesis of these developments has been that hull girder collapse would occur if the hull's maximum residual load-carrying capacity is insufficient to sustain the maximum hull girder loads applied [4]. Dominating hull girder loads are, usually considered, still-water bending moments that may increase as a consequence of flooding and wave bending moments that depend on the sea states in the time of the accident and exposure time to the sea waves [5]. Residual ultimate strength is assumed to be constant during a rescue. A possibility of crack appearance and propagation in damaged ship structure and consequent hull-girder failure because of the fluctuating wave loads are rarely studied. The number of cycles relevant for fatigue failure is the number of wave cycles encountered during the rescue of one week that could be around 60,000–80,000.

Therefore, the fatigue failure mode of the hull-girder needs to be investigated to clarify the relevance of the cyclic wave loading on the safety of a damaged ship structure.

Recently, the low-cycle fatigue (LCF) failure mode of a ship damaged in a collision has been investigated [6]. The LCF corresponds to the scenario where a ship encounters large wave amplitudes during the salvage, causing cyclic plasticity in some parts of the damaged hull structure. The obtained results indicated that LCF is not likely failure mode for damaged oil tankers, even for the most severe damage cases.

The crack propagation in a damaged ship caused by fluctuating stresses during rescue is studied at the conceptual level only [7–9]. In these studies, rather simple computational models were used that did not allow reaching a firm conclusion regarding the importance of fatigue crack propagation. The criterion for structural failure was the residual ultimate bending moment capacity that may be affected only by large fatigue crack. It has never been shown by credible computations that such large crack growth during the rescue period is feasible.

The aim of the present study is to propose the method for the fatigue crack propagation in ship damaged in collision during salvage. The present study is built upon previous research, where damage shape and size are modelled nearly realistic, taking bow shape and penetration depth of the striking ship into account [5,6]. Stress intensity factor (SIF) across the main deck of the struck ship, the most important parameter in crack propagation analysis, is determined numerically using Displacement Method and by using an expression based on the experiments of crack propagation in the stiffened panel [10]. Fluctuating wave–induced vertical wave bending moments (VWBM) of damaged ship are calculated by the seakeeping analysis, assuming stationary short-term sea conditions during salvage. Crack propagation is determined by the linear elastic fracture mechanics (LEFM), using Paris law, following recommendations of the Classification Societies for such kind of analysis [11]. Individual stress amplitudes for cycle-by-cycle crack propagation analysis are obtained by Monte Carlo (MC) simulation. The total crack increase during the salvage period is presented in the form of a histogram. Parametric analysis is then performed to investigate the influence of initial crack size and towing duration on crack propagation.

The outline of the paper is as follows: the basic theoretical background of crack propagation analysis is described firstly. The FE model of a damaged Aframax oil tanker, loading applied, and associated boundary conditions are described in Section 2, followed by a review of calculation methods for SIF and the procedure for cycle-by-cycle crack propagation. In Section 3, results are presented, and parametric analysis is performed with a modification of pertinent parameters to investigate their influence on crack propagation. A discussion regarding some of the assumptions made in this research and suggestions for practical implementation of the developed procedure are provided at the end of the paper.

2. Methodology

2.1. Fatigue Crack Propagation in a Damaged Ship

Fatigue crack propagation and failure assessment are based on the procedure and expressions provided in [11]. Fatigue life calculated from the crack propagation depends on the initial crack size, which for a ship in service can be measured during inspection or assumed. In any case, the initial surface crack size should not be less than 1 mm [11]. Normally, small surface cracks grow with a semi-elliptical shape into the material. After penetrating through the plate thickness, the 'through-thickness crack' propagates further in the transverse direction. The length of through-thickness crack is usually more than twice plate thickness. For a damaged ship, initial through-thickness crack size at the irregular boundary of damage in the ship hull may only be roughly assumed. In this study, a through-thickness crack size of 35 mm is assumed, representing the minimum possible through-thickness crack at the damage boundary. In the parametric study, a crack size of 70 mm is also assumed, representing a case when a larger crack appears after the sudden damage.

The main parameter governing the crack propagation calculation is the stress intensity factor (SIF), *K*. Based on the opening displacement of the crack lips, there are three principle crack opening modes: Mode I (tension), Mode II (shear), and Mode III (torsion) [11]. Mode I is the most common opening mode in practice [12]. The other two modes do not occur individually, but they may occur in combination with mode I, i.e., I-II, I-III or I-II-III. However, the majority of apparent combined mode cases are reduced to mode I by nature itself. This study is therefore based on Mode I only, and only SIF K_I is determined herein.

The SIF range (SIFR) is a parameter related to the dynamic stress range, and it is calculated by the following expression [13]:

$$\Delta K = Y \cdot \Delta\sigma \sqrt{\pi \cdot a}, \tag{1}$$

where *Y* is a geometry function, $\Delta\sigma$ is a tensile stress range, and *a* is a crack size. In case when $\Delta K \geq \Delta K_{th}$, the crack propagation is calculated by Paris's law:

$$\frac{da}{dN} = C \cdot \Delta K^m, \tag{2}$$

where ΔK_{th} is a threshold of ΔK, *C* and *m* are crack propagation law parameters; constants that depend on the material, load conditions, and corrosive environmental characteristics.

The crack growth is calculated cycle-by-cycle using the time history of stress cycles, where stress amplitudes are generated by MC simulation. The procedure is described in more detail in Sections 2.2 and 2.3. In addition, many such simulations are performed to cover different order of stress application, which generally leads to different extents of crack propagation.

2.2. Modelling of Damaged Ship Structure

The example ship used in the present study is an Aframax-class double-hull oil tanker. The main ship particulars are presented in Table 1. The ship structure is made of mild steel (R_{eH} = 235 N/mm^2), except for longitudinally effective deck and bottom structural elements, which are made of high-tensile steel (R_{eH} = 315 N/mm^2).

Table 1. Main particulars of Aframax-class double hull tanker.

Dimensions	Symbol	Value
Length between perp.	L_{pp} [m]	234
Breadth	B [m]	40
Depth	D [m]	21
Draught	T [m]	14
Deadweight	DWT [t]	105,000

Damage scenarios used for modelling damage shapes are described in [5,6]. Damage scenarios are based on the historical database of ship damages and accidents. Each collision scenario is described by non-dimensional vertical impact location (X_v/D), non-dimensional damage penetration depth (X_b/B), striking ship length, striking ship depth, and the bow shape of the striking ship. The FE ship model with collision damage is obtained from the intact ship model by removing damaged finite elements [5,6]. Collision scenarios are observed from the first contact between two ships until maximum penetration depth is reached. If damaged during penetration, finite elements are removed. Element elimination is done automatically by implementing the parametric striking ship model [14]. Removal of the elements is based upon overlap between geometry of striking and struck ship. If there is overlapping between geometry, elements (on struck model) are completely removed regardless if they are only deformed or totally damaged. Such modelling of damage shape is called "near realistic" and it is elaborated in detail in ref. [5,6]. More realistic approach would be to model damage by collision simulation and then to refine deformed elements on damage boundary for crack propagation analysis [15]. This approach is still considered as too complicated, especially as one of the goals of this research is to make fast calculation of crack propagation with available tools, mainly using linear elastic fracture mechanics. Problem of further crack initiation and bifurcation on deformed elements requires a plastic fracture mechanics and would require extensive calculations, beyond the scope of the present study.

A total of 50 random damage scenarios are defined in [5,6], while only one characteristic damage case, with parameters defined in Table 2, is used in the present study.

Table 2. Parameters of the characteristic collision scenario.

Vertical Impact Location (X_D/D)	Damage Penetration (X_B/B)	Striking Ship Length (L) [m]	Striking Ship Depth (D) [m]	Striking Ship Breadth (B) [m]
0.055	0.131	258.8	22.9	41.6

A detailed description of all damage scenarios and illustrative figures are given in [6] and therefore is not reproduced herein. The damage case described in Table 2 is considered representative of a major collision accident as there is a rupture of the main deck, side shell, and inner hull of the ship.

2.3. Computational Methods for Stress Intensity Factor (SIF)

FE analysis is carried out using FEMAP with NX Nastran software. During the collision, the forecastle of the striking ship penetrates through the deck of the struck ship. FE model of damaged struck oil tanker is shown in Figure 1.

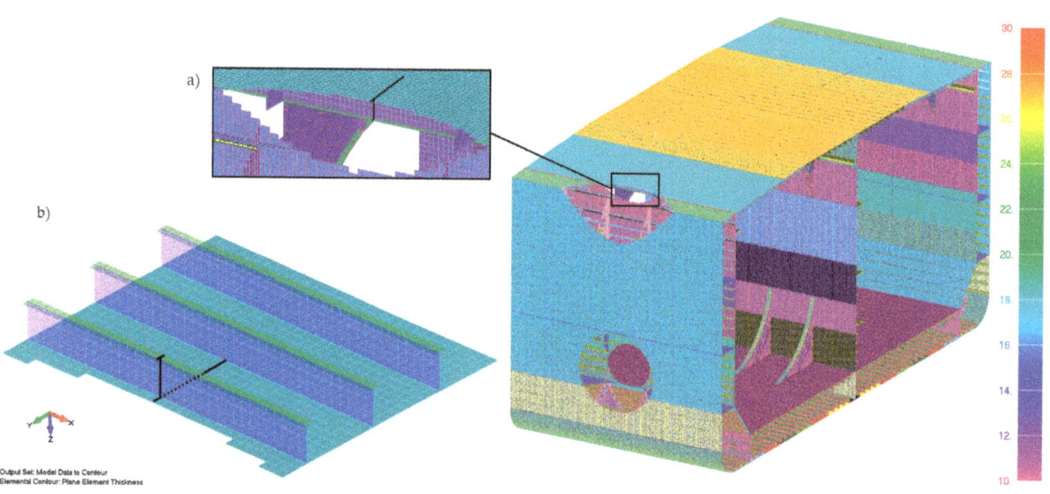

Figure 1. Damaged ship; (**a**) top view of assumed crack location; (**b**) bottom view of crack propagation path between two stiffeners.

As the concern of the present study is hull girder fatigue strength and fluctuating stresses due to vertical wave bending moments, the main attention is given to the deck structural area. Crack is assumed to originate at the farthest damage penetration point, across the main deck of the struck ship, as shown in Figure 1a. Crack is pre-determined according to the Mode I crack opening due to the symmetric boundary conditions and loading and because this crack path has the greatest contribution on overall structural integrity when considering vertical bending moment. For better visualization and understanding Figure 1b shows detailed crack path from origin point to second stiffener. The damaged deck section is rotated 90 degrees counterclockwise with stiffeners represented as transparent.

The analysis of SIF is performed with refined elements along the predicted crack propagation path, which is assumed to be in a direction perpendicular to the maximum principal stresses. Elements around the crack tip are refined in a box-like shape. With crack length increase, this "box" is shifted along the assumed crack path (Figures 2 and 3). The crack opening is simulated by the unzip feature in FEMAP, removing all connections between elements along the crack.

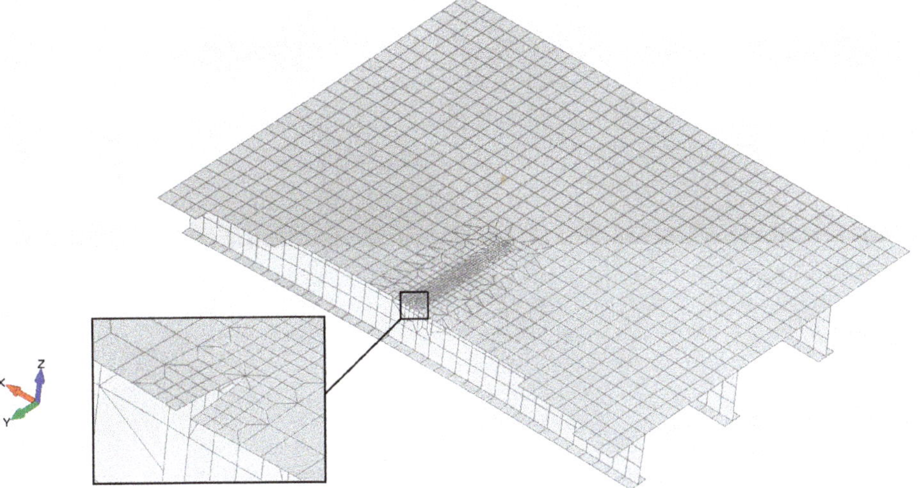

Figure 2. Crack till first stiffener.

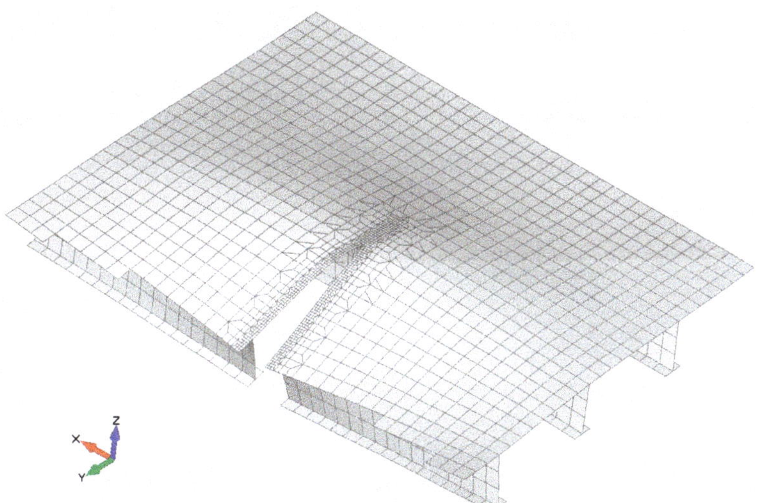

Figure 3. Crack till second stiffener.

Mesh element size in the refined model is approximately equal to the plate thickness, *t*, which is the standard element size in the fatigue analysis of ship structural details. CQUAD parabolic plate elements with four corner nodes and four mid nodes are employed for the refinement, except around the crack tip where TRIANGLE parabolic plate elements with three corner nodes and three mid nodes are used.

SIF calculation in a stiffened panel is generally possible through the application of analytical, numerical, and experimental methods. All methods for calculation of SIF attempt to somehow relate the SIF to certain physical quantities around the crack tip, such as displacement, nodal forces, or energy release. Using FE method, SIF calculation can be done with: (1) Displacement Method along crack face [16], (2)Weight Function Method where SIF equals integral of the product of weight function and the distribution of stress along assumed crack line [17], (3) Force Method which calculates forces on nodes along a ligament at a certain distance from the crack tip [18] and (4) Griffith Energy Method or J-integral method based on potential energy released when a crack propagates [19]. In the present study, the analysis is carried out using FEMAP with NX Nastran software by Displacement Method.

From the FE analysis, using the Displacement Method ΔK_I is calculated according to [16,20]:

$$\Delta K_I = \frac{E \, \Delta y_{qpn}}{4} \sqrt{\frac{2\pi}{r}}. \tag{3}$$

where E is the Young modulus, Δy_{qpn} is the displacement of the quarter-point node in the direction normal to the crack plane, and r is the distance from the crack tip to this quarter-point node. Based on ΔK_I from FE analysis, geometry function Y can be determined from Equation (1) as follows:

$$Y = \frac{\Delta K_I}{\Delta \sigma_n \cdot \sqrt{\pi a}}. \tag{4}$$

Vertical bending moment is applied to the FE model (Figure 1) using the standard assumption of linearly distributed strain along the cross-section in the vertical direction. Bending moment is imposed on the model boundary as enforced rotation around the control point, located at the neutral axis. The magnitude of the rotation is such that the nominal stress across the main deck in the intact condition reads 100 MPa. Based on such defined nominal stress range, SIFR is calculated by the described procedure using Displacement Method and Equation (3). Then, by employing Equation (4), geometry function Y is calculated along the crack path. The red line in Figure 4 shows the geometry function distribution along the crack propagation path. Results for geometry function are compared to other studies dealing with crack propagation in stiffened panels, e.g., [10,21,22], showing the same general trend of geometry function The sharp increase of Y when the crack passes the first stiffener is the consequence of the unrealistic assumption that the stiffener is immediately broken at this point. The experimental investigation has been performed, showing a smooth transition due to the partly intact stiffener over a certain time [10]. It was proposed in [10] that crack propagation through a stiffened panel may be approximated by crack propagation through the unstiffened plate, with a simple correction factor accounting for the effect of stiffeners.

SIFR for crack propagation through damaged main deck may therefore be approximated as propagation through a semi-infinite plate with stiffener correction factor:

$$\Delta K_I = \frac{2 \cdot b_s \cdot t + A_s}{2 \cdot b_s \cdot t} \cdot 1.122 \, \Delta \sigma_n \sqrt{\pi \cdot a}, \tag{5}$$

where the first term represents stiffener correction factor, while $1.122 \, \Delta \sigma_n \sqrt{\pi \cdot a}$ represents SIFR for edge crack in semi-infinite plate [23]; b_s is the distance between stiffeners; t is plate thickness while A_s is a cross-sectional area of one stiffener. Stiffener dimensions are provided in Table 3. The geometry function calculated according to Equation (5) reads 1.25. Compared with results obtained by the Displacement Method, it is found that this value represents approximately a mean value of the numerically obtained geometry function, and it is plotted as a blue line in Figure 4. This value of the geometry function is used in the further analysis of crack propagation as it is simpler and more realistic compared to the Y calculated by the Displacement Method.

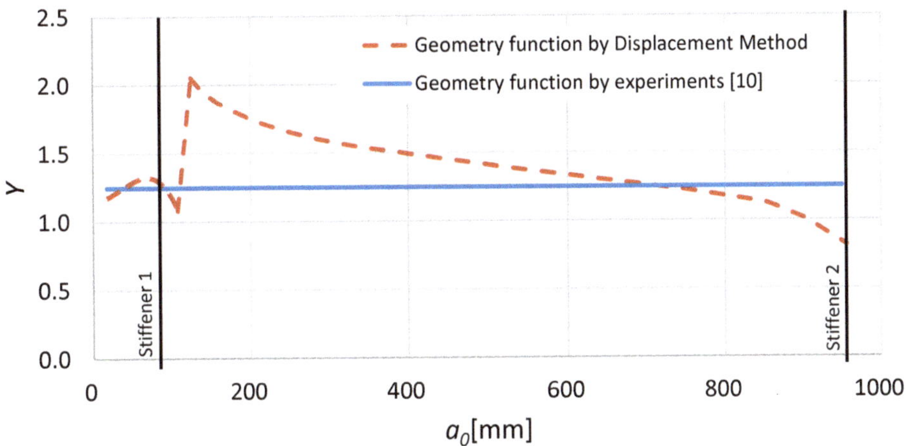

Figure 4. Geometry function Y.

Table 3. Characteristics of the plate and T-shaped stiffener $360 \times 12/140 \times 22$.

Parameter	Value
A_s [m^2]	0.0074
t [m]	0.0175
b_s [m]	0.85

2.4. Monte-Carlo Simulation of Wave-Induced Vertical Bending Moment (VWBM)

After a collision accident, the ship is subjected to some specific constant sea states until the salvage and rescue operation is completed. The salvage period during which the ship is towed to a safe port may vary from a few hours to few days, depending primarily on the location of the ship collision and weather conditions. According to Lee et al. [24], the design requirement for the damaged ship hull is to survive for four days in mean sea conditions. To apply the developed procedure on a ship subjected to such constant wave conditions, a careful definition of the sea states is necessary, in terms of significant wave height, zero crossing wave period, and wave angle. For this analysis, five different realistic wave scenarios are selected based on the wave scatter diagram of Sea Area 16 in the North Atlantic and Douglas sea scale [25,26]. Credible sea states for the analysis of the damaged ship are specified in Table 4, while only head waves are considered. Standard deviation s, for specific short-term condition, is obtained by the standard response prediction procedure by assuming the Pierson-Moskowitz wave spectrum. Crack propagation is assumed to be induced only by fluctuating global VWBM, while response amplitude operators (RAOs) of VWBM are calculated by closed-form expressions developed by Jensen and Mansour [24].

Table 4. The significant wave height (H_s) of each sea state and its corresponding zero crossing period (T_z) in Sea Area 16 for 4 days salvage period.

Sea State	H_s [m]	T_z [m]	n_c [cycles]	s [MNm]
3	1.25	7.5	46,080	43.9
4	2.5	8	43,200	113.1
5	4	8.5	40,659	199.3
6	6	9	38,400	319.3
7	9	10	34,560	512.2

The number of wave cycles n_c presented in Table 4 depends on the mean wave zero crossing period T_z and salvage period T_s as:

$$n_c = \frac{T_s}{T_z}.$$ (6)

The individual amplitudes of VWBM in each load cycle are obtained by using the MC simulation [27]. A random amplitude of VWBM in each wave cycle is simulated using the following expression obtained from Raleigh distribution:

$$M_{VBM} = s \cdot (-2 \ln c)^{\frac{1}{2}},$$ (7)

where c is the random number with uniform distribution in the interval [0, 1].

2.5. Analysis of Crack Propagation during a Salvage

As the results of crack propagation depend on the order of stress application, 5000 different simulations of wave cycles during the salvage period are performed. Each of the simulations (n_s) starts with the initial crack size a_0. The individual amplitudes of VWBM for each wave cycle are generated by MC simulation, as described in the previous Section, with parameters specified in Table 4.

The nominal applied stress amplitude σ_{nom} is calculated as the ratio of VWBM generated by MC simulation and section modulus W of the ship:

$$\sigma_{nom} = \frac{M_{VBM}}{W},$$ (8)

where section modulus W reads 32 m^3.

Nominal stress range $\Delta\sigma_{nom}$ is given as:

$$\Delta\sigma_{nom} = 2 \cdot \sigma_{nom}.$$ (9)

Using Equation (2) and corresponding parameters C, m, and a_0, crack propagation in one load cycle is calculated with the following expression:

$$\frac{\Delta a_i}{\Delta N_i} = C \cdot \left(Y_{i-1} \cdot \Delta\sigma_{nom,i} \cdot \sqrt{\pi \cdot a_{i-1}} \right)^m.$$ (10)

For small increments during one load cycle, $\Delta N = 1$, then crack growth Δa_i after one cycle, and new crack size a_i read:

$$\Delta a_i = C \cdot \left(Y_{i-1} \cdot \Delta\sigma_{nom,i} \cdot \sqrt{\pi \cdot a_{i-1}} \right)^m,$$ (11)

$$a_i = a_{i-1} + \Delta a_i.$$ (12)

A crack will propagate with each new cycle. When a total number of VWBM cycles in the salvage period is reached, the final crack size a_f is achieved. A procedure for calculation of the crack propagation during the salvage period is implemented using the MATLAB R2020b [28] programming language.

3. Results

The analysis is firstly performed for the "base case" combination of input parameters given in Table 5.

Table 5. Parameters for the "base case".

Parameter	Value
Salvage period [days]	4
C [MPa, m]	$7.27 \cdot 10^{-11}$ [11]
m	3
a_0 [m]	0.035

The analysis is performed for sea states from Table 4 and "base case" parameters given in Table 5. Ship towing in head waves is assumed. Results of MC simulations for crack propagation rate at the end of each simulation are presented as histograms in Table 5.

Figure 5 shows that the crack increase Δa for 4 days towing period is negligible except for SS7 when Δa reads between 0.073 and 0.083 m. For all Sea states, almost symmetrical bell-shaped histograms remained consistent. The average value and standard deviations of crack increase Δa during towing in different sea states are provided in Table 6.

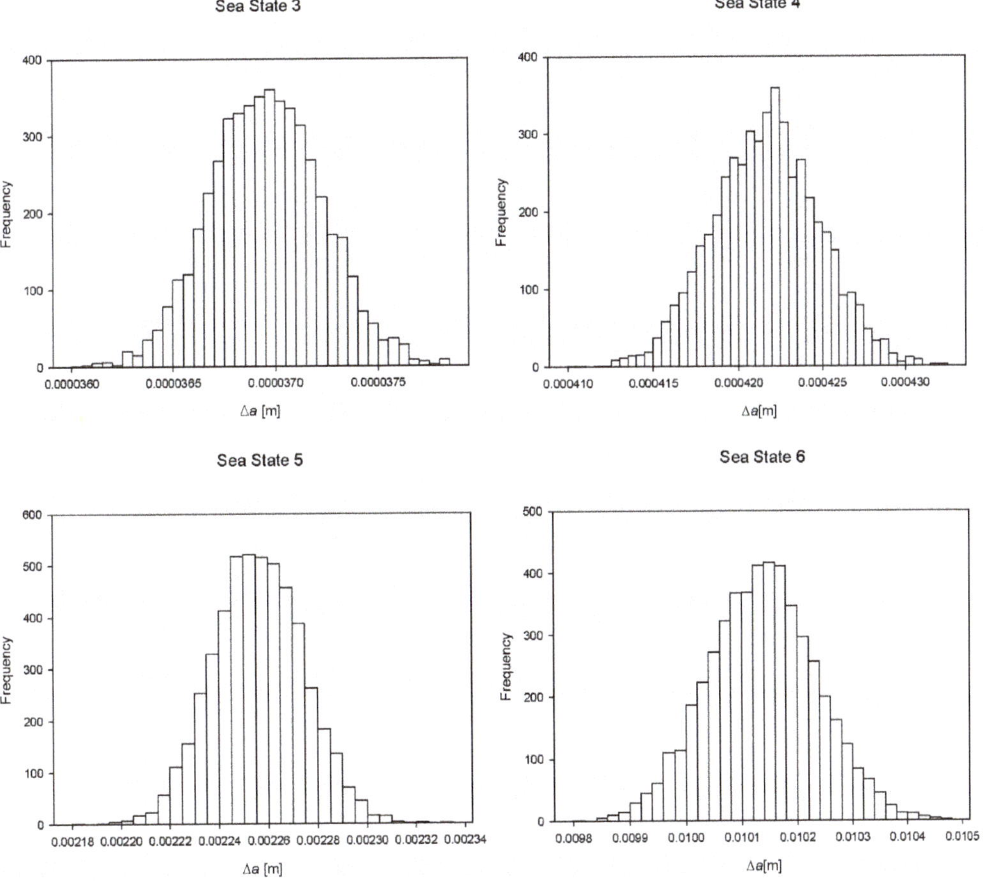

Figure 5. *Cont.*

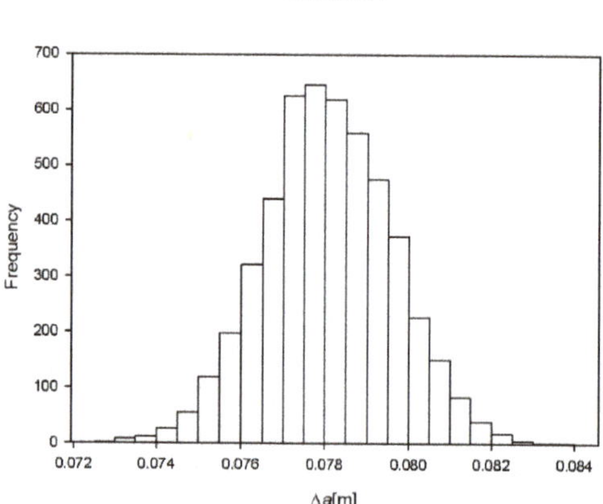

Figure 5. Histogram of 5000 simulations for each Sea state

Table 6. Mean value and standard deviation of Δa for 4 days towing period.

Sea State	Mean [m]	Standard Deviation [m]
3	$3.7 \cdot 10^{-5}$	$2.7 \cdot 10^{-7}$
4	$4.22 \cdot 10^{-4}$	$3.13 \cdot 10^{-6}$
5	$2.26 \cdot 10^{-3}$	$1.82 \cdot 10^{-5}$
6	$1.0 \cdot 10^{-2}$	$9.78 \cdot 10^{-5}$
7	$7.8 \cdot 10^{-2}$	$1.53 \cdot 10^{-3}$

Parametric Analysis

Parametric analysis of fatigue crack propagation by variation of salvage period and initial crack size a_0 is performed. Variation of parameters with respect to the "base case" is given in Table 7.

Table 7. Values of parameters for parametric analysis.

Parameter	Value
Salvage period [days]	7
a_0 [m]	0.07

As previously mentioned, the number of cycles is dependent on salvage period T_s and calculated by Equation (6). This is presented in Table 8.

Table 8. Number of cycles for each Sea state in Sea Area 16 for salvage period of 7 days.

Sea State	n_c [cycles]
3	80,640
4	75,600
5	71,153
6	67,200
7	60,480

Variation of parameters C and m have not been considered in the parametric analysis because the literature overview showed that credible values of these parameters were lower than the value set for the "base case". Variation using a lower value of these parameters would therefore result in smaller final crack lengths.

Results of the parametric analysis are presented in Figures 6 and 7 as crack propagation rate with the probability of exceedance of 1%, Δa (1%), for towing periods of 4 and 7 days, respectively. The crack propagation rate corresponding to 1% probability is calculated from the histogram. For the towing period of 4 days, there is almost no difference between crack increase for SS 3, 4, and 5, while it is very small for SS 6. However, for SS 7, this difference is observable, especially for a larger initial crack size of 70 mm. For the towing period of 7 days, as presented in Figure 7, crack propagation for $a_0 = 70$ mm becomes unstable and leads to infinity.

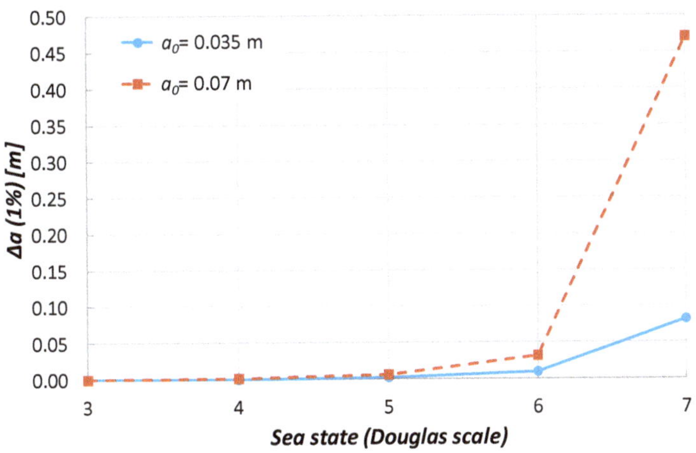

Figure 6. Crack propagation rate with the probability of exceedance of 1%, (Δa (1%)) for different SS and 4 days towing period.

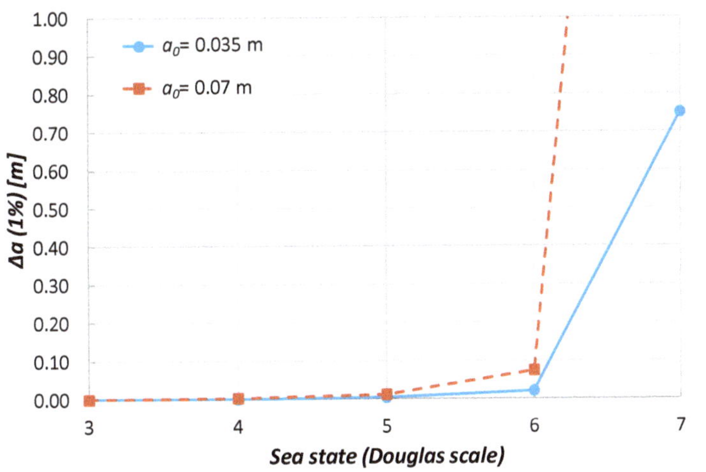

Figure 7. Crack propagation rate with the probability of exceedance of 1%, (Δa (1%)) for different SS and 7 days towing period.

4. Discussion

The procedure developed in this study may be used for the rapid calculation of crack propagation for other types of damaged ships and different damage shapes. If the FE model of the struck ship is available in practice, a pre-defined damage scenario can be employed in a similar way as described in the paper. Analysis of SIF by FE method requires fine meshing, but simplified expressions given by Equation (5) may be used as an alternative.

However, in practical application FE model may not be available, or it may be impractical to use in emergency response procedure, when computational and modelling time may be the limiting factor. In that case, a 2D sectional model may be used, as, for example, obtained by MARS software [29]. Such models are often available or can be quickly generated. The 2D sectional model with simplified collision damage is shown in Figure 8. A 2D sectional model enables the determination of the stress distribution in the main deck and the assessment of the crack propagation using the procedure described in Sections 2.2 and 2.3, where SIF, given by Equation (5), may be used.

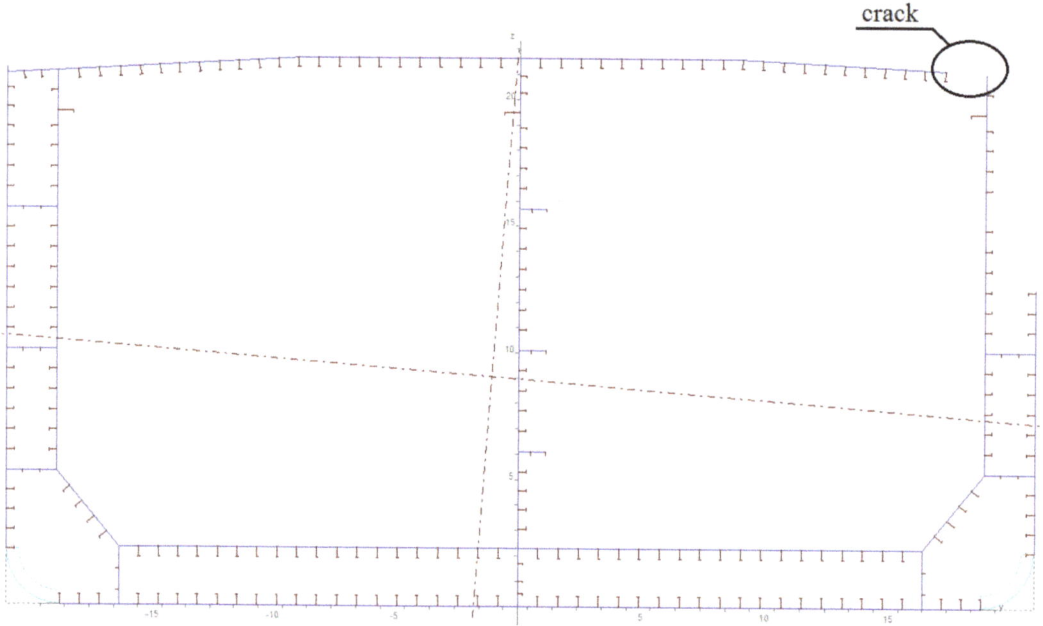

Figure 8. 2D sectional model with simplified collision damage.

The residual ultimate strength of the ship hull damaged in a collision is usually performed within emergency response procedure using progressive collapse analysis [3]. It is quite straightforward to extend such computation for the presence of fatigue crack if the crack propagation analysis shows that considerable crack during salvage may be developed. The effect of the crack on the residual strength may be simulated by deleting cracked elements, as shown in Figure 8.

Except for residual ultimate strength, the interaction between unstable fracture and collapse due to plastification needs to be considered using the Failure Assessment Diagram (FAD) [11]. This global failure mode could occur before the hull girder collapse.

Crack propagation, ultimate strength, and FAD depend not only on the dynamic (fluctuating) stress components but also on the still water stresses in a damaged condition. If the watertight integrity is compromised, effects of flooding of both cargo and void spaces and the corresponding oil outflow are to be considered [30]. The still water bending moment (SWBM) is the most important still water load component that can be significantly increased following a ship accident. In case of flooding of the cargo holds in the midship region, the tendency is to increase sagging SWBM at midship, as the weight is added in the mid-span of the ship hull girder. Such situation may decrease crack propagation in the main deck as compressive stresses would cause crack closure effect. On the contrary, tensile stresses in the bottom shell plating may increase and consequently enhance crack propagation. This indicates that the crack propagation in the bottom plating of a ship damaged in grounding may be dangerous. Practical methods for modelling bottom damage are described in [31].

For the considered crack location on the main deck at side, a combination of vertical and horizontal wave bending moment (HBM) may be considered as the critical load situation. If a damaged ship is found in quartering seas, the ratio of HBM to VWBM could be as large as 1.73 [32]. Even if the wave loading is only in the vertical plane, the presence of collision damage makes the ship hull structure asymmetrical and introduces asymmetrical bending loads. The neutral axis rotates and moves away from the damaged area [33]. These effects of the increase of stresses caused by combined bending moments are not considered herein.

The act of collision is complex, generating excessive deformation and residual stresses in the damaged area of the ship hull. Large plastification and bent plates are observed in the structures damaged by collision. This affects the initiation and propagation of cracks in a way that the prediction of cracks emanating from the damaged area is extremely difficult and uncertain. Experimental work and development of the numerical simulation of crashworthiness could help to resolve this issue in the future [15]. In the present study, however, the effect of large deformations around the damage opening is not considered. It should be mentioned that linear elastic fracture mechanics (LEFM) is used in the study, and plasticity zones resulting from the overload effect around crack tip are neglected, although according to [11], it plays a large role in the determination of the SIF. A closure effect from the plastic zone can reduce the SIF. The crack propagation rate decrease due to overloads is a very complex issue, and though it can provide somewhat safer results regarding the crack propagation analysis, it is typically not considered.

Only crack propagation in Mode I is considered herein that also deserves discussion. There are few reasons to consider only crack propagation in Mode I when calculating SIF. Firstly, shear stress in this specific case are rather small since the structural damage and fatigue loading are assumed as a symmetrical. This is a consequence of the assumption that the collision is perpendicular, i.e., that the striking and struck ship are at the right angle during the accident. Secondly, the analysis is in the first place intended for the analysis of the ultimate residual strength of damaged ship. For such an analysis, perpendicular crack, progressing in Mode I represent the "worst case" scenario. Finally, considering crack deflection because of Mode II and Mode III would introduce considerable complexity to the analysis.

Because of the damage shape complexity, there are many potential crack initiation locations. Consequently, multiple cracks could develop simultaneously, which could lead to the growth of the correlated cracks [34]. Such considerations are outside the scope of the present work.

Another aspect deserving attention is the consequences of the corrosion on the crack propagation. To get the impression of the corrosion influence, additional analysis is performed. The nominal stress range caused by fluctuating wave loads is increased by 10% because of the thickness diminution caused by the corrosion. This is a conservative assumption, as the section modulus for ship in service is normally reduced by less than 10%. Namely, the International Association of Classification Societies (IACS) requires keeping the longitudinal strength of an aging ship at the level higher than 90% of the initial state of the new building [35]. Other parameters in this study are taken as for the "base case" given in Table 5. Results of the analysis performed for each SS and for an initial crack size of 35 mm are shown in Figure 9 and Table 9. Significant increase of crack propagation because of corrosion can be noticed for higher sea states. Thus, the Δa (1%) is increased for SS 6 and SS 7 by 43% and 126%, respectively.

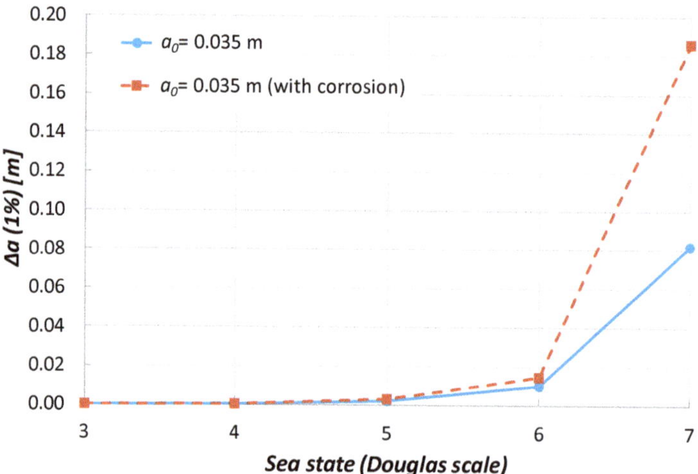

Figure 9. Comparison of crack propagation rate with probability of exceedance of 1%, (Δ*a* (1%)) for each Sea state and 4 days towing period, as-built and corroded condition.

Table 9. Increase rate of probability of exceedance Δ*a* (1%) due to corrosion.

Sea State	as-Built	Corrosion	Increase
	Δ*a* (1%) [m]	Δ*a* (1%) [m]	
3	$3.77 \cdot 10^{-5}$	$5.01 \cdot 10^{-5}$	33%
4	$4.29 \cdot 10^{-4}$	$5.74 \cdot 10^{-4}$	33.8%
5	$2.30 \cdot 10^{-3}$	$3.11 \cdot 10^{-3}$	35.2%
6	$1.04 \cdot 10^{-2}$	$1.49 \cdot 10^{-2}$	43.6%
7	$8.21 \cdot 10^{-2}$	$1.86 \cdot 10^{-1}$	126.6%

During rescue operations, sea conditions are not constant, especially for longer duration salvage instances. In such cases, it would be necessary to split total rescue duration in intervals with constant sea state properties. Crack propagation analysis could then be performed in principally same way as described herein. It should be mentioned that order of appearance of sea states is important, as the crack propagation is basically a non-linear process.

5. Conclusions

A method is proposed for the improvement of the emergency response prediction in case of a ship damaged in a collision, considering crack propagation caused by fluctuating wave loads during the salvage period. The shape and size of the damage are modelled using historical data of ship accidents. The crack propagation is based on the Paris law, while the stress intensity factor (SIF) is calculated both numerically and based on published experimental results on stiffened panels. Many Monte Carlo simulations are performed to cover different time-histories of the fatigue loading during salvage. Results of the analysis are presented as a histogram of the crack increase, where the influence of the variability of loading history is not found as very important. Parametric analysis is performed to investigate the effect of the sea state severity, initial crack size, and towing duration. Obtained results indicate that crack propagation can be significant for the Sea state 7, according to the Douglas scale (H_s = 9 m), and that crack may even become unstable for large initial crack size and long towing period. In case of damage of an aged ship, the crack could propagate faster because of the corrosion.

The proposed procedure is intended as a starting point for the improvement of a software tool for emergency response action during the salvage of damaged ships. Ideally, 2D sectional model of the ship hull could be used within such procedure. The influence of crack propagation on the residual ultimate strength of a damaged ship can be assessed using the progressive collapse analysis method, while the unstable fracture and collapse due to plastification could be investigated using the failure assessment diagram. However, it should be mentioned that further theoretical and experimental developments are necessary to model more realistically collision damage and post-accidental behavior of damaged structure regarding crack initiation and propagation.

Author Contributions: Conceptualization, I.G. and J.P.; Methodology, I.G. and J.P.; software, I.G. and A.M.; Formal analysis, I.G. and A.M.; Investigation, I.G.; Resources, I.G. and J.P.; Writing—original draft preparation, I.G. and J.P.; Writing—review and editing, I.G. and J.P.; Visualization, I.G.; Supervision, J.P.; Project administration, J.P.; Funding acquisition, J.P. All authors have read and agreed to the published version of the manuscript.

Funding: The work was fundedby the Croatian Science Foundation within the projects IP-2019-04-2085 and IP-2014-09-8658.

Institutional Review Board Statement: Not applicable.

Informed Consent Statement: Not applicable.

Data Availability Statement: The datasets analyzed or generated in this study are available from the corresponding author upon reasonable request.

Acknowledgments: The presented work was supported by the Croatian Science Foundation within the projects IP-2019-04-2085 and IP-2014-09-8658 and derived from I.G. doctoral thesis "Propagation of Damage in Ship Structure Caused by Collision or Grounding Accident" (in Croatian), under the mentorship of J.P., also supported by the Croatian Science Foundation.

Conflicts of Interest: The authors declare no conflict of interest. The funders had no role in the design of the study; in the collection, analyses, or interpretation of data; in the writing of the manuscript, or in the decision to publish the results.

References

1. Pedersen, P.T. Marine Structures: Future Trends and the Role of Universities. *Engineering* **2015**, *1*, 131–138. [CrossRef]
2. International Association of Classification Societies (IACS). Limit states. In *Common Structural Rules for Bulk Carriers and Oil Tankers (IACS CSR)*; International Association of Classification Societies: London, UK, 2014.
3. Wen, F. Rapid response damage assessment. In *Marine Technology (mt)*; Kelly, D., Ed.; SNAME: Alexandria, VA, USA, October 2017; pp. 40–47. Available online: http://sname.digitalwavepublishing.com/ (accessed on 8 June 2021).
4. Luis, R.M.; Teixeira, A.P.; Guedes Soares, C. Longitudinal strength reliability of a tanker hull accidentally grounded. *Struct. Saf.* **2008**, *31*, 224–233. [CrossRef]
5. Parunov, J.; Prebeg, P.; Rudan, S. Post-accidental structural reliability of double-hull oil tanker with near realistic collision damage shapes. *Ships Offshore Struct.* **2020**, *15* sup1, S190–S207. [CrossRef]
6. Gledić, I.; Parunov, J.; Prebeg, P.; Ćorak, M. Low-cycle fatigue of ship hull damaged in collision. *Eng. Fail. Anal.* **2019**, *96*, 436–454. [CrossRef]
7. Kwon, S.; Vassalos, D.; Mermiris, G. Adopting a risk-based design methodology for flooding survivability and structural integrity in collision/grounding accidents. In Proceedings of the 11th International Ship Stability Workshop, Wageningen, The Netherlands, 21–23 June 2010; pp. 1–8.
8. Sasa, K.; Incecik, A. New Evaluation on ship strength from the view point of stranded casualties in coastal areas under rough water. In Proceedings of the 28th International Conference on Ocean, Offshore and Arctic Engineering (OMAE), Honolulu, HI, USA, 31 May–5 June 2009; pp. 43–50.
9. Bardetsky, A. Fracture mechanics approach to assess the progressive structural failure of a damaged ship. In *Collision and Grounding of Ships and Offshore Structures*; Amdahl, J., Ehlers, S., Leira, B.J., Eds.; Taylor & Francis Group: London, UK, 2013; pp. 77–84.
10. Fricke, W.; Petershagen, H. Fatigue crack propagation in plate panels with welded stiffeners. In Proceedings of the Annual Assembly of International Institute of Welding, Hobart, Australia, July 1988; Volume: IIW-Doc. XIII-1272-88.
11. Bureau Veritas. *Guidelines for Fatigue Assessment of Steel Ships and Offshore Units*; Guidance Note NI 611 DT R00 E; Bureau Veritas: Neuilly sur Seine Cedex, France, 2016.
12. Broek, D. *The Practical Use of Fracture Mechanics*; Kluwer Academic Publishers: Dordrecht, The Netherlands, 1989; pp. 1–20.

13. British Standard. *Guide to Methods for Assessing the Acceptability of Flaws in Metallic Structures BS 7910*; British Standards Institution: London, UK, 2005.
14. Lützen, M. Ship Collision Damage. Ph.D. Thesis, Department of Mechanical Engineering, Technical University of Denmark, Kgs. Lyngby, Denmark, 2001.
15. Ringsberg, J.; Amdahl, J.; Chen, B.; Cho, S.; Ehlers, S.; Hu, Z.; Kubiczek, J.M.; Kõrgesaar, M.; Liu, B.; Marinatos, J.N.; et al. MARSTRUCT benchmark study on nonlinear FE simulation of an experiment of an indenter impact with a ship side-shell structure. *Mar. Struct.* **2018**, *59*, 142–157. [CrossRef]
16. Guinea, V.G.; Planas, J.; Elices, M. K_I evaluation by the displacement extrapolation technique. *Eng. Fract. Mech.* **2000**, *66*, 243–255. [CrossRef]
17. Carroll, L.B.; Tiku, S.; Dinovitzer, A.S. *Rapid Stress Intensity Factor Solution Estimation for Ship Structure Applications (SSC-429)*; Ship Structure Committee: Washington, DC, USA, 2003.
18. De Morais, A. Calculation of stress intensity factors by the force method. *Eng. Fract. Mech.* **2007**, *74*, 739–750. [CrossRef]
19. Han, Q.; Wang, Y.; Yin, Y.; Wang, D. Determination of stress intensity factor for mode I fatigue crack based on finite element analysis. *Eng. Fract. Mech.* **2015**, *138*, 118–126. [CrossRef]
20. Laird II, G.; Epstein, S.J. Fracture mechanics and finite element analysis. *Mech. Eng.* **1992**, *114*, 69–73.
21. Dexter, R.J.; Pilarski, P.J. Crack Propagation in Welded Stiffened Panels. *J. Constr. Steel Res.* **2002**, *58*, 1081–1102. [CrossRef]
22. Dexter, R.J.; Mahmoud, H.N.; Pilarski, P. Propagation of Long Cracks in Stiffened Box-sections under Bending and Stiffened Single Panels under Axial Tension. *Int. J. Steel Struct.* **2005**, *5*, 181–188.
23. Rooke, D.P.; Baratta, F.I.; Cartwright, D.J. Simple methods of determining stress intensity factors. *Eng. Fract. Mech.* **1981**, *14*, 397–426. [CrossRef]
24. Jensen, J.J.; Mansour, A.E. Estimation of Ship Long-Term Wave-Induced Bending Moment Using Closed-Form Expressions. *R. Inst. Nav. Archit.* **2002**, *W291*, 41–55.
25. Lee, Y.; Chan, H.S.; Pu, Y.; Incecik, A.; Dow, R.S. Global wave loads on damaged ship. *Ships Offshore Struct.* **2012**, *7*, 237–268. [CrossRef]
26. Sun, F.; Pu, Y.; Chan, H.S.; Dow, R.S.; Shahid, M.; Das, P.K. *Reliability-Based Performance Assessment of Damaged Ships*; Ship Structure Committee Report No. 459; Ship Structure Committee: Washington DC, USA, 2011.
27. Chen, N.Z. A stop-hole method for marine and offshore structures. *Int. J. Fatigue* **2016**, *88*, 49–57. [CrossRef]
28. MathWorks (Matlab Documentation). Available online: http://www.mathworks.com/help/index.html (accessed on 8 June 2021).
29. MARS. *User's Manual*; Bureau Veritas: Paris, France, 2020.
30. Bužančić Primorac, B.; Ćorak, M.; Parunov, J. Statistics of still water bending moment of damaged ship. In *Analysis and Design of Marine Structures*; Guedes Soares, C., Shenoi, R.A., Eds.; Taylor and Francis Group: London, UK, 2015; pp. 491–497.
31. Heinvee, M.; Tabri, K. A simplified method to predict grounding damage of double bottom tankers. *Mar. Struct.* **2015**, *43*, 22–43. [CrossRef]
32. Khan, I.A.; Das, P.K. Reliability Analysis of Intact and Damaged Ships Considering Combined Vertical and Horizontal Bending Moments. *Ships Offshore Struct.* **2008**, *3*, 371–384. [CrossRef]
33. Fujikubo, M.; Zubair, M.A.; Takemura, K.; Iijima, K.; Oka, S. Residual hull girder strength of asymmetrically damaged ships. *J. Jpn. Soc. Naval Arch. Ocean Eng.* **2012**, *16*, 131–140. [CrossRef]
34. Feng, G.Q.; Garbatov, Y.; Guedes Soares, C. Probabilistic model of the growth of correlated cracks in a stiffened panel. *Eng. Fract. Mech.* **2012**, *84*, 83–95. [CrossRef]
35. Paik, J.K.; Wang, G.; Thayamballi, A.K.; Lee, J.M.; Park, Y.I. Time-dependent risk assessment of aging ships accounting for general/pit corrosion, fatigue cracking and local denting damage. *Trans. Soc. Nav. Archit. Mar. Eng.* **2003**, *111*, 159–197.

Journal of
Marine Science and Engineering

Article

A Probabilistic Method for Estimating the Percentage of Corrosion Depth on the Inner Bottom Plates of Aging Bulk Carriers

Špiro Ivošević [1], Romeo Meštrović [1] and Nataša Kovač [2,*]

[1]   Faculty of Maritime Studies Kotor, University of Montenegro, Dobrota 36, 85330 Kotor, Montenegro;
     spiroi@ucg.ac.me (S.I.); romeo@ucg.ac.me (R.M.)
[2]   Faculty of Applied Sciences, University of Donja Gorica, Oktoih 1, Donja Gorica,
     81000 Podgorica, Montenegro
*   Correspondence: natasa.kovac@udg.edu.me

Received: 31 May 2020; Accepted: 14 June 2020; Published: 16 June 2020

Abstract: This paper presents an approach for the model estimating the probabilistic percent corrosion depth for inner bottom plates of fuel oil tanks located in the double bottom of aging bulk carriers. Assuming that corrosion begins after four years of exploitation, a statistical approach to investigations on the ratio of the corrosion rate and the average initial inner bottom plate's thickness of considered bulk carriers is given. We consider this ratio to be a random variable since it is included in the usual linear corrosion model. By applying adequate statistical tests to the available empirical dataset, three best fitted three-parameter distributions for estimating the cumulative density function and the probability density function of the random variable were obtained. These three distributions were further used to estimate the studied percentage of corrosion depth. Lastly, we present the corresponding numerical and graphical results concerning the obtained statistical and empirical results and give concluding remarks.

Keywords: bulk carrier; fuel oil tanks; inner bottom plates; corrosion rate; percentage of corrosion depth; probabilistic method; three-parameter distribution

---

1. Introduction

Corrosion is one of the most frequent degradation mechanisms that affects ship hulls as they age. Although there are numerous types of corrosion, such as galvanic corrosion, fretting corrosion, bio-corrosion, and pitting corrosion, the most widespread type that damages ship hull structures is general corrosion. The long periods of exploitation include operations with cargo. Additionally, the influential factors of the environment and maintenance all affect the corrosion rate over time. A high intensity of corrosion causes a faster wear of structural parts and requires a significant investment in maintenance. The maintenance and repairs delay the usage of ships, increase the expenses, and shorten the life expectancy of ships.

Previous studies have mainly focused on tankers, bulk carriers, and ships considered to be the most susceptible to the rapid decay caused by corrosion. Moreover, previously conducted research has been based on the analysis of corrosion mechanisms [1–3], influential factors of the environment [2,4–6], and the decay of particular structural parts of bulk carriers [4,7–14].

The findings reported so far have multiple benefits for shipbuilders, ship repair yards, ship owners, ship management companies, insurance companies, and classification societies. The quality standards and allowable wear limits are important for all stakeholders since each of the them has a specific role in the process of shipbuilding and exploitation.

Ship designers aim to create optimal ships with a flawless performance, and shipbuilders attempt to ensure a high quality of built-in materials, devices, and the services of suppliers and other stakeholders. The research conducted so far has investigated corrosion mechanisms over time in order to predict the corrosion margin that should be considered during the construction of steel plates. Furthermore, ship owners want to use their ships for a long period of time and postpone potential repairs and replacements of corroded surfaces. This can be achieved by choosing an appropriate maintenance method, which can significantly contribute to slowing down the corrosion process and reducing the number of repairs. Moreover, higher quality steels and adequate painting surface protection delay initial corrosion, which significantly extends the life cycle of steel ship structures. In this sense, the rules of classification societies regarding further exploitation and the usage of corroded surfaces represent a limiting factor. The rules clearly define the types of wastage of structural parts or entire structural areas.

This paper is organized in the following way. Section 2 explains the motivation for this study and relies on a survey investigating the analytic and probabilistic corrosion rate estimation models in relation to ship hull structures. The primary focus is the investigation of corrosion wastage of the inner bottom plates of bulk carriers. Additionally, the same section presents the data collection methods, a brief description of the input data set, and the probabilistic method developed in this study. The purposefully developed probabilistic model is a modified version of a frequently used linear probabilistic model that has been employed in previous studies and in our recent works [14–16]. In relation to this model, Section 3 describes the application of the statistical tests that introduced three best fitted three-parameter continuous distributions for the random variable $\frac{c_1}{d_0}$ into the considered class consisting of numerous three-parameter continuous distributions. Section 4 presents concluding remarks.

2. Materials and Methods

Currently, all stakeholders in the maritime industry are primarily concerned about the sustainability of ships in exploitation and the protection of the environment. However, the attitudes regarding ship sustainability vary among stakeholders. For example, ship owners and managers require the maximum exploitation of ships, while classification societies and flag states require compliance with the defined standards aimed at protecting people, material goods, and the environment. Likewise, in order to assess the need for repair and partial or complete replacement of the parts that are damaged by corrosion, classification societies analyze the allowed wear expressed as a percentage in relation to the original thickness of steel plates. A less significant wear percentage indicates minor corrosion and a longer period of exploitation, while excessive wear indicates notable corrosion and requires the replacement of a greater number of corroded surfaces. Classification societies prescribe allowable wear limits that ensure optimal security and protection of the environment, while ship owners want to maximally exploit their ships, up to the limits defined, with an intention to reduce the cost of maintenance and repairs.

Previous studies have examined the different types of corrosion such as general, pitting, cavitation, and galvanic corrosion. General corrosion was researched in relation to a reduction in the thickness of steel plates and expressed as the millimeters of wear [4,6,7,14,16]. Additionally, the thickness reduction caused by the pitting corrosion of steel plates has been analyzed in terms of the thickness wear percentage. Most authors have employed an analytic and probabilistic method to determine the millimeters of wear [12–17]. However, this paper established a corresponding methodology that suggested the best distributions for assessing the wear percentage of steel plates from a probabilistic and statistical perspective.

Monitoring the condition of ships in exploitation detected the beginning of corrosion. Based on previous research and the observation of an extensive database, it was assumed that the corrosion of susceptible areas starts after several years of exploitation [6,14,16] while, in less sensitive areas, the beginning of corrosion is postponed. For instance, some authors have assumed that corrosion starts after more than five years after construction (see References [10,11] where deck plates and ballast tanks are considered).

Previous research [12–14,16] has confirmed that inner bottom plates are significantly influenced by various factors that affect the rapid decay of plates. Regarding this, one side of the inner bottom plates was exposed to ballast water, dry space, and fuel oil tanks, while the other side was affected by cargo. More precisely, inner bottom plates were in constant contact with cargo in cargo holds and under the influence of handling equipment and maintenance processes such as cleaning before and after cargo operations. Therefore, this paper focuses on the plates constituting the parts of fuel oil tanks. Since fuel oil tanks become warm over time, the emergence of corrosion is naturally faster and requires early maintenance and repairs. This motivated research on corrosion rates expressed as the percentage of metal thickness reduction in relation to the original thickness.

In this regard, assuming that corrosion starts after four years of ship exploitation, this paper proposed a probabilistic method for estimating the corrosion depth percentage. The point at which the total wear reached 10% of the entire structural area of the fuel oil tank sheets was determined based on the estimations made and corresponded to the regulations. According to the estimations made, the structural area examined cannot be used throughout the 25 years of the projected life expectancy of ships.

2.1. Data Collecting Methodology

This research examined bulk carriers whose age varied between five and twenty-five years. Since previous research has proven that inner bottom plates and the bottom of cargo holds are the most susceptible to decay caused by corrosion, this study focused on the steel plates of fuel oil tanks. The steel plates are the sheets of fuel oil tanks and subsidiary cargo holds. Consequently, steel plates are exposed to different influences—ower sides are affected by warm fuel, while upper sides are under the influence of the atmosphere (if cargo holds are empty) or cargo (if cargo holds are full).

This research encompasses both sides of fuel oil tanks—the portside (P) and starboard (S) side. The data were collected by means of special surveys and regular measurements of each bulk carrier. The data obtained indicated the general corrosive decay expressed as the percentage of the reduction in the original steel thickness. The database does not include the damage caused by damages, cracks, or stress.

Since the fuel is located in the oil tanks (which are below inner bottom plates) in all bulk carriers, examined corrosion emerges from cargo holds, i.e., affects the top of sheets. The measuring of the thickness reduction was performed for cargo holds, but not for oil tanks.

Measuring ship hull structures required compliance with special procedures and standards with the aim of estimating a real condition of all structural parts. Although each special survey required the measuring of all structural elements of bulk carrier construction that are in contact with the sea, atmosphere, and cargo, this paper focused on the measured data regarding the wear of the sheets of fuel oil tanks located in double bottom areas. These steel plates are the inner bottom plates of cargo holds. Since the amount of measured data is precisely defined by classification societies, we selected the data in accordance with a predetermined order. Measuring the examined area only represents part of the extensive measuring of ships in exploitation. The selection included the data that objectively presented the condition of the structural area. In this sense, with a view toward ensuring a proper data analysis, this paper examined each sheet of fuel oil tanks as a separately measured surface. Each surface was transversally and longitudinally intersected by tin extending along the entire sheets of the double bottom of cargo holds.

The length of each fuel oil tank corresponded to the length of a cargo hold, except in three cases, when the lengths of fuel oil tanks equaled double the length of cargo holds. For that reason, only three tanks were in the way of two cargo holds and had 10 cross sections considered in this research. Each fuel oil tank in the way of a cargo hold was divided into five sections: two sections for after and before ends, and three sections at equal mutual distances in the middle, between the ends of tanks (Figure 1).

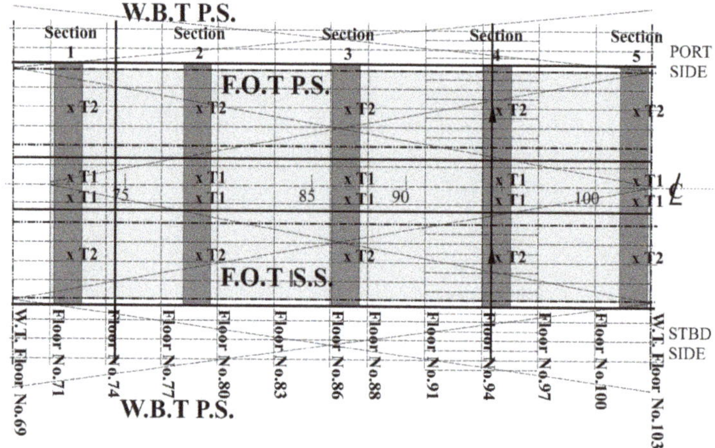

Figure 1. Collecting data scheme for inner bottom plates in the way of fuel oil tanks.

Each cross section included one measured point on each tin transversal on the fuel oil tanks that were included in the section, and is represented by an average value of the section. In this way, the average value of the section wear percentage represents several measured points. This established a database containing a total of 570 sections, which indicated the wear percentage of each section, i.e., the wear of inner bottom plates in relation to the original thickness of fuel oil tanks on the bulk carriers examined.

Considering that corrosion begins after four years of exploitation and in accordance with the statistical, graphical, and obtained numerical results, it can be assumed that three three-parameter continuous distributions are suitable for a probabilistic model for estimating the corrosion depth percentage of inner bottom plates of bulk carriers. The estimation concerned the percentage of thickness reduction of built-in structure plates of fuel oil tanks. Statistical tests confirmed this assumption with a high precision.

2.2. A Brief Description of the Input Data Set

As previously indicated, this paper only analyzed bulk carriers whose decay is, over time, caused by general corrosion and expressed as the percentage of the wear of steel plates. Based on previous research conducted by the authors of this paper [14,16,17], the examined database included 25 bulk carriers in exploitation whose age varied between five and twenty-five years. The measured data were gathered between 2005 and 2017, where certain bulk carriers were measured three times and others only once in special surveys. The analysis focused on four ships whose age was between five and ten years, seven ships that were 15 years old, 13 ships that were 20 years old, and 10 ships that were 25 years old. The measurements were conducted through 38 different special surveys, which established an input database for further analysis. Table 1 exhibits the database analyzed, including the number of ships in relation to their age, and the number of special surveys, fuel oil tanks, measured points, and sections examined. The last column of Table 1 presents the average values of wear, depending on the age intervals of the ships. A total of 110 fuel oil tanks were examined through 2810 measured points and 570 sections.

2.3. The Proposed Problem and Related Methodology

For a good survey of investigations of the analytic and probabilistic corrosion rate estimation model for different hull structural elements of bulk carriers, see the survey paper by Qin and Cui [18].

It is known that the corrosion wastage, $d(t)$ (at moment t), may be generally expressed as a power function of time (usually expressed in years) after the corrosion starts [18,19], i.e.,

$$d(t) = c_1(t - T_{cl})^{c_2}. \tag{1}$$

where $d(t)$ is the corrosion wastage, t is the elapsed time after the plate is used, T_{cl} is the life of the coating, and c_1 and c_2 are positive real coefficients. As noticed in Reference [18], in most of the related studies on the time-dependent reliability of ship structures [20–24], the effect of corrosion was represented by an uncertain but constant corrosion rate, which resulted in a linear decrease of plate thickness with time. Accordingly, for the subject of research of this paper, we use expression (1) with $c_2 = 1$ proposed by Paik, Kim, and Lee [4], and a related linear corrosion model is defined as

$$d(t) = c_1(t - T_{cl}), \tag{2}$$

where c_1 is the corrosion rate, usually expressed in mm/year.

Table 1. A database of the percentage of thickness reduction of inner bottom plates.

The Age of Ships (Years)	The Number of Ship Surveys	The Number of Tanks	The Number of Measured Points	The Number of Sections	The Average Values of Plate Thickness Reduction Caused by Corrosion (%)
0–5	4	9	230	45	0.5
5–10	4	10	266	55	2.8
10–15	7	19	500	100	9.8
15–20	13	43	998	220	11.7
20–25	10	29	816	150	17.7
SUM:	38	110	2810	570	

In the case of the investigation of the corrosion wastage of an arbitrary structural area of a group of bulk carriers, it can be of interest to estimate a percentage, $p(t)$ of corrosion wastage with respect to the average initial bulk carriers' structural area thickness. More precisely, if we denote the average initial structural area of the considered group of bulk carriers by $\overline{d_0}$, then it is natural to define a related value of $p(t)$ as:

$$p(t) = \frac{d(t)}{\overline{d_0}} \left(\text{with } t > T_{cl} \right). \tag{3}$$

Dividing Equation (2) by $\overline{d_0}$, we can obtain the following.

$$p(t) = \frac{d(t)}{\overline{d_0}} = \frac{c_1}{\overline{d_0}} (t - T_{cl}) \ (\text{with } t > T_{cl}). \tag{4}$$

Since the effect of corrosion is generally of an uncertain nature, the coefficient c_1 can be considered as a continuous random variable. Since $\overline{d_0}$ is a real constant, this is also true for the ratio $\frac{c_1}{\overline{d_0}}$. Hence, the determination of the values $p(t)$ is closely related to the probabilistic estimation of a continuous random variable $\frac{c_1}{\overline{d_0}}$ for determining the best fitted three-parameter continuous distributions for the probability density function (PDF) and the cumulative density function (CDF) for $\frac{c_1}{\overline{d_0}}$. To the best of our knowledge, three-parameter continuous distributions have very rarely been considered in related literature. Very recently, in Reference [25], the suggested theoretical distribution for a related initial distribution approach is the three-parameter-Weibull distribution. Applying a Kolmogorov-Smirnov test, the probabilistic estimation model for the random variable $\frac{c_1}{d_0}$ and related applications are given in the next section.

3. Results

Assuming that corrosion starts after four years of exploitation and, based on the application of Equation (2) from Section 2.3, this section proposes a probabilistic method for estimating the random variable $\frac{c_1}{d_0}$ that was defined in Section 2.3 for the inner bottom plates of the investigated fuel oil tanks on 38 aging bulk carriers. Namely, the section suggests a statistical approach to approximating the CDF of the variable $\frac{c_1}{d_0}$. The relevant estimations were only detected among three-parameter continuous distributions, which defined the three best-fitted distributions for the random variable $\frac{c_1}{d_0}$. Expression (4) where $T_{cl} = 4$, i.e.,

$$\frac{c_1}{d_0} = \frac{d(t)}{(t-4)\overline{d_0}} \tag{5}$$

was used for these purposes.

Figure 2 presents a total of 570 measurements of plate thickness expressed as the percentage of the reduction of the original thickness due to corrosion on the inner bottom plates of 38 ships surveyed. The data set includes all measured points from each plate in corresponding transversal sections. The following text presents different multi-parameter functions that were validated by corresponding tests. Based on the tests considered, the paper analyzes the average values of the obtained parameters in order to determine the linear dependence of the values examined and the point at which the average values exceeded the allowable limits defined by classification societies.

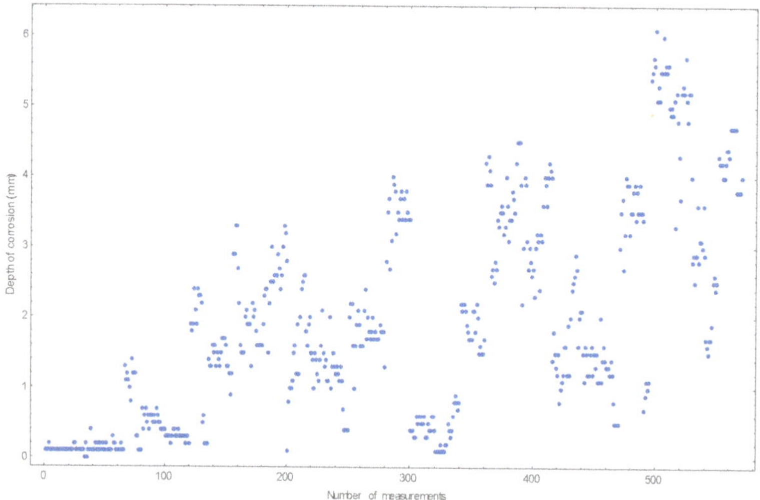

Figure 2. The data of 570 thickness measurements of inner bottom plates obtained through 38 ship surveys.

3.1. Appropriate Statistical Analysis Related to Measurements of Inner Bottom Plates

Table 2 presents the numerical characteristics of 465 empirical data that are expressed as the percentage of corrosion wastage and examined in this paper. Table 2 contains the standard statistical parameters that describe the empirical data (first and third quartiles are denoted as Q1 and Q3 respectively). These parameters describe the shape of the data as well as some of the related percentiles.

Table 2. The descriptive statistics of the empirical data for $T_{cl} = 4$ years.

Statistic	Value	Percentile	Value
Sample Size	465	Min	0.00129
Range	0.02041	5%	0.00197
Mean	0.00788	10%	0.00284
Variance	1.5617×10^{-5}	25% (Q1)	0.00448
Standard Deviation	0.00395	50% (Median)	0.00746
Coefficient of Variation	0.5017	75% (Q3)	0.01111
Standard Error	1.8326×10^{-4}	90%	0.01336
Skewness	0.22245	95%	0.01424
Excess Kurtosis	−0.83477	Max	0.0217

As previously indicated, a total of 570 measurements of corrosion wear were performed in the structural area of inner bottom plates (IBP) of bulk carriers and all obtained values are expressed as millimeters. The papers by Ivošević et al. [14] and Ivošević et al. [16] examined the same dataset consisting of 570 measured data on plate thickness reduction due to the corrosion of IBP (i.e., related values $d(t)$ on the depth of corrosion, expressed as millimeters), which were measurements carried out on 38 bulk carriers. It is known that, for practical reasons, a difference between the measured steal thickness and initial thickness of less than 0.3 mm can be considered negligibly small (no corrosion due to steel under the paint), and, hence, those measurements were eliminated from our further statistical analysis. In this sense, the database relevant for the paper consists of 465 measurements on the corrosive wear of the IBP area. This section statistically analyzes the corrosion wear of steel plates expressed as the wear percentage in relation to the average values of the original thickness of inner bottom plates. Therefore, the 465 sets of data expressed as millimeters were first converted into equivalent percentages. Since the initially measured steel wear caused by corrosion was measured in millimeters, the examined data set had to be expressed as a percentage, i.e., as the ratio $(\frac{d(t)}{\overline{d_0}})$ of corrosion wear at a particular moment t ($d(t)$) to the original thickness of inner bottom plates ($\overline{d_0}$) (see Section 2.3). The IBP area was constructed of steel plates whose thickness varied. For the purposes of further analysis, the steel plates were grouped into the following categories, depending on the original thickness: [16,17], [17,18], [18,19], [19,20], [20,21], [21,22], [22,23] (where all values are expressed as mm). The ordinal numbers of intervals are denoted by i, where $i \in \{1, 2, \ldots, 7\}$, and the average value of an interval is denoted by d_i, where $d_i \in \{16.5, 17.5, 18.5, 19.5, 20.5, 21.5, 22.5\}$. Since the original thickness of the steel examined was considered when measuring, the entire data set (i.e., all measured points) can be related to one of the previously formed intervals. In this way, it is possible to determine the total number of points that belong to the formed interval i ($i \in \{1, 2, \ldots, 7\}$). The average value of the original thickness of inner bottom plates ($\overline{d_0}$) was calculated in accordance with the grouped data in the following way.

$$\overline{d_0} = \frac{1}{N} \sum_{i=1}^{7} d_i n(d_i) \tag{6}$$

where $n(d_i)$ denotes the frequency of d_i ($i \in \{1, 2, \ldots, 7\}$).

Since N is the total number of samples (measured points) based on the dataset used for the statistical analysis, the calculated value of the original average thickness is $\overline{d_0} = 18.43455$ mm. Taking into account the average value obtained for the original thickness of inner bottom plates ($\overline{d_0}$), the data expressed as mm were easily converted into a percentage. A probabilistic model that can describe the wear percentage caused by corrosion of the steel plates of the IBP was developed based on the percentage data listed in Table 2. An initial assumption in the development of this model was that $T_{cl} = 4$ years and that the ratio $\frac{c_1}{\overline{d_0}}$ is a continuous random variable (see Section 2.3).

The fitting of theoretical distributions based on a set of empirical data includes three basic phases.

1. The choice of an adequate model, i.e., the choice of optimal theoretical distributions,
2. The determination of the optimal values for the parameters that characterize theoretical distributions,
3. The determination of the significance level and the quality of the distributions fitted.

Each theoretical distribution depends on the set of initial parameters that describe its characteristics. The most frequent parameters of theoretical distributions are the location, scale, shape, and threshold parameters. The location parameter describes the position of the graph of the theoretical distribution in relation to the horizontal axis (x-axis). The scale parameter defines the degree of dispersion. The shape parameter describes the tendency of a distribution to skew to the left, while the threshold parameter describes the lowest value that a distribution reaches in relation to the x-axis. A distribution does not necessarily have all the parameters listed. There are two-parameter and multi-parameter distributions.

This paper focuses on the theoretical distributions that are characterized by three parameters, i.e., three-parameter distributions. The optimal set of parameters that characterizes relevant distributions should be determined during the distribution fitting process in order to determine the best distributions that adequately describe the set of empirical percentage data. For this purpose, this paper relies on the maximum likelihood estimation method (e.g., see Hays, [26]), which is one of the most commonly used methods for determining the unknown parameters of theoretical distributions. This method is based on a procedure that detects the parameters that maximize the likelihood of the observed data that occur with respect to a theoretical model. In order to determine the goodness of fit (e.g., see Stephens, [27]), the last phase applies the Kolmogorov–Smirnov test. This test compares empirical and theoretical models by computing the maximum absolute difference between empirical and theoretical distribution functions. The first column of Table 3 exhibits an overview of the most prominent three-parameter distributions that successfully fit into the set of the 465 empirical data previously described. The second column of Table 3 presents the parameters that describe each theoretical distribution, along with the values obtained by means of the maximum likelihood estimation method. The theoretical distributions in Table 3 are not sorted in accordance with the goodness of fit, but in alphabetical order.

Table 3. Fitted three-parameter distributions of $\frac{c_1}{d_0}$ for IBP where $T_{cl} = 4$ years.

Distribution	Parameters
Burr	$k = 4.2273E + 13\ \alpha = 0.93196\ \beta = 3.4010x10^{12}$
Dagum	$k = 0.08817\ \alpha = 15.238\ \beta = 0.01399$
Erlang	$m = 7\ \beta = 0.0016\ \gamma = -0.00262$
Error	$k = 4.1549\ \sigma = 0.00395\ \mu = 0.00788$
Fatigue Life	$\alpha = 0.21019\ \beta = 0.01867\ \gamma = -0.0112$
Frechet	$\alpha = 1.0402E + 8\ \beta = 3.4816x10^{5}\ \gamma = -3.4816x10^{5}$
Gamma	$\alpha = 6.5552\ \beta = 0.0016\ \gamma = -0.00262$
Gen. Extreme Value	$k = -0.1979\ \sigma = 0.00381\ \mu = 0.00631$
Gen. Gamma	$k = 0.94793\ \alpha = 3.6606\ \beta = 0.00198$
Gen. Logistic	$k = 0.04888\ \sigma = 0.00226\ \mu = 0.00769$
Gen. Pareto	$k = -0.81361\ \sigma = 0.01158\ \mu = 0.00149$
Inv. Gaussian	$\lambda = 0.45761\ \mu = 0.0195\ \gamma = -0.01163$
Log-Logistic	$\alpha = 7.6305\ \beta = 0.01799\ \gamma = -0.01047$
Log-Pearson3	$\alpha = 6.1211\ \beta = -0.2504\ \gamma = -3.4722$
Lognormal	$\sigma = 0.18192\ \mu = -3.8403\ \gamma = -0.01397$
Pearson 5	$\alpha = 57.607\ \beta = 1.6874\ \gamma = -0.02194$
Pearson 6	$\alpha 1 = 3.4686\ \alpha 2 = 6330.4\ \beta = 14.369$
Pert	$m = 0.00571\ a = 7.0282x10^{-4}\ b = 0.0232$
Power Function	$\alpha = 0.57172\ a = 0.00129\ b = 0.0217$
Triangular	$m = 0.00129\ a = 0.00128\ b = 0.02175$
Weibull	$\alpha = 1.9729\ \beta = 0.00845\ \gamma = 3.8091x10^{-4}$

All three-parameter distributions were rated based on the Kolmogorov–Smirnov test, which identified the three best fitted three-parameter distributions, i.e., the three distributions that best described the empirical data of the percentage of corrosion wear. Based on the distribution of the random variable $\frac{c_1}{d_0}$ for IBP where $T_{cl} = 4$ years, the Kolmogorov–Smirnov test indicated that the three best fitted three-parameter distributions were Generalized Pareto, Error, and Log-Pearson 3. These three distributions and corresponding fitted parameters are presented in Table 4.

Table 4. The three best fitted three-parameter distributions of $\frac{c_1}{d_0}$ for IBP with $T_{cl} = 4$ years.

Distribution	Parameters
Gen. Pareto	$k = -0.8\ \sigma = 0.01\ \mu = 0.00149$
Error	$k = 7.0\ \sigma = 0.08\ \mu = 0.16$
Log-Pearson 3	$\alpha = 5.7\ \beta = -0.26\ \gamma = -0.54$

The second column of Table 4 shows the parameters and corresponding values that characterize the best fitted three-parameter distributions. These parameters are defined as follows.

(1) For the Generalized Pareto distribution:

k—continuous shape parameter $(-\infty < \sigma < +\infty)$;

σ—continuous scale parameter $(0 < \sigma < +\infty)$;

μ—continuous location parameter $(-\infty < \mu < +\infty)$.

The PDF, $f_{GP}(x)$, of the Generalized Pareto distribution with its domain $-\infty < x \leq \mu$ when $k \geq \mu$ and $\mu \leq x \leq \mu - \rho/k$ when $k < 0$ is defined as:

$$f_{GP}(x) = \frac{1}{\sigma}\left(1 + \frac{k(x-\mu)}{\rho}\right)^{-\left(\frac{1}{\xi}-1\right)} \tag{7}$$

The CDF of the Generalized Pareto distribution, $F_{GP}(x)$, is defined as:

$$F_{GP}(x) = \begin{cases} 1 - \left(1 + k\frac{(x-\mu)}{\sigma}\right)^{-1/k} & k \neq 0 \\ 1 - exp\left(-\frac{(x-\mu)}{\sigma}\right) & k = 0 \end{cases} \tag{8}$$

(2) For the Error distribution:

k—continuous shape parameter,

σ—continuous scale parameter $(\sigma > 0)$,

μ—continuous location parameter.

The PDF, $f_E(x)$, of the Error Distribution for its domain $-\infty < x < +\infty$ is defined as:

$$f_E(z) = \frac{c_1}{\sigma}exp(-|c_0 z|^k) \tag{9}$$

where $c_0 = \left(\frac{\Gamma\left(\frac{3}{k}\right)}{\Gamma\left(\frac{1}{k}\right)}\right)^{1/2}$, $c_1 = \frac{kc_0}{2\Gamma\left(\frac{1}{k}\right)}$, and $z = \frac{x-\mu}{\sigma}$.

The CDF, $F_E(x)$, of the Error Distribution for its domain $-\infty < x < +\infty$ is defined as:

$$F_E(x) = 0.5\left(1 + \frac{\Gamma_{|c_0 x^k|}(1/k)}{\Gamma(1/k)}\right) \text{ when } x \geq \mu, \text{ and} \tag{10}$$

$$F_E(x) = 0.5\left(1 - \frac{\Gamma_{|c_0 x^k|}(1/k)}{\Gamma(1/k)}\right) \text{ when } x < \mu. \tag{11}$$

(3) For the Log-Pearson 3 Distribution:

α—continuous shape parameter $(0 < \alpha < +\infty)$,
β—continuous scale parameter $(\beta \neq 0)$,
γ—continuous location parameter.

The PDF, $f_{LP}(x)$, of the Log-Pearson 3 Distribution for its domain $e \leq x < +\infty$ when $\beta < 0$ and $e \leq x < +\infty$ when $\beta \geq 0$, is defined as:

$$f_{LP}(x) = \frac{1}{\left(x|\beta|\Gamma(\alpha)\right)\left(\frac{lnx-\gamma}{\beta}\right)^{\alpha-1}} e^{-\left(\frac{lnx-\gamma}{\beta}\right)} \tag{12}$$

The CDF, $F_{LP}(x)$, of the Log-Pearson 3 Distribution is defined as:

$$F_{LP}(x) = \frac{\Gamma_{(\ln(x)-\gamma)/\beta}(\alpha)}{\Gamma(\alpha)} \tag{13}$$

The data from Table 4, together with the expressions (1)–(4), imply the following statistical results, given in Tables 5 and 6. In these tables, $f_{GP}(x)$, $f_E(x)$, and $f_{LP}(x)$, $(F_{GP}(x)$, $F_E(x)$, and $F_{LP}(x))$ denote the probability density functions of the best fitted Generalized Pareto, Error, and Log-Pearson 3 distributions (the cumulative distribution functions) for $\frac{c_1}{d_0}$, respectively.

Table 5. The empirical values and the values of PDF for the three best fitted three-parameter distributions for $\frac{c_1}{d_0}$ with respect to IBP.

		IBP with T_{cl} = 4 Years			
Lower Bound	Upper Bound	Empirical PDF of $\frac{c_1}{d_0}$	$f_{GP}(x)$	$f_E(x)$	$f_{LP}(x)$
0	0.00227	0.073	0.06681	0.08237	0.05704
0.00227	0.00454	0.187	0.18958	0.15123	0.18019
0.00454	0.00681	0.191	0.18092	0.18039	0.21897
0.00681	0.009079	0.155	0.17050	0.18290	0.19497
0.009079	0.011349	0.157	0.15752	0.17981	0.14648
0.011349	0.013619	0.161	0.13914	0.14778	0.09680
0.013619	0.015889	0.067	0.09552	0.06668	0.05684
0.015889	0.018159	0.006	0.00001	0.00870	0.02941
0.018159	0.020429	0.000	0	0.00014	0.01306
0.020429	0.022699	0.002	0	0	0.00474

Table 6. The empirical values and the values of cumulative density functions (CDF) for the three best fitted distributions for $\frac{c_1}{d_0}$ with respect to IBP.

		IBP T_{cl} = 4 Years			
Lower Bound	Upper Bound	Empirical CDF of $\frac{c_1}{d_0}$	$F_{GP}(x)$	$F_E(x)$	$F_{LP}(x)$
0	0.00227	0.073	0.06681	0.08237	0.05704
0.00227	0.00454	0.260	0.25639	0.23360	0.23723
0.00454	0.00681	0.451	0.43731	0.41399	0.45620
0.00681	0.009079	0.606	0.60781	0.59689	0.65117
0.009079	0.011349	0.763	0.76533	0.77670	0.79765
0.011349	0.013619	0.924	0.90447	0.92448	0.89445
0.013619	0.015889	0.991	0.99999	0.99116	0.95129
0.015889	0.018159	0.997	1	0.99986	0.98070
0.018159	0.020429	0.999	1	1	0.99376
0.020429	0.022699	0.999	1	1	0.99850

Figures 3 and 4 present the graphs of PDF and CDF of the three best fitted three-parameter distributions of $\frac{c_1}{d_0}$ for IBP where $T_{cl} = 4$ years, as given in Table 4. Figure 5 shows the probability difference graph of the three best fitted three-parameter distributions for $\frac{c_1}{d_0}$ for IBP where $T_{cl} = 4$ years. The probability difference graph was formed in order to show the quality of the fitted distributions by means of a comparison of the corresponding theoretical CDF and empirical CDF (see Figure 5). Additionally, the probability difference graph enables a comparison of the goodness- of-fit of the three best fitted distributions presented in Table 4. The graph in Figure 5 represents the difference between the empirical CDF and theoretical CDF and illustrates the quality of the fitted distributions, bearing in mind that the difference does not exceed the interval $(-0.06, 0.06)$.

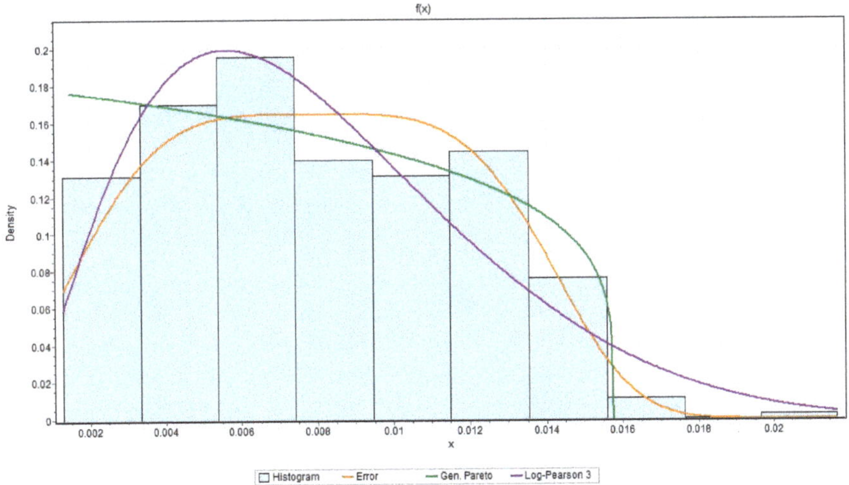

Figure 3. The empirical PDF and the PDF of the three best fitted three-parameter distributions of $\frac{c_1}{d_0}$ for IBP with $T_{cl} = 4$ years.

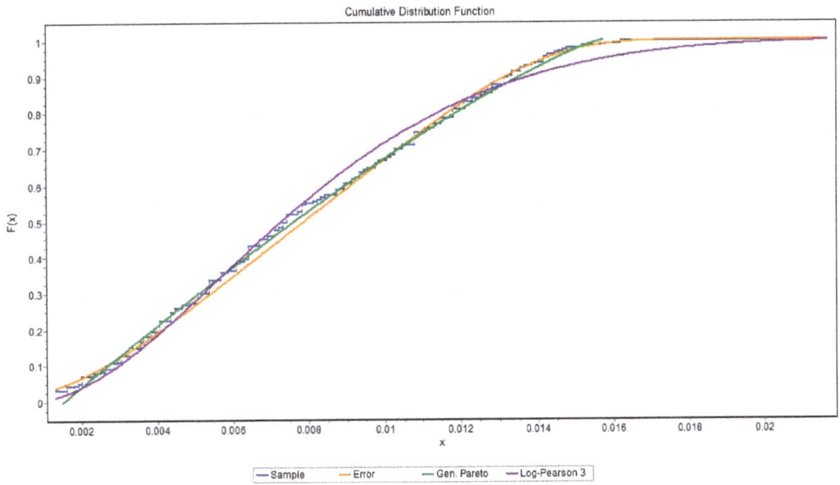

Figure 4. The empirical CDF and the CDF of the three best fitted three-parameter distributions of $\frac{c_1}{d_0}$ for IBP with $T_{cl} = 4$ years.

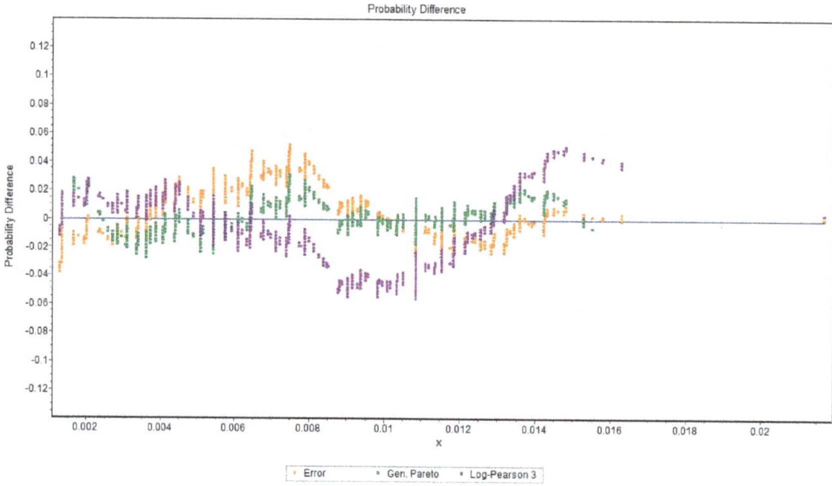

Figure 5. The probability difference graph of the three best fitted three-parameter distributions of $\frac{c_1}{d_0}$ for IBP with $T_{cl} = 4$ years.

Furthermore, Figure 6 presents probability-probability (P-P) plots of the three best fitted three-parameter distributions of $\frac{c_1}{d_0}$ for IBP where $T_{cl} = 4$ years given in Table 4. According to Figure 6, it is evident that the Generalized Pareto, Error, and Log-Pearson 3 distributions properly describe the tendencies of the empirical data since the deviations from the line shown in the graph are negligible. On the basis of the linearity observed in Figure 6, it can be concluded that the three theoretical distributions adequately shape all aspects of the empirical data including the location and scope.

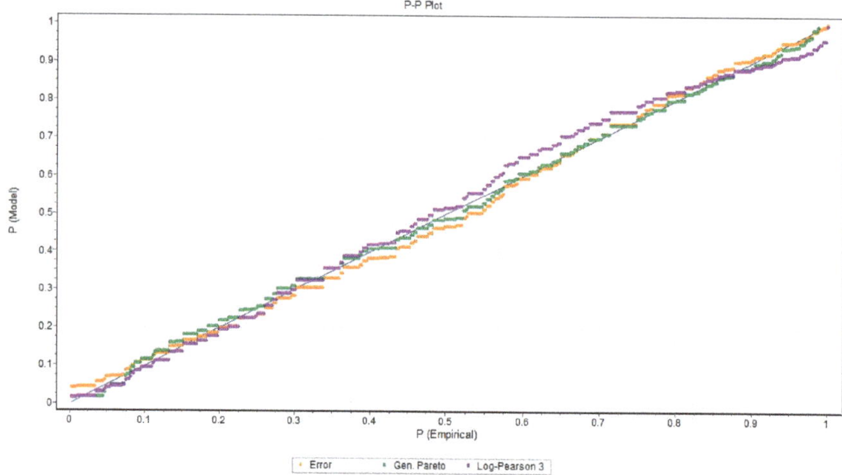

Figure 6. Probability-probability (P-P) plots of the three best fitted three-parameter distributions of $\frac{c_1}{d_0}$ for IBP with $T_{cl} = 4$ years.

Table 7 shows the statistical values and corresponding p-values in the case of the three best fitted three-parameter distributions that were previously presented in Table 4. Since the three calculated p-values are below the value chosen for the significance level (α), it is evident that the three theoretical distributions properly follow the empirical data. The same conclusion is still relevant for a lower significance level (e.g., 0.02 or 0.01).

Table 7. Test statistical values for the best fitted three-parameter distributions.

Distribution	Statistic	*p*-Value
Gen. Pareto	0.03075	0.75958
Error	0.05153	0.16339
Log-Pearson 3	0.05805	0.08375

The second column of Table 8 shows the average values of the wear percentage of steel, which were obtained on the basis of empirical data on the structural area of the IBP on the examined bulk carriers that were 5, 10, 15, 20, and 25 years old. The equation $\frac{d(t)}{d_0} = 0.00793523\,(t-4)$ was obtained upon fitting a line based on these data. More precisely, the solution of the equation $\frac{d(t)}{d_0} = 0.10$ is $t = 16.602$ years, while the solution of the equation $\frac{d(t)}{d_0} = 0.15$ is $t = 22.903$ years. These facts are presented in Figure 7.

Table 8. Average values of $\frac{d(t)}{d_0}$ for IBP.

Years	$\dfrac{d(t)}{d_0}$
5	0.0216986
10	0.0399617
15	0.104289
20	0.116896
25	0.166935

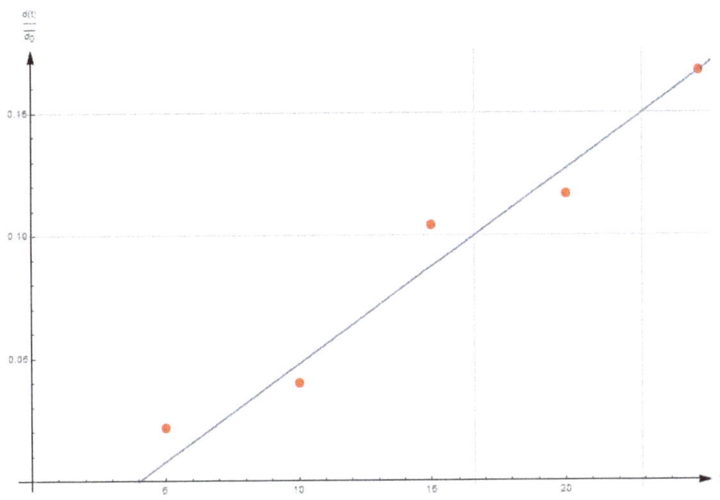

Figure 7. Estimates of $\frac{d(t)}{d_0}$ for IBP as a function of t.

The key parameter of the estimation of the further exploitation of the structural area of bulk carriers (bottom and side shell plating, deck plating, and inner bottom plating) is based on the percentage of the allowable wear of the entire structural area. Although particular cases of allowable wear of inner bottom plates vary between 15% and 30% (depending on the ship size and classification society), classification societies usually allow the wear of 10% for the entire structural area. In that regard, the data presented in Figure 7 are in accordance with the values of wear percentage obtained for the entire inner bottom area.

We conducted an additional analysis in order to verify whether (in statistical terms) it can be confirmed that the proposed methodology follows the empirical data. It is well-known that the expected value of the Generalized Pareto distribution is equal to (see Reference [28]):

$$E(GP) = \mu + \frac{\sigma}{1-k} \tag{14}$$

Bearing in mind the results presented in Table 4 and the previous formula, it can be concluded that the mean of the best fitted three-parameter distribution of variable $\frac{c_1}{d_0}$ for IBP with $T_{cl} = 4$ years is equal to $E(GP) = 0.00704556$.

In order to estimate the percentage values of corrosion $p(t)$ (with $t > 4$ years) given by Equation (4), we applied the usual statistical assumption that the mean value of $\frac{c_1}{d_0}$ is equal to its expected value, i.e., for the General Pareto distribution of $\frac{c_1}{d_0}$, its expected value is $E(GP) = 0.00704556$/year. Substituting this "probability approximation" into Equation (5), we found that:

$$p(t) = \frac{d(t)}{d_0} = \frac{c_1}{d_0}(t-4) \approx E(GP)(t-4) = 0.00704556(t-4) \tag{15}$$

and it follows that:

$$p(t) = \frac{d(t)}{d_0} = \frac{c_1}{d_0}(t-4) \approx E(GP)(t-4) = 0.00704556(t-4) = 0.7046(t-4)\%. \tag{16}$$

The numerical results given in Table 9 present the estimated values of $p(t)$. These estimations were obtained by applying Equation (16) and varying variable T_{cl} from 5 to 24 years.

Table 9. Estimated values of $p(t) = \frac{d(t)}{d_0}$ for IBP with $T_{cl} > 4$ years (expressed as a percentage) in accordance with Equation (16).

T_{cl}	$p(t)$	T_{cl}	$p(t)$	T_{cl}	$p(t)$	T_{cl}	$p(t)$
5	0.704556	10	4.22734	15	7.75012	20	11.2729
6	1.40911	11	4.93189	16	8.45467	21	11.9775
7	2.11367	12	5.63645	17	9.15923	22	12.682
8	2.81822	13	6.341	18	9.86378	23	13.3866
9	3.52278	14	7.04556	19	10.5683	24	14.0911

From Table 9, it can be seen that the calculated numerical values are very close to the related empirical data presented in Figure 7. The correctness of the proposed methodology is confirmed by this analysis and the fact that the numerical values follow the empirical data set well.

As can be seen from Figure 7, a linear curve intersects the value of 10% of the wear of the structural area after 16.602 years while the value of 15% of the total wear is achieved after 22.903 years. This proves that the corrosion processes on the inner bottom plates of fuel oil tanks are so intensive that the entire structural area should be replaced after 16.602 years. Considering the fact that a projected life expectancy of ships is between 20 and 25 years, the structural area examined in this paper and presented in Figure 7 cannot withstand the projected life expectancy due to excessive corrosive processes.

4. Conclusions

This paper is an extension of the authors' investigations on the probabilistic corrosion rate estimation model for inner bottom plates of bulk carriers. It has presented a probabilistic method for estimating the percentage of corrosion depth on the inner bottom plates of a considered group of aging bulk carriers. By applying the usual linear corrosion model, we have focused our attention on a statistical approach toward estimating the ratio of the corrosion rate and the average initial inner bottom plates thickness of considered bulk carriers. Motivated by several earlier investigations and the fact that many factors influence the corrosion wastage of ship hull structures and are of an uncertain nature, the previously defined ratio was considered as a continuous random variable.

Applying the Kolmogorov-Smirnov test at a significance level of 0.05, it was confirmed that, of the three-parameter continuous distributions, the Generalized Pareto distribution, Error distribution, and Log-Pearson 3 distribution were the best fitted distributions for the defined ratio. We have also given related statistical, numerical, and graphical results. These numerical and graphical results show that, in a statistical sense, the three best-fitted distributions for the defined ratio closely match the corresponding empirical data and follow all of their important characteristics well. In particular, the numerical results related to the Generalized Pareto distribution confirm the fact that the values calculated for estimating the studied percentage of corrosion depth are very close to the corresponding empirical data.

The proposed methodology clearly shows the dependence of the corrosion depth as a function of time. This methodology based on statistical analysis allows us to determine the final limits of utilizing the inner bottom plating of fuel oil tanks and to compare them with those defined by the rules of classification societies.

The proposed statistical methodology could be applied for related investigations involving other ship hull structure members of bulk carriers. Clearly, related statistical analysis can be performed for a large class of two-parameter continuous distributions.

Author Contributions: Conceptualization, Š.I.; Data curation, Š.I.; Formal analysis, R.M. and N.K.; Investigation, Š.I.; Methodology, Š.I., R.M. and N.K.; Project administration, Š.I.; Resources, Š.I.; Software, N.K.; Supervision, R.M.; Validation, Š.I., R.M. and N.K.; Visualization, N.K.; Writing—original draft, Š.I., R.M. and N.K.; Writing—review & editing, R.M. All authors have read and agreed to the published version of the manuscript.

Funding: This research received no external funding.

Acknowledgments: The approved Thickness Measurement Company - INVAR-Ivošević Company supported this research. More information about the company can be found at URL: http://www.invar.me/index.html. Namely, the data collected and systematized during the last 25 years by the company operators and experts have been included in the above simulation and probabilistic analysis of the corrosion effects for the analyzed group of 10 aged bulk carriers fuel tanks. It should be pointed out that the INVAR-Ivošević Company provides its customers with marine services of ultrasonic thickness measurements of vessels' hull structures and it has valid certificates issued by recognized classification societies: LR, BV, DNV, GL, RINA, and ClassNK. Currently, more than 300 vessels, mainly aged bulk carriers, are being inspected by the company.

Conflicts of Interest: The authors declare no conflicts of interest

References

1. Paik, J.F.; Brennan, F.; Carlsen, C.A.; Daley, C.; Garbatov, Y.; Ivanov, L.; Rizzo, C.; Simonsen, B.C.; Yamamoto, N.; Zhuang, H.Z. Report of Committee V.6 Condition Assessment of Aging Ships. In Proceedings of the 16th International Ship and Offshore Structures Congress, Southampton, UK, 20–25 August 2006.

2. Roberts, S.E.; Marlow, P.B. Casualties in dry bulk shipping (1963–1996). *Mar. Policy* **2002**, *26*, 437–450. [CrossRef]

3. Bulk carrier casualty report, IMO, MSC 83/INF.6, 3 July 2007. Available online: http://www.safedor.org/resources/MSC_83-INF-8.pdf (accessed on 16 June 2020).

4. Paik, J.K.; Kim, S.K.; Lee, S.K. A probabilistic corrosion rate estimation model for longitudinal strength members of bulk carriers. *Ocean Eng.* **1998**, *25*, 837–860. [CrossRef]

5. Guedes Soares, C.; Garbatov, Y.; Zayed, A. Effect of environmental factors on steel plate corrosion under marine immersion conditions. *Corros. Eng. Sci. Technol.* **2011**, *46*, 524–541. [CrossRef]

6. Paik, J.K.; Thayamballi, A.K.; Park, Y.I.; Hwang, J.S. A time-dependent corrosion wastage model for seawater ballast tank structures of ships. *Corros. Sci.* **2004**, *46*, 471–486. [CrossRef]

7. Yamamoto, N.; Ikagaki, K. A Study on the Degradation of Coating and Corrosion on Ship's Hull Based on the Probabilistic Approach. *J. Offshore Mech. Arct. Eng.* **1998**, *120*, 121–128. [CrossRef]

8. Guo, J.; Wang, G.; Ivanov, L.; Perakis, A.N. Time-varying ultimate strength of aging tanker deck plate considering corrosion effect. *Mar. Struct.* **2008**, *21*, 402–419. [CrossRef]

9. Garbatov, Y.; Guedes Soares, C.; Wang, G. Non-linear time dependent corrosion wastage of deck plates of ballast and cargo tanks of tankers. In Proceedings of the 22nd International Conference on Offshore Mechanics and Arctic Engineering, OMAE 2005–67579, Halkidiki, Greece, 12–17 June 2005.

10. Jurišić, P.; Parunov, J.; Garbatov, Y. Aging effects on Ship Structural integrity. *Brodogradnja/Shipbuilding* **2017**, *68*, 15–28. [CrossRef]

11. Jurišić, P.; Parunov, J.; Garbatov, Y. Comparative analysis based on two nonlinear corrosion models commonly used for prediction of structural degradation of oil tankers. *Trans. FAMENA* **2014**, *38*, 21–30.

12. Paik, J.K.; Lee, J.M.; Park, Y.I.; Hwang, J.S.; Kim, C.W. Time–variant ultimate longitudinal strength of corroded bulk carriers. *Mar. Struct.* **2003**, *16*, 567–600. [CrossRef]

13. Paik, J.K.; Thayamballi, A.K.; Park, Y.I.; Hwang, J.S. A time-dependent corrosion wastage model for bulk carrier structures. *Int. J. Marit. Eng.* **2003**, *145 Pt A2*, 61–87.

14. Ivošević, Š.; Meštrović, R.; Kovač, N. Probabilistic estimates of corrosion rate of fuel tank structures of aging bulk carriers. *Int. J. Naval Arch. Ocean Eng.* **2019**, *11*, 165–177. [CrossRef]

15. Ivošević, Š. Analysis of Ships Hull Structural Degradation. Ph.D. Thesis, University of Montenegro, Maritime Faculty Kotor, Kotor, Montenegro, 2012.

16. Ivošević, Š.; Meštrović, R.; Kovač, N. An approach to the probabilistic corrosion rate estimation model for inner bottom plates of bulk carriers. *Brodogradnja/Shipbuilding* **2017**, *68*, 57–70. [CrossRef]

17. Ivošević, Š.; Meštrović, R.; Kovač, N. A comparison of some multi-parameter distributions related to estimation of corrosion rate of aging bulk carriers. In Proceedings of the 7th International Conference on Marine Structures, Dubrovnik, Croatia, 6–8 May 2019; CRC Press: Boca Raton, FL, USA, 2019; pp. 403–410.

18. Qin, S.; Cui, W. Effect of corrosion models on the time-dependent reliability of steel plated elements. *Mar. Struct.* **2003**, *16*, 15–34. [CrossRef]

19. Paik, J.K.; Thayamballi, A.K. Ultimate strength of aging ships. *J. Eng. Marit. Environ.* **2002**, *1*, 57–77.

20. Guedes Soares, C.; Garbatov, Y. Reliability of maintained ship hulls subjected to corrosion. *J. Ship Res.* **1996**, *40*, 235–243.

21. Guedes Soares, C.; Garbatov, Y. Reliability of maintained ship hull girders subjected to corrosion and fatigue. *Struct. Saf.* **1998**, *20*, 201–219. [CrossRef]

22. Guedes Soares, C.; Garbatov, Y. Reliability of plate elements subjected to compressive loads and accounting for corrosion and repair. In *Structural Safety and Reliability*; Shiraishi, N., Shinozuka, M., Wen, Y.K., Eds.; Balkema: Rotterdam, The Netherlands, 1998; Volume 3, pp. 2013–2020.

23. Guedes Soares, C.; Garbatov, Y. Reliability of corrosion protected and maintained ship hulls subjected to corrosion and fatigue. *J. Ship Res.* **1998**, *43*, 65–78.

24. Guedes Soares, C.; Garbatov, Y. Reliability of maintained ship hulls subjected to corrosion and fatigue under combined loading. *J. Constr. Steel Res.* **1999**, *52*, 93–115. [CrossRef]

25. Katalinić, M.; Parunov, J. Uncertainties of Estimating Extreme SignificantWave Height for Engineering Applications Depending on the Approach and Fitting Technique—Adriatic Sea Case Study. *J. Mar. Sci. Eng.* **2019**, *8*, 259. [CrossRef]

26. Hays, W.L. *Statistics for the Social Sciences*, 2nd ed.; Rinehart & Winston: New York, NY, USA, 1973.

27. Stephens, M.A. Introduction to Kolmogorov (1033) on the empirical determination of a distribution. In *Breakthroughs in Statistics*; Springer: New York, NY, USA, 1992.

28. De Smith, M.J. *Statistical Analysis Handbook a Comprehensive Handbook of Statistical Concepts, Techniques and Software Tools*; The Winchelsea Press: London, UK, 2018.

Journal of
Marine Science and Engineering

MDPI

Article

Long-Term Marine Environment Exposure Effect on Butt-Welded Shipbuilding Steel

Goran Vukelic [1,2,*], Goran Vizentin [1,2], Josip Brnic [3], Marino Brcic [3] and Florian Sedmak [4]

[1] Marine Engineering Department, Faculty of Maritime Studies, University of Rijeka, Studentska 2, 51000 Rijeka, Croatia; vizentin@pfri.hr
[2] Center for Marine Technologies, Faculty of Maritime Studies, University of Rijeka, M. Baraca 19, 51000 Rijeka, Croatia
[3] Department of Engineering Mechanics, Faculty of Engineering, University of Rijeka, Vukovarska 58, 51000 Rijeka, Croatia; mbrcic@riteh.hr (M.B.); brnic@riteh.hr (J.B.)
[4] 3. Maj Shipyard, Liburnijska 3, 51000 Rijeka, Croatia; florian.sedmak@uljanik.hr
* Correspondence: gvukelic@pfri.hr

Abstract: Extreme environments, such as marine environments, have negative impacts on welded steel structures, causing corrosion, reduced structural integrity and, consequently, failures. That is why it is necessary to perform an experimental research sea exposure effect on such structures and materials. Research presented in this paper deals with the mechanical behavior of butt-welded specimens made of AH36 shipbuilding steel when they are exposed to a natural marine environment (water, seawater, sea splash) for prolonged periods (3, 6, 12, 24, and 36 months). The usual approach to such research is to perform accelerated tests in a simulated laboratory environment. Here, relative mass change due to corrosion over time is given along with calculated corrosion rates. Corroded surfaces of specimens were inspected using optical and scanning electron microscopy and comparison, based on the numbers and dimensions of the corrosion pits (diameter and depth) in the observed area. As a result, it can be concluded that exposure between 3 and 6 months shows significant influence on mass loss of specimens. Further, sea splash generally has the most negative impact on corrosion rate due to the combined chemical and mechanical degradation of material. Pit density is the highest at the base metal area of the specimen. The diameters of the corrosion pits grow over the time of exposure as the pits coalesce and join. Pit depths are generally greatest in the heat affected zone area of the specimen.

Keywords: marine environment effect; corrosion; pitting; AH 36 steel

Citation: Vukelic, G.; Vizentin, G.; Brnic, J.; Brcic, M.; Sedmak, F. Long-Term Marine Environment Exposure Effect on Butt-Welded Shipbuilding Steel. *J. Mar. Sci. Eng.* **2021**, *9*, 491. https://doi.org/10.3390/jmse9050491

Academic Editors: Joško Parunov and Yordan Garbatov

Received: 9 April 2021
Accepted: 29 April 2021
Published: 1 May 2021

Publisher's Note: MDPI stays neutral with regard to jurisdictional claims in published maps and institutional affiliations.

1. Introduction

Structural integrity of marine steel structures can be greatly reduced by corrosive environments in which they are set. Corrosion can be caused by outer (e.g., water, seawater, sea atmosphere, tides) or inner factors (e.g., ballast water, fuel oil, aggressive cargo) [1,2]. Consequently, loss of structural integrity can lead to failure of steel structures. Marine propulsion systems are generally susceptible to corrosion, leading to a series of possible mechanical failures [3], which is why it is important to gain insight into their fracture behaviors, even if it is the case of marine shaft steels or marine exhaust steels with improved corrosion resistance [4,5]. Damage of masts on the training vessel was caused by deficient welds at the foundation of the masts [6] that accumulated water, enabled corrosion, and reduced the integrity of the structure. A cargo ship [7] sunk due to corrosive influence of the cargo where leaking acid accelerated corrosion of the floor panels, leading to failure. A numerical study showed that corrosion defects in pipelines significantly reduce failure pressure, leading to pipe bursts [8]. Finite element analysis revealed that a barge midship section can decrease for more than a third compared to its initial measure, due to corrosion [9].

Pitting corrosion can be considered one of the most damaging forms of corrosion. Small holes, or pits, are caused by corrosive medium that attacks limited areas on the surface of metal. Stochastic nature of pit distribution poses a threat to the integrity of corroded steel structures. Thus, characterization and quantification of size and distribution of pits is necessary to estimate the structural integrity loss. Different techniques, such as 3D profile measurement [10], X-ray tomography [11], scanning electron microscopy [12] or white light interferometry [13] can be used for such task. As for the effects of pitting corrosion on the marine structures, a recent experimental study revealed that surface machining considerable affects corrosion resistance of steel in seawater environments, even provoking stress-corrosion cracking that has origins in corrosion pits [14]. Further, a study on pitting corrosion resistance of the austenitic stainless steel welded joint revealed that the weld zone is the most critical for pit initiation [15]. Moreover, localized pitting occurs in marine steel-reinforced concrete structures at air-voids in concrete-steel interfaces, provoking further corrosion [16]. In regards to corrosion of steel reinforcement in concrete, a comprehensive model referencing field and laboratory observations, built into the bi-modal solution, was developed [17]. Moreover, a comprehensive study with the aim of developing an advanced technique to predict time-dependent damage caused by the corrosion of marine structures was developed with a corrosion depth formulating as a function of time [18], and a technique applied to determine damage of the ship ballast tank [19].

When dealing with steel welded joints exposed to a corrosive environment, the problem of hydrogen embrittlement should be considered. It was found that cathodic protection can lead to hydrogen embrittlement and that the welds are especially endangered [20]. Moreover, there is a challenge in overcoming indirect environmental effects on the quality of welded joints, in regards to the influence of the storage environment on welding consumables [21].

Additionally, when trying to resolve a problem of corrosion propagation over time, several authors proposed different models of time-variant or time-dependent corrosion wastage models. Just by chronologically selecting the efforts made in the last two decades, a mathematical model that provides statistical characteristics of corrosion wastage as a function of time [22] can be emphasized. Further, a corrosion wastage model was proposed based on a standard non-linear time dependent corrosion model modified by the effects of the different environmental factors contained in the crude oil tank atmosphere [23] or marine atmosphere [24]. As the statistical scatter of corrosion wastage is usually wide, and the reliability of the model relies on it, a model with a formulated Weibull function was developed to account that matter [25]. A similar approach was used to propose a time-dependent pit depth corrosion model for sub-sea gas pipelines [26]. Application of a time-dependent corrosion analysis was used to assess the corrosion impact on automotive structures [27] as they are also prone to corrosion damage and failures [28]. As for the corrosion wastage models for steel bars in concrete, effort was made to develop a model that considers the effects of longitudinal cracking on the chloride and oxygen diffusion [29] or integrate chemo–physical–mechanical, and electrochemical, modeling techniques in an integrated framework [30].

Regarding previously performed studies of the marine environment's effect on ship-building steel corrosion, most of the research was performed in laboratory settings simulating real environments [31] in an accelerated manner [10]. Obviously, there is a lack of long-term experiments in the natural environment [32] focused on the steel corrosion effects. The results of usual laboratory experiments that are performed under controlled conditions and in an accelerated manner prove useful, but they may not relate adequately to the natural environment when investigating corrosion mechanisms [33].

Questions from previous findings served as motivation for the research presented here. The goal of this research was to investigate the prolonged influences of different types of marine environments on butt-welded AH36 steel. Results can be subsequently used in the material selection stage of the structural design process.

2. Materials and Methods

AH36 steel butt-welded specimens were exposed to a real marine environment (water, seawater, sea splash) for prolonged periods (3, 6, 12, 24, and 36 months) in order to determine the influence of these factors on corrosive behavior of such materials. Once extracted, relative mass change of specimens due to corrosion over time was measured. Results are presented in correlation to period of exposure and the type of environment. Moreover, corrosion rates were calculated. Specimen surfaces were inspected using optical and scanning electron microscopy and profilometry. Comparison, based on the numbers and dimensions of corrosion pits in the observed area, is given for specimens exposed to different environments and for different locations on specimens (base metal, heat affected zone, weld metal).

2.1. Material

The presented research was performed on AH36 shipbuilding steel. It is a low-alloy high-strength steel under ASTM A131 standard and a common choice for building large commercial ships, bulk carriers, ferries, and marine and offshore structures, such as pipelines, power plants, port facilities, and oil platforms in general [34]. Minimum yield strength of AH36 should be 355 MPa, while ultimate tensile strength should be in the 490–620 MPa range at room temperature. The most conventional welding methods can be used on AH36 as it has excellent weldability [35].

Chemical composition of tested steel is given in Table 1, along with equivalent carbon content (CE) calculated according to [36]:

$$CE_q = C + \frac{Si}{25} + \frac{Mn + Cu}{16} + \frac{Ni}{40} + \frac{Cr}{10} + \frac{Mo}{15} + \frac{V}{10} \tag{1}$$

Table 1. Chemical composition of AH36 steel (w_t %).

C	Si	Mn	P	S	Cr	Cu	Al	Ti	V	Mo	Ni	CE_q
0.157	0.392	1.501	0.014	0.003	0.03	0.015	0.042	0.003	0.003	0.08	0.01	0.2763

2.2. Specimen Preparation and Marine Environment Exposure

In order to prepare specimens for the experiment, AH36 plates of 12 mm thickness were first machined into 320 × 110 mm rectangular weld blanks, with one side prepared for a single V-groove welded joint. Two blanks were welded with complete joint penetration using the metal active gas (MAG) 136 welding technique. Other details regarding the welding process were: 1.2 mm diameter of welding wire AWS 5.20: E71T-1C; C1 (100% CO_2) shielding gas; 15–18 L/min flow rate; Lincoln DC 400, LF 37 welding machine; vertical up welding position; 4 passes; 140–150 A, 21–22 V root parameters, 190–200 A, 24–25 V finish parameters. Butt-welded specimens were cut out parallel to the rolling direction of the plates using water jet technology, Figure 1.

Specimens were exposed to three different types of the environment, Figure 2:

- Tap water in laboratory;
- Seawater, 10 m below sea surface, location: the northern Adriatic, in front of the city of Rijeka in Croatia;
- Sea splash, same geographical location, but with specimens at the sea surface exposed to daily change of tides and splashing of waves.

Specimens were divided into 12 groups, each group containing five specimens, giving a total of 60 specimens. A control group of specimens was kept at room atmosphere. Groups of specimens were exposed to a corrosive environment for 3, 6, 12, 24, and 36 months.

Tap water at room temperature was changed daily to ensure that specimens were in contact with fresh water. Specimens exposed in the sea were affected by changes of sea temperature, salinity, and pH value. At the location of the experiment, sea temperature varied between 10 °C and 14 °C annually, salinity changed between 37.8 and 38.3‰, while the pH value was between 8.22 and 8.29 [37]. After being exposed for 3, 6, 12, 24, and 36 months, the specimens were retrieved from the corrosive environment. Corrosion products were removed from the surface mechanically with a soft brush in running water; specimens were washed with distilled water and dried at room temperature. The procedure was chosen in order to not remove the base material, just corrosion products, but several other procedures can also be applied [38].

Figure 1. Welded blanks with cut out specimens.

(a) (b) (c)

Figure 2. Exposure of specimens to corrosive environment: (**a**) water, (**b**) seawater, (**c**) sea splash.

2.3. Testing Procedures

Identical digital balance Ohaus 3000 was used before and after the exposure to weigh the specimens in order to determine mass loss.

An 11 MP resolution Sony digital camera was used to obtain digital macro images to investigate localized corrosion on the surface. An area of 200 mm^2 (10 mm × 20 mm) was selected on the images to measure the number of corrosion points with ImageJ digital image processing software.

Changes in surface morphology caused by localized corrosion were detected by Olympus SZX10 stereo optical microscope. Moreover, the diameter of corrosion pits was measured this way. Pit depth was measured by analogue Somet pit gauge with 0.01 mm precision. Morphology of the corroded was studied using scanning electron microscope (SEM) FEI Quanta 250.

3. Results

3.1. Mass Loss

Based on measured values before and after the exposure, the average percentage of mass loss for a group of specimens exposed to different environments was calculated and put in relation to exposure time; Figure 3. Along with measured data, dependency of the average mass loss percentage on the exposure time is represented by approximation curves connecting the experimental values. Additionally, coefficient of determination R^2 is displayed.

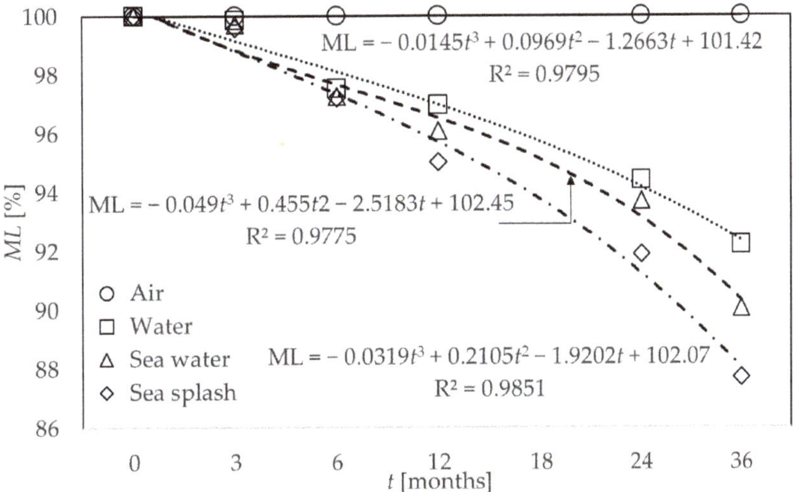

Figure 3. Average percentage of mass loss (*ML*) for AH36 steel specimens exposed over time (*t*) to air, water, seawater, and sea splash for 3, 6, 12, 24, and 36 months.

Average corrosion rate *CR* is calculated according to Equation (2) [39] and given in Figure 4:

$$CR = \frac{KW}{ATD}\left[\frac{\text{mg}}{\text{dm}^2\text{day}}\right],\tag{2}$$

where constant $K = 2.4 \times 10^6 \cdot D$, material density $D = 7.8$ kg/cm^3, W = measured mass loss [g], A = specimen area [cm^2], T = time of exposure [h]. Corrosion rate describes the material wastage caused by corrosion and it can help in assessing the remaining operational life of material under the influence of the corrosive environment.

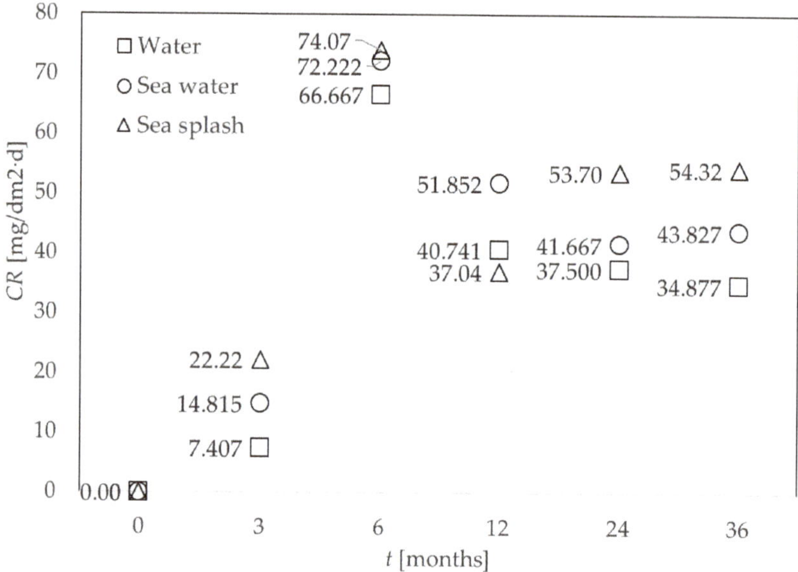

Figure 4. Average corrosion rate (*CR*) for AH36 steel specimens exposed over time (*t*) to air, water, seawater, and sea splash for 3, 6, 12, 24, and 36 months.

3.2. Pit Density and Dimensions

Optical microscope was used to assess the corroded surface at three different areas of specimens: base metal (BM), heat affected zone (HAZ), and weld metal (WM), Figure 5. Visual tests were performed prior to cleaning the specimens, but due to heavily corroded surfaces, priority was given to microscope observations that are presented here. Typical images obtained by optical microscope are presented in Figure 6.

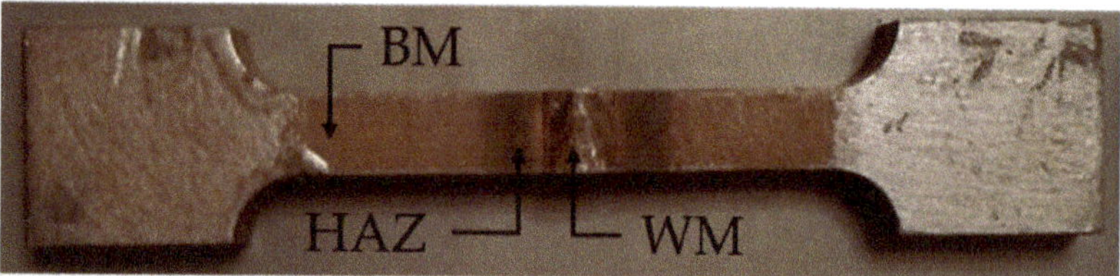

Figure 5. Areas of specimens observed by microscope: base metal—BM, heat affected zone—HAZ and weld metal—WM.

In Table 2, the pit density obtained by processing digital images of corroded surfaces is presented.

Values of mean pit diameters (*d*) obtained by optical microscopy along with mean pit depths (*h*) obtained by measuring actual pit depths with pit depth gauge are presented in Figure 7. Since every group of examined specimens consisted of five items, there was a dissipation of results. Thus, error bars representing standard deviation are also given.

3.3. Corroded Surface Morphology

Scanning electron microscope (SEM) was used to assess the corroded surface at three different areas of specimen: base metal (BM), heat affected zone (HAZ), and weld metal (WM); Figure 5. Besides this technique, alternates were used by other authors, such as roughness meters [40]. Typical images obtained by SEM are presented in Figure 8.

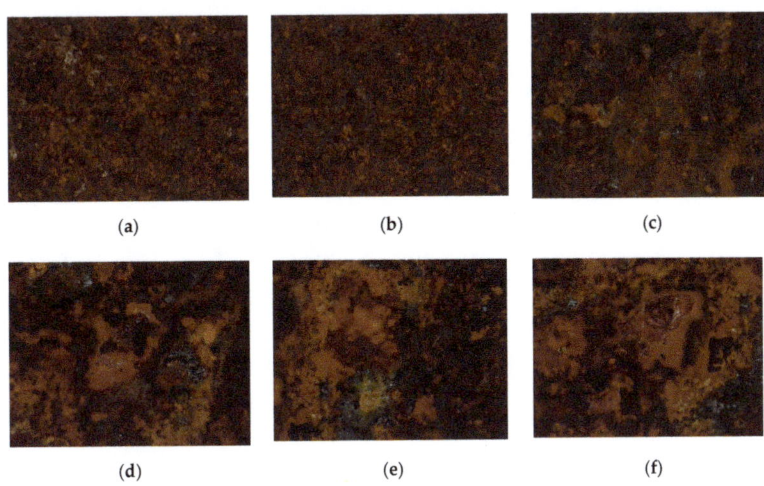

Figure 6. Optical microscope images of specimens exposed to seawater at magnification of 32× for: (**a**) BM, 6 months exposure; (**b**) HAZ, 6 mt.; (**c**) WM, 6 mt.; (**d**) BM, 12 mt.; (**e**) HAZ, 12 mt.; (**f**) WM, 12 mt.

Table 2. Pit density.

Environment:	Water														
Location:	BM					HAZ					WM				
Exposure Time (months):	3	6	12	24	36	3	6	12	24	36	3	6	12	24	36
Pit density (10 mm^{-2}):	3 ± 1	4 ± 1	5 ± 1	4 ± 1	5 ± 1	2 ± 1	3 ± 1	5 ± 2	4 ± 1	5 ± 1	2 ± 1	2 ± 1	3 ± 1	3 ± 1	4 ± 1
Environment:	Seawater														
Location:	BM					HAZ					WM				
Exposure Time (months):	3	6	12	24	36	3	6	12	24	36	3	6	12	24	36
Pit Density (10 mm^{-2}):	6 ± 1	9 ± 3	6 ± 2	5 ± 2	5 ± 1	2 ± 1	3 ± 1	3 ± 1	2 ± 1	2 ± 2	4 ± 1	7 ± 2	6 ± 2	6 ± 1	5 ± 1
Environment:	Sea splash														
Location:	BM					HAZ					WM				
Exposure Time (months):	3	6	12	24	36	3	6	12	24	36	3	6	12	24	36
Pit Density (10 mm^{-2}):	4 ± 1	6 ± 2	6 ± 2	5 ± 1	5 ± 1	3 ± 2	5 ± 2	5 ± 2	4 ± 1	5 ± 1	2 ± 1	4 ± 1	3 ± 1	3 ± 1	2 ± 1

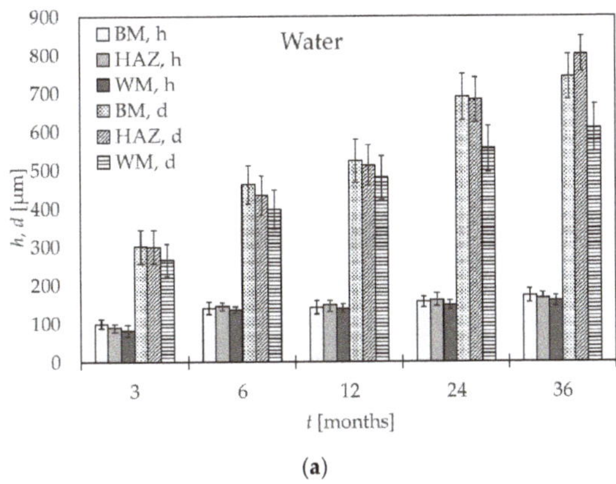

(**a**)

Figure 7. *Cont.*

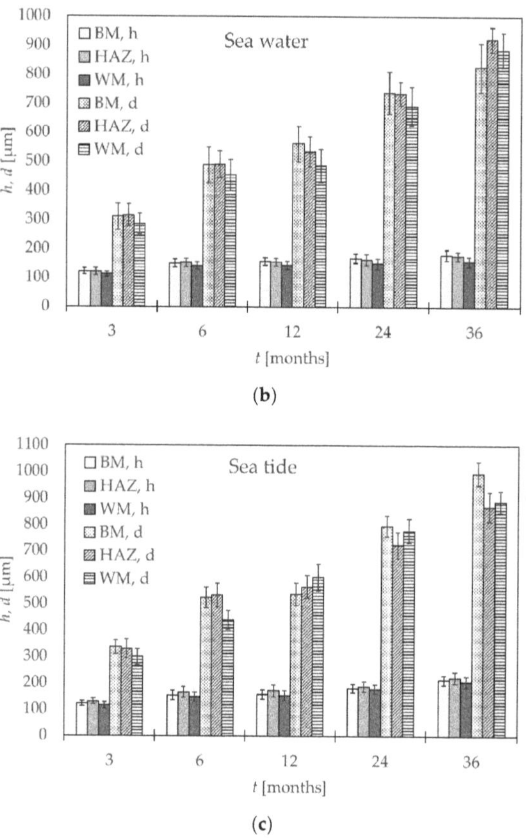

(b)

(c)

Figure 7. Values of average pit depth (*h*) and pit diameter (*d*) of specimens over exposed time (*t*). (**a**) water, (**b**) sea water, (**c**) sea tide.

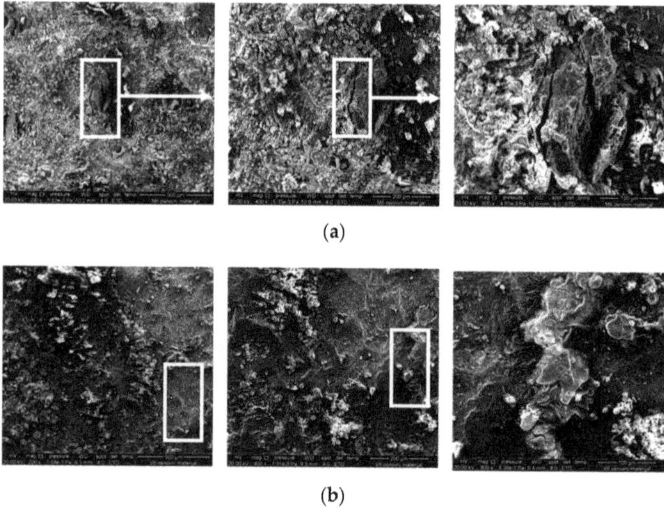

(a)

(b)

Figure 8. SEM images at magnification of 200×, 400×, and 800× for BM specimen area exposed for 6 months to: (**a**) sea water, (**b**) sea splash.

4. Discussion

Percentage of mass loss for specimens exposed to different corrosive environments, Figure 3, shows that the significant mass loss occurs after the three-month exposure period. Up to three months, mass loss is negligible. After a six-month exposure period, there is a gain in mass loss, but the rate at which it occurs is somewhat alleviated. This can be contributed to the formation of a protective rust layer on the surface of the metal [41]. However, taking into account the whole 36-month period, one can observe significant mass loss, reaching more than 15% of initial mass. The greatest effect on mass loss was the exposure to sea splash, contributing to combined influence of the corrosive environment (sea) and mechanical loads (splash) [2]. Equations on approximation curves, Figure 3, can help in assessing the mass loss for steel AH36 over the exposure time.

Consistent with previous data, the highest impact on corrosion rate can be observed between the third and sixth month of exposure; Figure 4. Moreover, sea splash generally has the most negative impact on the corrosion rate.

When comparing images of corroded surfaces obtained by optical microscope at three different areas of the specimen exposed to seawater, Figure 6a–c, it can be noted that the number of corrosion pits is highest at the BM area and lowest at the HAZ area. These findings are consistent with pit density values presented in Table 2 and in correlation with previous findings of pit density in austenitic stainless steel welded joints [42].

When comparing images of specimens exposed to the same corrosive environment for two periods (6 and 12 months), Figure 6a–c with Figure 6d–f, it can be noted that pits tend to be larger at specimens exposed for longer periods. On the other hand, pit density tends to be lower for longer periods; Table 2. This is due to fact that pits grow in time and coalesce, making them larger in size, but smaller in number [43].

When assessing a localized corrosion attack, measured pit depth tends to be a better indicator than mass loss. Mean pit depths obtained experimentally, Figure 7, show that the most severe pitting occurs during the first six months of exposure to a corrosive environment. After that, the rate of pit depth growth slows. Moreover, the greatest values of pit depths are measured in the heat-affected zone of the specimens, followed by base material and weld metal. When comparing influence of different corrosive environments, the most negative impact on the surface of the metal has a combined influence of sea and waves (sea splash).

SEM images at three different magnifications ($200\times$, $400\times$, $800\times$) for BM of specimens exposed for 6 months to seawater and sea splash, Figure 8, show differences in the shape of the pits at the corroded surfaces. Pits for the metal exposed to the sea have clear and sharp edges, while ones at the metal exposed to the continuous influence of waves have round and smooth edges. These types of smoothed pits present a lower risk of evolving into cracks that could cause failure of a structure [44]. However, generally, greater dimensions of pits at specimens exposed to sea splash than seawater, Figure 8, annulated this fact and bought significant risk of structural integrity loss to the steel structure exposed to the tides.

As for the future direction of the research, it would be useful to study the change of mechanical properties of steel exposed to different types of marine environments for prolonged periods, and to compare them to properties determined at room temperature. Research was conducted on similar steel, various temperatures, and types of environments [45–47], but further investigations that are carried out for prolonged periods of exposure, real marine environments, and welded joints could shed additional light on this interesting topic. An educated guess would be that a marine environment could adversely impact the mechanical properties of the steel. Moreover, insight into the change of mechanical properties, due to the change of ambient temperature, would be helpful. Further, the types of experimental data presented in this paper could prove useful in building a numerical model of steel specimen [48] with stochastically distributed corrosion pits over it, performing numerical design optimization [49]. Although there were studies on that matter [10], they did not consider material dissimilarities over BM, HAZ, and WM.

5. Conclusions

To avoid failures, and to help in the preparation of inspection procedures [50], the study of surface conditions of butt-welded specimens made of AH36 shipbuilding steel, exposed to a corrosive marine environment, is presented here. Studying previous similar investigations by other authors, this research has several specific characteristics that need to be emphasized:

- Material was exposed to a natural corrosive environment, which was not the case for most of the previous research studies based in laboratory experiments in simulated environments.
- Exposure to a corrosive environment was performed for prolonged periods, contrary to usual accelerated laboratory tests.
- Findings can be summarized as:
- Exposure between three and six months shows significant influence on mass loss of specimens.
- Sea splash generally has the most negative impact on the corrosion rate due to combined chemical and mechanical degradation of material.
- Pit density is the highest at the BM area of the specimen and lowest at the HAZ area of the specimen.
- The diameter of corrosion pits grows over the time of exposure as the pits coalesce and join.
- Sea splash impacts the shape of the pits, making them rounder and smoother compared to specimens exposed to the sea.
- Pit depths grow fastest during the first six months of exposure to a corrosive environment; after that, the rate of growth tends to slow down.
- Pit depths are generally greatest in the HAZ area of the specimen, followed by BM and WM.

Author Contributions: Conceptualization, G.V. (Goran Vukelic); methodology, G.V. (Goran Vukelic), G.V. (Goran Vizentin); validation J.B.; investigation, G.V. (Goran Vizentin), F.S., M.B.; resources, F.S., M.B.; data curation, G.V. (Goran Vizentin), F.S.; writing—original draft preparation, G.V. (Goran Vukelic); writing—review and editing, J.B., M.B.; supervision, G.V. (Goran Vukelic), J.B.; project administration, G.V. (Goran Vizentin); funding acquisition, G.V. (Goran Vukelic). All authors have read and agreed to the published version of the manuscript.

Funding: This research was funded by the University of Rijeka, grant numbers uniri-technic-18-200 and uniri-technic-18-42. The APC was funded by University of Rijeka.

Institutional Review Board Statement: Not applicable.

Informed Consent Statement: Not applicable.

Acknowledgments: The authors would like to thank Anel Kudic and Branimir Mihaljec, former students of University of Rijeka, Faculty of Maritime Studies, for their technical support.

Conflicts of Interest: The authors declare no conflict of interest. The funders had no role in the design of the study; in the collection, analyses, or interpretation of data; in the writing of the manuscript, or in the decision to publish the results.

References

1. Ivošević, Š.; Meštrović, R.; Kovač, N. Probabilistic estimates of corrosion rate of fuel tank structures of aging bulk carriers. *Int. J. Nav. Archit. Ocean Eng.* **2019**, *11*, 165–177. [CrossRef]
2. Ivošević, Š.; Meštrović, R.; Kovač, N. A Probabilistic Method for Estimating the Percentage of Corrosion Depth on the Inner Bottom Plates of Aging Bulk Carriers. *J. Mar. Sci. Eng.* **2020**, *8*, 442. [CrossRef]
3. Vizentin, G.; Vukelic, G.; Murawski, L.; Recho, N.; Orovic, J. Marine propulsion system failures—A review. *J. Mar. Sci. Eng.* **2020**, *8*, 662. [CrossRef]
4. Vukelic, G.; Brnic, J. Predicted fracture behavior of shaft steels with improved corrosion resistance. *Metals* **2016**, *6*, 40. [CrossRef]

5. Vukelic, G.; Brnic, J. Using experimental and numerical characterization in comparing marine exhaust system stainless steels. In *Proceedings of the 6th. European Conference on Computational Mechanics (Solids, Structures and Coupled Problems) ECCM 6 and 7th. European Conference on Computational Fluid Dynamics ECFD 7*; Owen, R., de Borst, R., Reese, J., Pearce, C., Eds.; CIMNE: Glasgow, UK, 2018; pp. 4423–4431.

6. Kozak, J.; Tarełko, W. Case study of masts damage of the sail training vessel POGORIA. *Eng. Fail. Anal.* **2011**, *18*, 819–827. [CrossRef]

7. Zunkel, A.; Tiebe, C.; Schlischka, J. "Stolt Rotterdam"–The sinking of an acid freighter. *Eng. Fail. Anal.* **2014**, *43*, 221–231. [CrossRef]

8. Lo, M.; Karuppanan, S.; Ovinis, M. Failure Pressure Prediction of a Corroded Pipeline with Longitudinally Interacting Corrosion Defects Subjected to Combined Loadings Using FEM and ANN. *J. Mar. Sci. Eng.* **2021**, *9*, 281. [CrossRef]

9. Salazar-Domínguez, C.M.; Hernández-Hernández, J.; Rosas-Huerta, E.D.; Iturbe-Rosas, G.E.; Herrera-May, A.L. Structural Analysis of a Barge Midship Section Considering the Still Water and Wave Load Effects. *J. Mar. Sci. Eng.* **2021**, *9*, 99. [CrossRef]

10. Xu, S.; Wang, H.; Li, A.; Wang, Y.; Su, L. Effects of corrosion on surface characterization and mechanical properties of butt-welded joints. *J. Constr. Steel Res.* **2016**, *126*, 50–62. [CrossRef]

11. Česen, A.; Kosec, T.; Legat, A. Characterization of steel corrosion in mortar by various electrochemical and physical techniques. *Corros. Sci.* **2013**, *75*, 47–57. [CrossRef]

12. Reiser, D.B.; Alkire, R.C. The measurement of shape change during early stages of corrosion pit growth. *Corros. Sci.* **1984**, *24*, 579–585. [CrossRef]

13. Holme, B.; Lunder, O. Characterisation of pitting corrosion by white light interferometry. *Corros. Sci.* **2007**, *49*, 391–401. [CrossRef]

14. Ren, Z.; Ernst, F. Stress–Corrosion Cracking of AISI 316L Stainless Steel in Seawater Environments: Effect of Surface Machining. *Metals* **2020**, *10*, 1324. [CrossRef]

15. He, J.; Xu, S.; Ti, W.; Han, Y.; Mei, J.; Wang, X. The Pitting Corrosion Behavior of the Austenitic Stainless Steel 308L-316L Welded Joint. *Metals* **2020**, *10*, 1258. [CrossRef]

16. Melchers, R.E. Long-Term Durability of Marine Reinforced Concrete Structures. *J. Mar. Sci. Eng.* **2020**, *8*, 290. [CrossRef]

17. Melchers, R.E. Experience-Based Physico-Chemical Models for Long-Term Reinforcement Corrosion. *Corros. Mater. Degrad.* **2021**, *2*, 6. [CrossRef]

18. Kim, D.K.; Wong, E.W.C.; Cho, N.K. An advanced technique to predict time-dependent corrosion damage of onshore, offshore, nearshore and ship structures: Part I = generalisation. *Int. J. Nav. Archit. Ocean Eng.* **2020**, *12*, 657–666. [CrossRef]

19. Kim, D.K.; Lim, H.L.; Cho, N.K. An advanced technique to predict time-dependent corrosion damage of onshore, offshore, nearshore and ship structures: Part II = Application to the ship's ballast tank. *Int. J. Nav. Archit. Ocean Eng.* **2020**, *12*, 645–656. [CrossRef]

20. Świerczyńska, A.; Fydrych, D.; Landowski, M.; Rogalski, G.; Łabanowski, J. Hydrogen embrittlement of X2CrNiMoCuN25-6-3 super duplex stainless steel welded joints under cathodic protection. *Constr. Build. Mater.* **2020**, *238*, 117697. [CrossRef]

21. Świerczyńska, A.; Landowski, M. Plasticity of Bead-on-Plate Welds Made with the Use of Stored Flux-Cored Wires for Offshore Applications. *Materials* **2020**, *13*, 3888. [CrossRef] [PubMed]

22. Paik, J.K.; Thayamballi, A.K.; Park, Y.I. Hwang, J.S. A time-dependent corrosion wastage model for seawater ballast tank structures of ships. *Corros. Sci.* **2004**, *46*, 471–486. [CrossRef]

23. Guedes Soares, C.; Garbatov, Y.; Zayed, A.; Wang, G. Corrosion wastage model for ship crude oil tanks. *Corros. Sci.* **2008**, *50*, 3095–3106. [CrossRef]

24. Soares, C.G.; Garbatov, Y.; Zayed, A.; Wang, G. Influence of environmental factors on corrosion of ship structures in marine atmosphere. *Corros. Sci.* **2009**, *51*, 2014–2026. [CrossRef]

25. Paik, J.K.; Kim, D.K. Advanced method for the development of an empirical model to predict time-dependent corrosion wastage. *Corros. Sci.* **2012**, *63*, 51–58. [CrossRef]

26. Mohd, M.H.; Kim, D.K.; Kim, D.W.; Paik, J.K. A time-variant corrosion wastage model for subsea gas pipelines. *Ships Offshore Struct.* **2014**, *9*, 161–176. [CrossRef]

27. Saeed, A.; Khan, Z.A.; Nazir, M.H. Time dependent surface corrosion analysis and modelling of automotive steel under a simplistic model of variations in environmental parameters. *Mater. Chem. Phys.* **2016**, *178*, 65–73. [CrossRef]

28. Pastorcic, D.; Vukelic, G.; Bozic, Z. Coil spring failure and fatigue analysis. *Eng. Fail. Anal.* **2019**. [CrossRef]

29. Chen, J.; Zhang, W.; Gu, X. Modeling time-dependent circumferential non-uniform corrosion of steel bars in concrete considering corrosion-induced cracking effects. *Eng. Struct.* **2019**, *201*, 109766. [CrossRef]

30. Yu, Y.; Gao, W.; Castel, A.; Chen, X.; Liu, A. An integrated framework for modelling time-dependent corrosion propagation in offshore concrete structures. *Eng. Struct.* **2021**, *228*, 111482. [CrossRef]

31. Choi, Y.Y.; Lee, S.H.; Park, J.C.; Choi, D.J.; Yoon, Y.S. The impact of corrosion on marine vapour recovery systems by VOC generated from ships. *Int. J. Nav. Archit. Ocean Eng.* **2019**, *11*, 52–58. [CrossRef]

32. Schoefs, F.; Boéro, J.; Capra, B. Long-Term Stochastic Modeling of Sheet Pile Corrosion in Coastal Environment from On-Site Measurements. *J. Mar. Sci. Eng.* **2020**, *8*, 70. [CrossRef]

33. Duan, T.; Peng, W.; Ding, K.; Guo, W.; Hou, J.; Cheng, W.; Liu, S.; Xu, L. Long-term field exposure corrosion behavior investigation of 316L stainless steel in the deep sea environment. *Ocean Eng.* **2019**, *189*, 106405. [CrossRef]

34. Martins, K.L.; Pinto, V.T.; Fragassa, C.; Real, M.V.; Rocha, L.A.O.; Isoldi, L.A.; dos Santos, E.D. A Simplified Numerical Method for the Design and Analysis of FPSO Platform Brackets Subjected to Operational Conditions. *J. Mar. Sci. Eng.* **2020**, *8*, 929. [CrossRef]

35. Kim, J.D.; Myoung, G.H.; Suh, J. Butt weldability of shipbuilding steel AH36 using laser-arc hybrid welding. *Trans. Korean Soc. Mech. Eng. A* **2016**, *40*, 901–906. [CrossRef]

36. Dueren, C.F. Prediction of the hardness in the HAZ of HSLA steels by means of the C-equivalent. In Proceedings of the Conference on Hardenability of Steels 1990, Derby, UK, 17 May 1990.

37. Institute of Oceanography and Fisheries. *Početna Procjena Stanja i Opterećenja Morskog Okoliša Hrvatskog Dijela Jadrana*; Institute of Oceanography and Fisheries: Split, Croatia, 2012.

38. Garbatov, Y.; Saad-Eldeen, S.; Guedes Soares, C.; Parunov, J.; Kodvanj, J. Tensile test analysis of corroded cleaned aged steel specimens. *Corros. Eng. Sci. Technol.* **2019**, *54*, 154–162. [CrossRef]

39. *American Society for Testing and Materials, ASTM G1-03 Standard Practice for Preparing, Cleaning and Evaluating Corrosion Test Specimens*; West Conshohocken: Montgomery County, PA, USA, 2011.

40. Florescu, S.N.; Gheonea, M.C.; Mihailescu, D.; Teodor, V. Influence of marine corrosion on the roughness of MAG welded joint surfaces. In *Proceedings of the IOP Conference Series: Materials Science and Engineering*; IOP Publishing Ltd.: Bristol, UK, 2020; Volume 968, p. 012008.

41. Muthanna, B.G.N.; Amara, M.; Meliani, M.H.; Mettai, B.; Božić, Ž.; Suleiman, R.; Sorour, A.A. Inspection of internal erosion-corrosion of elbow pipe in the desalination station. *Eng. Fail. Anal.* **2019**. [CrossRef]

42. Luo, L.; Huang, Y.; Weng, S.; Xuan, F.Z. Mechanism-related modelling of pit evaluation in the CrNiMoV steel in simulated environment of low pressure nuclear steam turbine. *Mater. Des.* **2016**, *105*, 240–250. [CrossRef]

43. Li, S.X.; Akid, R. Corrosion fatigue life prediction of a steel shaft material in seawater. *Eng. Fail. Anal.* **2013**, *34*, 324–334. [CrossRef]

44. Zerbst, U.; Madia, M.; Klinger, C.; Bettge, D.; Murakami, Y. Defects as a root cause of fatigue failure of metallic components. III: Cavities, dents, corrosion pits, scratches. *Eng. Fail. Anal.* **2019**, *97*, 759–776. [CrossRef]

45. Rajput, A.; Park, J.H.; Hwan Noh, S.; Kee Paik, J. Fresh and sea water immersion corrosion testing on marine structural steel at low temperature. *Ships Offshore Struct.* **2020**, *15*, 661–669. [CrossRef]

46. Rajput, A.; Paik, J.K. Effects of naturally-progressed corrosion on the chemical and mechanical properties of structural steels. *Structures* **2021**, *29*, 2120–2138. [CrossRef]

47. Tavares, S.S.M.; Batista, R.T.; Landim, R.V.; Velasco, J.A.C.; Senna, L.F. Investigation of the effect of low temperature aging on the mechanical properties and susceptibility to sulfide stress corrosion cracking of 22%Cr duplex stainless steel. *Eng. Fail. Anal.* **2020**, *113*, 104553. [CrossRef]

48. Kodvanj, J.; Garbatov, Y.; Guedes Soares, C.; Parunov, J. Numerical Analysis of Stress Concentration in Non-uniformly Corroded Small-Scale Specimens. *J. Mar. Sci. Appl.* **2020**, 1–9. [CrossRef]

49. Vukelic, G.; Vizentin, G.; Masar, A. Hydraulic torque wrench adapter failure analysis. *Eng. Fail. Anal.* **2019**, *96*, 530–537. [CrossRef]

50. Poggi, L.; Gaggero, T.; Gaiotti, M.; Ravina, E.; Rizzo, C.M. Recent developments in remote inspections of ship structures. *Int. J. Nav. Archit. Ocean Eng.* **2020**, *12*, 881–891. [CrossRef]

MDPI

St. Alban-Anlage 66

4052 Basel

Switzerland

Tel. +41 61 683 77 34

Fax +41 61 302 89 18

www.mdpi.com

Journal of Marine Science and Engineering Editorial Office

E-mail: jmse@mdpi.com

www.mdpi.com/journal/jmse